CORROSION AND DEGRADATION OF IMPLANT MATERIALS: SECOND SYMPOSIUM

A symposium
sponsored by ASTM
Committees F-4 on
Medical and Surgical Materials
and Deviced and G-1 on
Corrosion of Metals
Louisville, KY, 9–10 May 1983

ASTM SPECIAL TECHNICAL PUBLICATION 859
Anna C. Fraker, National Bureau
of Standards, and
Charles D. Griffin, Carbomedics, Inc.
editors

ASTM Publication Code Number (PCN)
04-859000-27

 1916 Race Street, Philadelphia, PA 19103

Library of Congress Cataloging in Publication Data

Corrosion and degradation of implant materials.

(ASTM special technical publication: 859)
Papers presented at the Second International Symposium on Corrosion and Degradation of Implant Materials.
Includes bibliographies and index.
"ASTM publication code number (PCN) 04-859000-27."
1. Implants, Artificial—Materials—Corrosion—Congresses. 2. Implants, Artificial—Materials—Deterioration—Congresses. 3. Biomedical materials—Corrosion—Congresses. 4. Biomedical materials—Deterioration—Congresses. I. Fraker, Anna C. II. Griffin, Charles D. III. International Symposium on Corrosion and Degradation of Implant Materials (2nd: 1983: Louisville, KY) IV. ASTM Committee F-4 on Medical and Surgical Materials and Devices. V. American Society for Testing and Materials. Committee G-1 on Corrosion of Metals. VI. Series.
RD132.C59 1985   617'.95   84-70337
ISBN 0-8031-0427-8

NOTE

The Society is not responsible, as a body,
for the statements and opinions
advanced in this publication.

Printed in Ann Arbor, MI
June 1985

Foreword

The Second Symposium on Corrosion and Degradation of Implant Materials was held in Louisville, Kentucky, 9–10 May 1983. The symposium was sponsored by ASTM Committees F-4 on Medical and Surgical Materials and Devices and G-1 on Corrosion of Metals. Anna C. Fraker, National Bureau of Standards, and Charles D. Griffin, Carbomedics, Inc., presided as symposium chairmen and editors of this publication.

Related
ASTM Publications

Medical Devices: Measurements, Quality Assurance, and Standards, STP 800 (1983), 04-800000-54

Titanium Alloys in Surgical Implants, STP 796 (1983), 04-796000-54

Corrosion and Degradation of Implant Materials, STP 684 04-684000-27

A Note of Appreciation
to Reviewers

The quality of the papers that appear in this publication reflects not only the obvious efforts of the authors but also the unheralded, though essential, work of the reviewers. On behalf of ASTM we acknowledge with appreciation their dedication to high professional standards and their sacrifice of time and effort.

ASTM Committee on Publications

ASTM Editorial Staff

Helen M. Hoersch
Janet R. Schroeder
Kathleen A. Greene
Bill Benzing

Contents

SUMMARY

Introduction

The Second International Symposium on the Corrosion and Degradation of Implant Materials was held in May 1983 in Louisville, Kentucky. The symposium was sponsored by the ASTM Joint F-4/G-1 Section on the Corrosion of Implant Materials. Members of this group come from ASTM Committee F-4 on Medical and Surgical Materials and Devices and ASTM Committee G-1 on Corrosion of Metals. This was the second symposium on the topic sponsored by this joint group and was a five-year follow-up of the original meeting. The first meeting, held in Kansas City, Kansas, in 1978, resulted in *Corrosion and Degradation of Implant Materials, ASTM STP 684*. This is an excellent and much used book, but due to the growth in the field and advances in technology since 1978, a re-examination of the theme was considered necessary. At a meeting of the ASTM Joint F-4/G-1 Section on Corrosion of Implant Materials in 1981, planning began that resulted in a successful update to the original symposium.

The introduction to *ASTM STP 684* refers to man's efforts to repair the body over the centuries and the role of physicians, scientists, and engineers in making vast progress in the synthetic implant field. More history mentioned in the introduction to *Titanium Alloys in Surgical Implants, ASTM STP 796*, cites the importance of antiseptic techniques in surgery and indicates the level of demand on the implant material after these techniques extended the life of implants by reducing infection and other complications. Today, we are in an age of high technology where metal alloys, polymers, ceramics, and other materials are being modified and new materials are being developed. As these creations and innovations occur, it is important to regularly assess the biocompatibility and durability of the materials in the human body. As material technology and design improvements extend the life and success of medical implants, these basic questions become more important than ever.

This Second International Symposium was intended to cover a number of different implant materials as well as topics dealing with standards, retrieval analysis, and legal considerations. The papers in this book represent a contribution from the authors, reviewers, ASTM Staff, and the editors. It is hoped that the papers will be useful in advancing the science of biomaterials and in providing better implant materials for surgical use.

Anna C. Fraker
National Bureau of Standards, Washington, DC 20234; symposium chairman and editor.

Charles D. Griffin
Carbomedics, Inc., Austin, TX 78752; symposium chairman and editor.

Metallic Materials: *In Vitro* Studies

Norbert D. Greene[1]

Corrosion of Surgical Implant Alloys: A Few Basic Ideas

REFERENCE: Greene, N. D., "**Corrosion of Surgical Implant Alloys: A Few Basic Ideas,**" *Corrosion and Degradation of Implant Materials: Second Symposium, ASTM STP 859*, A. C. Fraker and C. D. Griffin, Eds., American Society for Testing and Materials, Philadelphia, 1985, pp. 5–10.

ABSTRACT: The corrosion processes occurring on surgically implanted alloys are reviewed. Crevice corrosion and other forms of attack observed on surgical implants are described. The theory, techniques and results of linear polarization corrosion rate measurements are discussed.

KEY WORDS: implant materials, biological degradation, fatigue (materials), corrosion, corrosion tests, corrosion reactions, implants, passive, surgical implants, polarization, linear polarization, *in vivo* tests, stainless steels

Corrosion Processes

Corrosion Reactions

The chemical reactions that occur on the surface of a surgically implanted alloy are identical to those observed during exposure to seawater (namely, aerated sodium chloride). The metallic components of the alloy are oxidized to their ionic forms and dissolved oxygen is reduced to hydroxide ions

$$M \rightarrow M^{n+} + ne \quad \text{(oxidation)} \tag{1}$$

$$O_2 + 2H_2O + 4e \rightarrow 4OH^- \quad \text{(reduction)} \tag{2}$$

Equation 1 is the general reaction for the oxidation of a metal atom. The electrons that are released are conducted through the metal to another site where they are consumed in the reduction process (Eq 2). For a chromium-nickel alloy

[1]Professor of Metallurgy, Department of Metallurgy, Institute of Materials Science, University of Connecticut, Storrs, CT 06268.

the reactions are

$$Cr \rightarrow Cr^{+3} + 3e \text{ (oxidation)} \tag{3a}$$

$$Ni \rightarrow Ni^{+2} + 2e \text{ (oxidation)} \tag{3b}$$

$$O_2 + 2H_2O + 4e \rightarrow 4OH^- \text{ (reduction)} \tag{4}$$

During a corrosion reaction, the total rates of oxidation and reduction, expressed as electron production and consumption, must be equal. Thus, the overall reaction rate is controlled by the slowest of these two processes. Most metals and alloys used for surgical implants are passive in saline solutions and plasma. That is, their surfaces are rapidly covered by a very thin, oxygen-rich protective layer. Although the mechanism is unclear, this layer markedly decreases oxidation reaction rates. Decreases of 10^6 or more are common. Thus, the corrosion of surgical implants is controlled by the rate of metal oxidation and changes in dissolved oxygen concentration over a wide range have no effect on the rate of corrosion.

There are a number of ways to reduce corrosion [1].[2] Altering the environment is a common method used for industrial applications. Changing the electrolyte concentration, reducing the temperature, or adding chemical inhibitors or all are used to decrease corrosion. Unfortunately, these techniques cannot be used for surgical implants since the environment is fixed and cannot be altered without destructive biological effects. Coatings are also widely used to retard corrosion. They have only limited use for protecting implants since many are subjected to abrasion and wear, especially orthopaedic devices. Applied current protection methods (anodic and cathodic protection) are too clumsy and impractical for use with surgically implanted devices. The only general useful way to reduce the corrosion of surgical implants is by appropriate alloy selection.

Forms of Corrosion

There are many forms of corrosion damage [2] and several of these occur on alloys surgically implanted in humans and animals. General or uniform corrosion is present on all implanted metals and alloys. The rate of attack is usually very low due to the presence of passive surface films. Metal losses of less than 0.1 μm per year are typical. This cannot be detected by visual inspection or weight loss measurements. However, the corrosion rate can be determined by electrochemical techniques described later.

Crevice attack refers to corrosion at shielded sites such as screw/plate interfaces and under washers. This is often observed on stainless steels and other passive alloys in the presence of chlorides. If the passive film within a narrow crevice is damaged, it reforms slowly due to oxygen depletion in these areas. As a result, rapid attack occurs. Stainless steel (18Cr-8Ni and 18Cr-8Ni-2Mo)

---

[2]The italic numbers in brackets refer to the list of references appended to this paper.

implants are very susceptible to this type of corrosion. Over 90% of crevice-containing stainless steel implants (for example, screw-plate assemblies) show evidence of crevice corrosion [3].

Pitting and stress corrosion cracking, although usually associated with stainless steels in chloride media, are not observed on recovered surgical implants. Implants often exhibit cracks and surface pitting, but these are most likely the result of improper manufacture rather than corrosion [3,4].

Fatigue refers to fracture failures caused by cyclic loading. The presence of a corrosive (for example, chloride) often accelerates this process. This is termed corrosion fatigue and it may occur on implants that are repeatedly stressed at high levels.

Corrosion Rate Measurements

Theory

Obviously, an implant alloy should corrode very slowly during its service life. If this is the case, mechanical damage will be minimal. More importantly, the amount of metal compounds released into the system will be small.

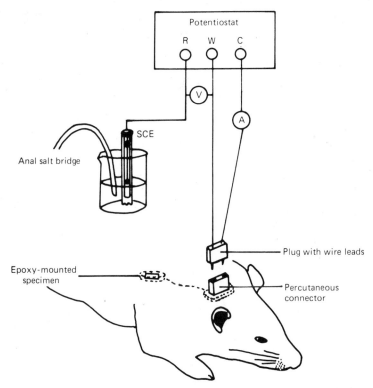

FIG. 1—*Apparatus for* in vivo *electrochemical measurements in a rat* [7].

Most present and proposed implant alloys corrode at very low rates. These rates are too small to be detected by dimensional changes or weight loss. Tafel extrapolation and linear polarization are two electrochemical methods that have been used to measure low corrosion rates *in situ* [5]. Of the two, linear polarization is the more suitable for studies of implant alloys. It is rapid, utilizes very low applied currents and is applicable to a variety of corroding systems.

Linear polarization measurements are accomplished by applying small potentials to the test specimen and measuring the resulting currents. If the offset potential is 10 mV or less, there is a linear relationship between current and potential. Theoretically and practically, some curvature is present. However, it is so small that the normal experimental procedure obscures it. The ratio of potential/current has the units of resistance, R, ohms. For iron, nickel, cobalt, and their alloys, it can be shown that corrosion rate may be approximated by [6].

$$C = \frac{330\ 200}{RA}$$

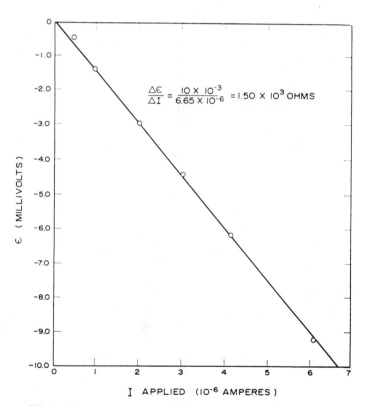

FIG. 2—*Linear polarization of a 1018 carbon steel specimen in a dog* [8].

Where C is the corrosion rate in microns penetration per year, A is the sample area in cm², and R is the resistance in ohms.

Techniques

Electrochemical methods are easily applied to both *in vitro* and *in vivo* studies of implant corrosion. In either case, only three electrical connections are required; to the specimen, to a counter electrode, and to a potential reference electrode. Animal experiments require more care than conventional, *in vitro* laboratory tests, but are relatively simple. Figure 1 shows a schematic diagram of a system used for studies in rats [7]. The plug that connects to the specimen and counter electrode (contained in the connector) may be removed between tests. Before each measurement, the rat is anesthetized, the plug is connected, and an anal catheter connected to a saturated calomel electrode (SCE) is inserted.

Results

Figures 2 and 3 show some results obtained using a sample of low carbon steel implanted in a dog [8]. Although the corrosion rate of this alloy is too

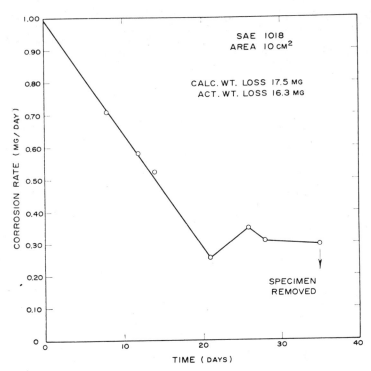

FIG. 3—*Corrosion of a 1018 carbon steel specimen in a dog as determined by repeated linear polarization measurements* [8]. *Weight loss after removal is compared to that calculated by integrating this curve.*

large to be of use as an implant material, it allows the linear polarization data to be confirmed by weight loss measurements (Fig. 2). Using linear polarization, it is possible to measure corrosion rates below 0.1 μm metal loss per year.

Summary

The corrosion processes occurring on alloy implants are identical to those that are observed during exposure to aerated saline solutions. The reaction rates are usually very slow due to the presence of passive films. Although it is usually impossible to measure these corrosion rates by weight loss, they can be accurately determined by linear polarization methods.

References

[1] Fontana, M. G. and Greene, N. D., *Corrosion Engineering, 2nd ed.*, McGraw-Hill, New York, 1978, pp. 194–222.
[2] See Ref *1*, 28–115.
[3] Colangelo, V. J. and Greene, N. D., *Journal of Biomedical Materials Research*, Vol. 3, 1969, pp. 247–265.
[4] Cahoon, J. R. and Paxton, H. W., *Journal of Biomedical Materials Research*, Vol. 2, 1968, pp. 1–22.
[5] See Ref *1*, pp. 342–345.
[6] Greene, N. D. and Jones, D. A., *Journal of Materials*, Vol. 1, No. 2, 1966, pp. 345–353.
[7] Abru, E. and Greene, N. D., unpublished results. See also Ref *1*, pp. 390–392.
[8] Colangelo, V. J., Greene, N. D., Kettlekamp, D. B., Alexander, H., and Campbell, C. J., *Journal of Biomedical Materials Research*, Vol. 1, 1967, pp. 405–414.

B. J. Edwards,[1] *M. R. Louthan, Jr.,*[1] *and R. D. Sisson, Jr.*[2]

Hydrogen Embrittlement of Zimaloy: A Cobalt-Chromium-Molybdenum Orthopedic Implant Alloy

REFERENCE: Edwards, B. J., Louthan, M. R., Jr., and Sisson, R. D., Jr., "**Hydrogen Embrittlement of Zimaloy: A Cobalt-Chromium-Molybdenum Orthopedic Implant Alloy,**" *Corrosion and Degradation of Implant Materials: Second Symposium, ASTM STP 859,* A. C. Fraker and C. D. Griffin, Eds., American Society for Testing and Materials, Philadelphia, 1985, pp. 11–29.

ABSTRACT: Controlled potential, slow strain rate tests of Zimaloy (a cobalt-based orthopedic implant alloy) in Ringer's solution (a physiological saline solution) at 37°C show that hydrogen absorption may degrade the mechanical properties of that alloy. Potentials were controlled so that the tensile sample was either cathodic or anodic with respect to the metal's free corrosion potential. Hydrogen was generated on the sample surface when the specimen was cathodic, and dissolution of the sample was encouraged when the sample was anodic. The results of these controlled potential tests showed no susceptibility of this alloy to stress corrosion cracking at anodic potentials.

However, when samples were tested at a potential of 150 mV cathodic to the free corrosion potential, the difficulty was decreased by 37%. The 150-mV shift in potential was selected because that was the potential observed when this alloy, in a creviced condition, was placed in Ringer's solution for two months. Support for this reduction in ductility during the cathodic exposures was obtained by testing the hydrogen compatibility of the alloy. Samples charged in gaseous hydrogen showed significant losses in ductility. The most severe hydrogen charging (207 MPa at 300°C for ten days) reduced the tensile ductility to 0% reduction in the area at fracture while less severe charging (6.9 MPa at 200°C for seven days) caused a 50% decrease in ductility. Results of these tests indicate that this cobalt-based alloy suffers a significant loss in ductility under conditions that may arise if a creviced sample is exposed to saline solutions.

KEY WORDS: hydrogen embrittlement, polarization, tension tests, crevice potential, Zimaloy, cobalt-chromium-molybdenum alloy, implant materials, fatigue (materials), degradation

Cast cobalt-chromium-molybdenum (Co-Cr-Mo) alloys have been successfully used for surgical implant devices since the 1930's [1].[3] Recent innovations in alloy processing have improved both mechanical properties and formability. One

[1]Graduate student and professor, respectively, Materials Engineering Department, Virginia Polytechnic Institute and State University, Blacksburg, VA 24061.

[2]Professor, Mechanical Engineering Department, Worcester Polytechnic Institute, Worcester, MA 01609.

[3]The italic numbers in brackets refer to the list of references appended to this paper.

such technique consolidates metal powders at high temperatures and pressures and is known as hot isostatic pressing (HIP). A hot isostatic pressed Co-Cr-Mo alloy, known as HIP Zimaloy, has the same chemical composition as cast alloy [2,3], but has significantly improved mechanical properties [1].

The mechanical and corrosive properties of cast Co-Cr-Mo alloys both in saline solutions and in-service conditions are well known [4–7]. However, to date, *in vitro* testing of HIP Zimaloy has been limited to corrosion and stress corrosion studies [2]. The objective of the present study was to determine the effects of corrosion potential on the slow strain rate mechanical properties and failure mode of HIP Zimaloy in Ringer's solution at 37°C and to determine the effect of hydrogen charging on these properties.

Background

Cast Cobalt Alloys

In-service failure of the cast Co-Cr-Mo material may become an increasing problem because as surgical technique is refined and long-term replacement operations become acceptable for younger patients, the expected service life of implant materials increases. Clearly, fatigue failures may become more prevalent and numerous studies have shown fatigue to be a potential problem [1,5,6,8,9].

Cobalt-based implant alloys have excellent corrosion resistance and Scales [10] reported in a 1956 study that no *in vivo* corrosion of this metal had been observed. However, between 1962 and 1965, Scales [10] reported a study of over 600 patients where 22% of the Co-Cr-Mo implants were corroded or fretted. Other isolated cases of corrosion were reported in 1969 [11] and 1972 [12]. It has also been shown that at least some *in vivo* corrosion occurs in all implanted materials, regardless of their corrosion resistance [13,14].

Laboratory tests have shown that simple saline solution can drastically reduce the fatigue life of cast Co-Cr-Mo implants [7,15], even though it is difficult to distinguish between corrosion fatigue and fatigue for *in vivo* failure [7]. This difficulty suggests that many failures that were cited as fatigue failures may have been aided by corrosion. Some authors have suggested there may be confusion in distinguishing between fatigue, corrosion fatigue, and stress corrosion cracking [16,17], as the latter two processes require the development of similar passive films and subsequent rupture of those films. It is therefore suggested that stress corrosion cracking or environmentally enhanced fatigue or both could be included as possible failure modes for Co-Cr-Mo implant materials.

Corrosion Potential Shifts

Slow strain rate tests at the free corrosion potential of the cast Co-Cr-Mo in a physiological solution similar to Ringer's showed no evidence of environment-assisted cracking [15]. However, it has been shown that the corrosion potential can shift with time (weeks and months) and that the direction of the shift depends on the condition of the material [18,19]. The potential of cast Co-Cr-Mo in saline solution has been shown to change up to 300 mV in three weeks, [19] and up to 150 mV after ten weeks [4]. These changes were in the anodic direction. The effect of crevices on potential shifts has also been studied in stainless steel and titanium where a crevice caused a potential shift after one month [18] of up to 150 mV cathodic with respect to the uncreviced potential. An effect of straining on the corrosion potential of cast Co-Cr-Mo was seen in fully-reversed torsion fatigue tests [7] that resulted in failure after 6 h. The corrosion potential changed 200 mV in the cathodic direction with the onset of plastic strain. This change in potential reached 600 mV just before the specimen failed [7]. Such potential changes could cause hydrogen to be generated at the surface of the cobalt-based implant materials.

Hydrogen Uptake

The uptake and transport of hydrogen in Zimaloy (or similar cobalt alloys) have not been determined experimentally. However, the diffusivity and permeability of hydrogen in pure cobalt have been measured [20]. These data for pure cobalt are used to estimate the behavior of hydrogen in Zimaloy.

Hydrogen diffusivity was calculated from experimentally measured permeation rates and the rate of rise to the steady-state permeation rate. This diffusivity (D) is given by

$$D = 3.4 \times 10^{-2} \exp(-13\ 600/RT) \text{ cm}^2/\text{s}$$

Using this diffusivity equation, the time required to saturate a 0.08-cm-thick sample can be calculated [21] from

$$\left(\frac{Dt}{\ell^2}\right) \sim (1.2)^2$$

where t is the time of exposure and ℓ is the half sample thickness. Such calculations suggest that an implant in the human body can saturate with hydrogen during the typical 10 to 20-year life.

Procedure

The material used in this investigation was hot isostatically pressed (HIP) Zimaloy.[4] A typical composition of this material is given in Table 1. An electrolytic etch with 5% hydrochloric acid revealed the grain structure shown in Fig. 1. Carbide particles were found to be nonuniformly distributed. The dark particles at grain boundaries are most likely $M_{23}C_6$ while the light particles within

---

[4]Supplied by Zimmer, Inc., Warsaw, IN 46580.

TABLE 1—*Chemical analysis of HIP Zimaloy* [2].

	HIP Zimaloy	ASTM F 75-82[a]
Carbon	0.24	0.35 max
Phosphorus	0.011	...
Sulfur	0.012	...
Manganese	0.41	1.0 max
Silicon	0.75	1.0 max
Chromium	29.1	27.0 to 30.0
Nickel	0.18	2.50 max
Molybdenum	6.28	5.0 to 7.0
Iron	0.10	0.75 max
Tungsten	0.054	...
Copper	0.015	...
Cobalt	remainder	remainder

[a]ASTM Specification for Cast Cobalt-Chromium-Molybdenum Alloy for Surgical Implant Applications (F 75-82).

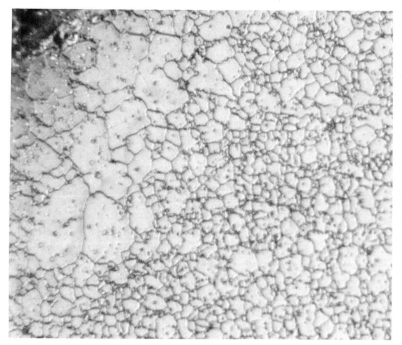

FIG. 1—*Optical photomicrograph of HIP Zimaloy showing duplex fine grain structure (original magnification ×500).*

the grains are probably M_7C_3. A point count analysis revealed that the typical carbide fraction varied by factors of two times for $M_{23}C_6$ and four times for M_7C_3 particles throughout. These variations were based on point count analysis of metallographic cross sections magnified $\times 500$.

Anodic Polarization

The anodic polarization curves were determined for samples that were polished through 600 grit finish using silicon carbide paper. Each sample was polished, rinsed in distilled water, and dried in room air. Specimens were prepared so that drying took place about 1½ h before the start of the polarization test. After immersion of the sample in the solution for polarization studies, the sample was allowed to reach a stable corrosion potential. This was defined as the potential reached when no more than 1 mV change occurred over a 5 min period. When the corrosion potential was achieved, usually after 2 h, the polarization test was initiated. The initial level of potential control was 150 mV cathodic with respect to the corrosion potential. The potential was then increased in the anodic direction 50 mV every 5 min. The test was ended when the current density reached the value of 100 $\mu A/cm^2$. All potentials listed in this paper, unless otherwise noted, are given relative to a saturated calomel reference electrode (SCE).

An nonaerated (exposed to room air) buffered Ringer's solution at 37 $\pm$ 1°C was used in the polarization tests. Solution pH, checked at the beginning of each test, was 7.4. Results (Fig. 2) agreed with Zimmer's polarization data for HIP Zimaloy [3] as well as with data collected on the cast cobalt alloy [22]. The controlled potential slow strain rate tension tests were done under the same conditions as the anodic polarization tests.

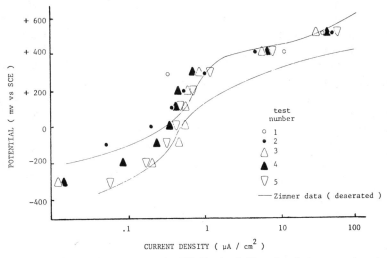

FIG. 2—*Anodic polarization curves for HIP Zimaloy in Ringer's solution exposed to air.*

Hydrogen in Zimaloy

The effect of hydrogen on the mechanical properties of HIP Zimaloy was determined by tension tests of a standard tensile specimen that was charged in 107 MPa (30 000 psi) gaseous hydrogen at 300°C for ten days. The test sample was then pulled to failure in 207 MPa hydrogen at room temperature using test techniques identical to those described in Ref *23*. Another sample was charged in 6.89 MPa (1000 psi) hydrogen at 200°C for seven days before being pulled to failure in Ringer's solution.

Hydrogen could be generated at the surface of HIP Zimaloy implants because of the electrochemical corrosion reaction under certain localized conditions. Such conditions include a crevice or other differential aeration cell where hydrogen reduction replaces oxygen reduction as the cathodic reaction.

A creviced condition has also been shown to cause cathodic shifts in the corrosion potential of other materials. The potential shift associated with exposure of this alloy to a creviced condition was experimentally determined. Two flat pieces of HIP Zimaloy, each 3.175 mm (⅛ in.) thick, 6.35 mm (¼ in.) wide, and 25.4 mm (1 in.) long, were placed with the 6.35 by 25.4-mm (¼ by 1-in.) faces together. The two pieces were separated by a thin coat of lacquer except for the bottom 1.59 mm (1/16 in.) of the interface. An electrical connection was spot welded between the pieces at the top. That junction and all outer surfaces of the two pieces were then coated with lacquer down to 1.59 mm (1/16 in.) from the bottom. The resulting crevice was 0.127 mm (0.005 in.) wide and 1.59 mm (1/16 in.) deep. The bottom half of the assembly was immersed in solution and the potential was monitored for two months.

Controlled Potential Slow Strain Rate Tests

Tension tests at a controlled strain rate of 1.6×10^{-4}/s were made on samples whose corrosion potentials were controlled at: (1) the material's corrosion potential, (2) +100 mV, (3) +350 mV, (4) −480 mV, and (5) −1370 mV with respect to the SCE. (A test was also run to determine the change in sample corrosion potential with strain as the sample was pulled to failure.) The positive potentials (+350 and +100) were chosen to maintain the sample in the passive region (see Fig. 2). The negative potential of −480 mV was chosen to be 150 mV cathodic with respect to the free corrosion potential developed during the tension test. The potential of −1370 mV was the potential at which gas bubbles generated by the cathodic current first became visible on the sample surface. The current through each sample during the controlled strain rate test was recorded so that abrupt changes could be detected. The potentiostat was used to impress the desired potential on the sample while simultaneously monitoring the current as the sample was strained to failure.

The fractured sample was rinsed in distilled water, examined under a low-power (×10) microscope, and measured with dial calipers, also under the

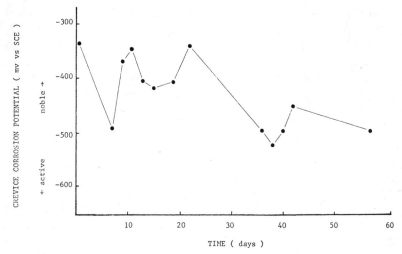

FIG. 3—*Corrosion potential of creviced HIP Zimaloy as a function of time.*

microscope. Ultimate strength and percent reduction in area were determined, and selected samples were examined under a scanning electron microscope (SEM).

Results

Corrosion Potential Shifts

Corrosion potential of creviced HIP Zimaloy shifted generally in the cathodic direction as shown in Fig. 3. As a result of these findings, some controlled

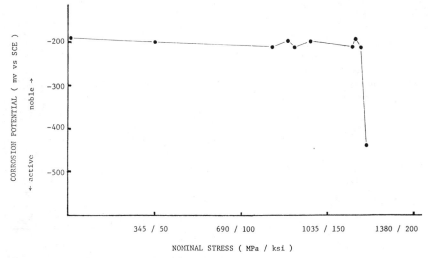

FIG. 4—*Free corrosion potential of HIP Zimaloy as a function of nominal stress measured during a slow strain rate ($1.6 \times 10^{-4}/s$) tension test.*

potential slow strain rate tests were performed at various potentials that were cathodic with respect to the corrosion potential.

The shift in the corrosion potential of cast Co-Cr-Mo alloys due to strain has been discussed previously [7]. However, in this study as the specimen was strained to failure, the corrosion potential showed a change of no more than 10 mV until the failure stress was reached. A graph of corrosion potential versus load is given in Fig. 4. The change was gradual and was in the cathodic direction. Immediately before failure, the corrosion potential decreased by 250 mV.

TABLE 2—*Summary of results of controlled potential slow strain rate tests of HIP Zimaloy.*

Sample Number	Test Environment	Ultimate Tensile Strength, MPa/ksi	Reduction in Area, %	Fracture Surface
15 Y	air	1500/218	26	Fig. 6
26 W	In Ringer's; monitor potential	1130/161	23	Fig. 5
32 Y	In Ringer's; potential set at corrosion potential	1360/197	24	not shown
33 Y	In Ringer's; potential = +100 mV	1320/191	21	Fig. 13
34 W	Hydrogen charged 6.9 MPa in Ringer's; potential set at corrosion potential	1180/171	12	Fig. 7
34 Y	Hydrogen charged 6.9 MPa in Ringer's; potential = +100 mV	1180/171	18	not shown
35 W	In Ringer's; potential = −480 mV	1430/207	16	Fig. 9
35 Y	In Ringer's; potential = +350 mV	1360/197	22	Fig. 14
36 W	In Ringer's; generated hydrogen for 1 h, potential = −1370 mV	1200/171	15	Fig. 10 Fig. 11 Fig. 12
37 W	Hydrogen charged[a]	NA[b]	0	Fig. 8

NOTES:—All tests done at strain rate = $1.6 \times 10/s$; temperature = 37°C.
[a]207 MPa failed in 207 MPa.
[b]NA = not available.

Hydrogen's Effect on Zimaloy

Hydrogen was shown to reduce the ductility of Zimaloy to varying degrees depending on the charging condition (Table 2). The percent reduction in area decreased and became zero with sufficient hydrogen charging. The SEM fractographs of the failure surfaces showed evidence of intergranular cracking. This cracking increased with the severity of charging and little evidence of microscopic ductility was found in the severely charged samples.

This contrasts with the fracture surfaces of both the uncharged sample that was failed in Ringer's solution (Fig. 5) and the sample fractured in air (Fig. 6). Fracture under both of these test conditions showed a mixed-mode failure consisting of microvoid coalescence and twin boundary fracture. The sample charged at 6.89 MPa (1000 psi) hydrogen for one week (Fig. 7) showed larger, more sharply defined areas of twin boundary fracture and the microvoid coalescence mode was replaced by smaller facets and cracking typical of intergranular fractures. The sample charged in 207 MPa (30 000 psi) hydrogen for ten days showed a predominantly intergranular fracture surface with limited areas of twin boundary fracture (Fig. 8).

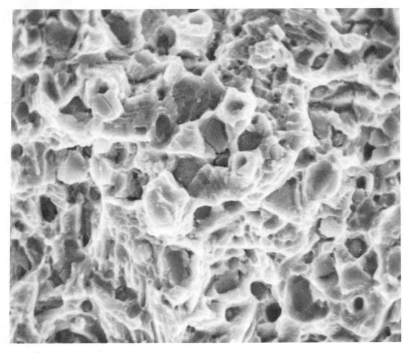

FIG. 5—*Fractograph of HIP Zimaloy tested in Ringer's solution at the free corrosion potential (original magnification ×3000).*

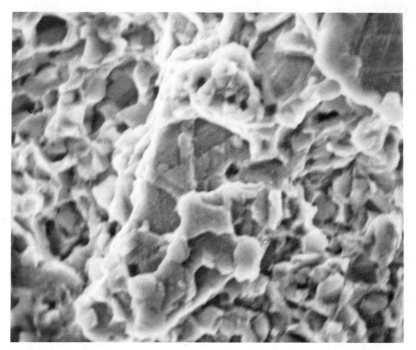

FIG. 6—*Fractograph of HIP Zimaloy tested in air (original magnification ×3000).*

The sample charged in 207 MPa hydrogen failed with no measurable plastic strain. By comparison, samples tested in air and in Ringer's solution at the free corrosion potentials experienced an average of 24% reduction in area. The sample charged in 6.89 MPa (1000 psi) hydrogen and tested at its corrosion potential experienced 12% reduction in area. These data are summarized in Table 2.

Fracture surfaces of samples tested under cathodic control (Figs. 9 and 10) were different from the surfaces of samples tested at free corrosion potential (Fig. 5) only in localized areas. Microvoid coalescence was mixed with small areas of twin boundary fracture and intergranular cracking. The corner of the sample tested at −1370 mV showed that failure was initiated by a mix of twin boundary and intergranular fracture (Figs. 11 and 12). This hydrogen-effected crack propagated across the sample until overload fracture by microvoid coalescence caused rupture of the remainder of the specimen. The hydrogen-induced crack initiation sites caused samples tested at cathodic potentials to show significant losses in ductility despite the ductile appearance of much of their fracture surfaces.

Tests at anodic potentials of +100 mV and +350 mV should cause a change in crack morphology if this material was susceptible to stress corrosion cracking.

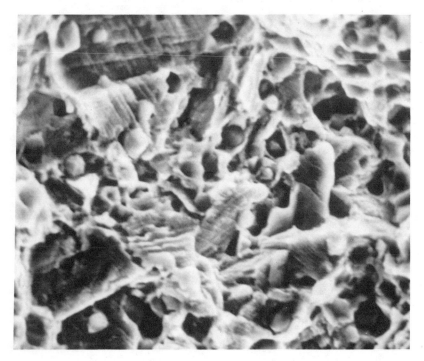

FIG. 7—*Fractograph of HIP Zimaloy charged in hydrogen (6.9 MPa at 200°C for seven days) and tested in Ringer's solution at the free corrosion potential (original magnification ×3000).*

Fractographs of anodically controlled samples (Figs. 13 and 14) differ very little from samples tested at the free corrosion potential. Predominant regions of microvoid coalescence are connected by small areas of twin boundary fracture. Furthermore, the anodic controlled samples showed no significant losses in ductility. Thus, no indications of susceptibility to stress corrosion were observed.

Tension tests of samples controlled at the corrosion potential showed a small but steady increase in current as the sample strained. Additionally, spikes in the current were observed at stresses of 480 MPa (70 ksi) and 950 MPa (138 ksi). The yield stress of Zimaloy is 916 MPa (133 ksi) [2]. A sample tested at +100 mV showed a similar, though more rapid increase in current and even more pronounced disturbances at nearly the same two stress levels. A comparison of these two behaviors is shown in Fig. 15.

Hydrogen-charged samples exhibited a generally less stable current during all stages of the test (Fig. 16). The hydrogen-charged sample tested at +100 mV showed a significant current spike near the yield stress. Furthermore, after the yield stress had been exceeded, the hydrogen-charged samples showed a series of small and frequent current spikes.

FIG. 8—*Fractograph of HIP Zimaloy charged in hydrogen (207 MPa at 300°C for ten days) and tested in 207 MPa hydrogen at room temperature (original magnification ×3000).*

Discussion

There are several possible failure modes for Co-Cr-Mo implants in the body. Corrosion has been seen to act in combination with fretting [10] and fatigue [15]. *In vivo* fatigue failures have also been identified [1,5,6,8,9], but it is not certain that the biological environment did not aid in these failures. Hydrogen embrittlement was an environmentally induced failure mode found in this study of HIP Zimaloy. Although hydrogen effects on cobalt alloys have not been discussed in the literature, this form of embrittlement may have contributed to some of the observed service failures of the cast alloy.

The fact that short time exposure to cathodic potentials (test times were from 1 to 2 h at 37°C) affected ductility of the HIP Zimaloy as much as charging in 6.9 MPa gaseous hydrogen for one week (168 h at 200°C) was surprising. The diffusion of hydrogen in cobalt at 37°C is two orders of magnitude slower than at 200°C, thus hydrogen uptake by diffusion would be limited in the low temperature tests. However, hydrogen can be transported by dislocation motion [24]. Evidence for this transport is an increase in hydrogen release rate when a hydrogen-charged sample was pulled in tension. These increases correlated with increases in the amount of dislocation movement [24]. During the tension test

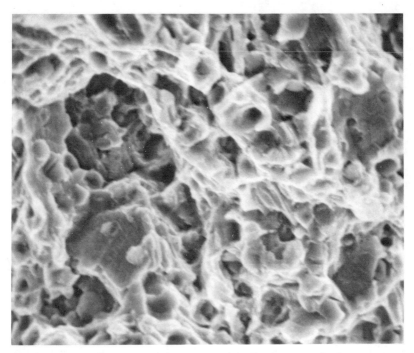

FIG. 9—*Fractograph of HIP Zimaloy tested in Ringer's solution at a controlled potential of −480 mV versus SCE, near center of sample (original magnification ×3000).*

of HIP Zimaloy charged in hydrogen, several small bursts of anodic current were observed. This effect can be seen in Fig. 16, which shows stress versus current in hydrogen-charged samples tested in Ringer's solution. After yielding, the specimen current required to maintain the sample at the control potential went through a period of small frequent spiking. Such behavior was not seen in the uncharged samples (Fig. 15). A hydrogen atom transported to the surface would go into solution as a hydrogen ion plus an electron, increasing the anodic current. Therefore, this spiking is attributed to dislocation transport of hydrogen to the metal surface.

If hydrogen release rate is affected by dislocation motion in Zimaloy, it is reasonable to assume hydrogen uptake also is affected. Thus, hydrogen generated electrolytically on samples may be transported to the advancing crack by dislocation motion causing the significant losses in ductility observed in the tests at 37°C.

Once straining has progressed far enough to cause the formation of a crack or crevice in the passive film, a condition for localized hydrogen generation and concentration is set up. As the crack is first opened, the briefly unprotected metal within corrodes. These metal ions generated may then combine with water

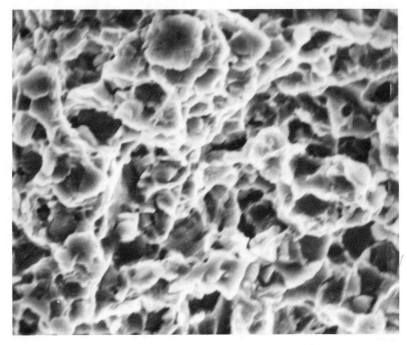

FIG. 10—*Fractograph of HIP Zimaloy tested in Ringer's solution at a controlled potential of −1370 mV versus SCE, center of sample (original magnification ×3000).*

in the following reaction, similar to a reaction for iron ions discussed by Kruger [25].

$$M^{+n} + nH_2O \rightarrow M(OH)_n + nH^+$$

Because this is occurring within the crack, the products of reaction cannot readily escape. The buildup of hydrogen ions lowers the pH within a crack. The cathodic reaction within the crack will most likely be [25].

$$H^+ + e^- \rightarrow H$$

If this occurs, hydrogen atoms are continuously being generated within the crack and the possibility exists for hydrogen embrittlement.

Conclusions

This study has demonstrated that a cobalt-chromium-molybdenum orthopedic implant alloy is very susceptible to hydrogen embrittlement. Controlled potential, slow strain rate tests of samples exposed in synthetic body fluids at cathodic potentials were shown to degrade the tensile properties of the alloy. Test potentials were selected by experimental determination of the potential that developed

FIG. 11—*Fractograph of HIP Zimaloy tested in Ringer's solution at a controlled potential of −1370 mV versus SCE, stress concentration and electrochemical attack combined to initiate fracture at the corner (original magnification ×500).*

in a creviced sample exposed to Ringer's solution at 37°C. Combination of these two experimental observations indicates that hydrogen-enhanced crack growth may be a potential mechanism for degradation of cobalt-based orthopedic implant materials.

Acknowledgments

The authors wish to thank Zimmer U.S.A., for providing the material used in this study and the College of Engineering, Virginia Polytechnic Institute, for financial support.

DISCUSSION

J. Lemons[1] *(written discussion)*—The specimen design and geometry of crevices are quite critical to the test results. Please describe the specimen size and geometry, especially for the crevice situation.

[1]University of Alabama at Birmingham, Birmingham, AL 35294.

FIG. 12—*Fractograph of HIP Zimaloy tested in Ringer's solution at a controlled potential of −1370 mV versus SCE, near corner of sample (original magnification ×3000).*

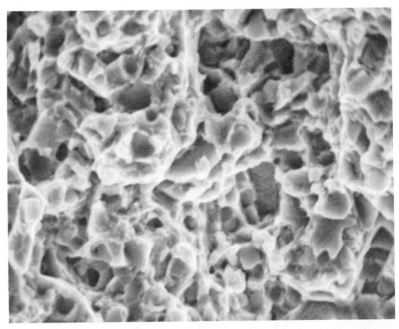

FIG. 13—*Fractograph of HIP Zimaloy tested in Ringer's solution at a controlled potential of +100 mV versus SCE (original magnification ×3000).*

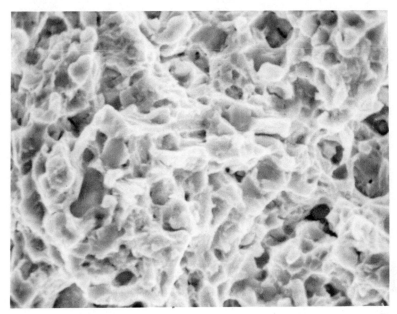

FIG. 14—*Fractograph of HIP Zimaloy tested in Ringer's solution at a controlled potential of +350 mV versus SCE (original magnification ×3000).*

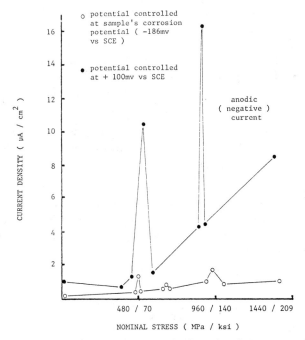

FIG. 15—*Corrosion current of HIP Zimaloy tested in Ringer's solution at controlled potentials of −186 mV and +100 mV versus SCE as a function of nominal stress measured during a slow strain rate (1.6 × 10⁻⁴/s) tension test.*

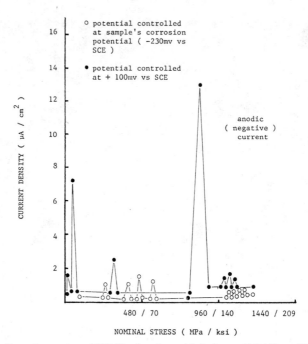

FIG. 16—*Corrosion current of HIP Zimaloy charged in hydrogen (6.9 MPa at 200°C for seven days) and tested in Ringer's solution at controlled potentials of −230 mV and +100 mV versus SCE as a function of nominal stress measured during a slow strain rate (1.6 × 10⁻⁴/s) tension test.*

B. J. Edwards, M. R. Louthan, Jr., and R. D. Sisson, Jr. (authors' closure)— The geometry and dimensions were not described in the presentation but have been included in the Procedure section under Hydrogen in Zimaloy.

Jonathan Black[2] (*written discussion*)—Effects at low cathodic potentials are more likely to be intergranular oxidation effects rather than hydrogen embrittlement. Shape of *I-V* curve suggests that these experiments were run under maximum oxygen diffusion conditions.

B. J. Edwards, M. R. Louthan, Jr., and R. D. Sisson, Jr., (authors' closure)— This is true for the general sample surface since the solution was nonaerated. Our argument for hydrogen's action in the uncharged samples depends on formation of a crack or other defect in the passive film, and hence, the opportunity for very localized hydrogen accumulation.

References

[1] Bardos, D. I., *Biomaterials, Medical Devices, and Artificial Organs,* Vol. 7, No. 1, 1979, p. 73.
[2] "Micrograin Zimaloy," Zimmer Technical Report, 1978.

²University of Pennsylvania, Philadelphia, PA 19104.

[3] Greer, K. W., Schobert, C. M., and Bardos, D. I., paper presented at the Sixth International Congress on Metallic Corrosion, Sydney, Australia, Dec. 1975.

[4] Hoar, T. P. and Mears, D. C., *Proceedings,* Royal Society, Vol. A294, 1966, p. 486.

[5] Ducheyne, P., Meester, P. D., Aernoudt, E., Martens, M., and Mulier, J. C., *Journal of Biomedical Materials Research,* Vol. 9, No. 6, 1975, p. 199.

[6] Miller, H. L., Rostoker, W., Galante, J. O., *Journal of Biomedical Materials Research,* Vol. 10, 1976, p. 399.

[7] Imam, M. A., Fraker, A. C., and Gilmore, C. M., *Corrosion and Degradation of Implant Materials, ASTM STP 684,* B. C. Syrett and A. Acharya, Eds., American Society for Testing and Materials, 1979, p. 128.

[8] Charnley, J., *Clinical Orthopaedics and Related Research,* Vol. 111, 1975, p. 105.

[9] Galante, J. O., Rostoker, W., and Doyle, J. M., *Journal of Bone and Joint Surgery,* Vol. 57-A, 1975, p. 230.

[10] Scales, J. T., *Journal of Bone and Joint Surgery,* Vol. 53-B, 1971, p. 311.

[11] Weightman, B. O., Zarek, J. M., and Bingold, A. C., *Medical and Biological Engineering,* Vol. 7, 1969, p. 679.

[12] Rose, R. M., Schiller, A. L., and Radin, E. L., *Journal of Bone and Joint Surgery,* Vol. 54-A, 1972, p. 851.

[13] Ferguson, A. B., Laing, P. G., and Hodge, E. S., *Journal of Bone and Joint Surgery,* Vol. 42-A, 1960, p. 77.

[14] Cohen, J., *Journal of Materials,* Vol. 1, 1966, p. 351.

[15] Jones, R. L., Wing, S. S., and Syrett, B. C., *Corrosion,* Vol. 34, No. 7, 1978, p. 226.

[16] Lisagor, W. B., *ASTM Standardization News,* Vol. 20, 1975, p. 21.

[17] Cohen, J., *Journal of Biomedical Materials Research,* Vol. 4, 1970, p. 245.

[18] Levine, D. L. and Staehle, R. W., *Journal of Biomedical Materials Research,* Vol. 11, No. 1, 1977, p. 553.

[19] Syrett, B. C. and Wing, S. S., *Corrosion,* Vol. 31, No. 4, 1978, p. 138.

[20] Caskey, G. R., Jr., Derrick, R. G., and Louthan, M. R., Jr., *Scripta Metallurgica,* Vol. 8, 1974, p. 481.

[21] Darken, L. S. and Gurry, R. W., *Physical Chemistry of Metals,* McGraw-Hill, New York, 1953.

[22] Brettle, J., *Injury,* Vol. 2, 1970, p. 26.

[23] West, A. J. and Louthan, M. R., Jr., *Metals Transactions,* Vol. 13A, 1982, p. 2049.

[24] Louthan, M. R., Jr., Caskey, G. R., Jr., Donovan, J., and Rawl, D. E., *Materials Science Engineering,* Vol. 10, 1972, p. 357.

[25] Kruger, J., *Corrosion and Degradation of Implant Materials, ASTM STP 684,* B. C. Syrett and A. Acharya, Eds., American Society for Testing and Materials, 1979, p. 107.

Prabhat Kumar,[1] *Anthony J. Hickl,*[1] *Aziz I. Asphahani,*[1] *and Alan Lawley*[2]

Properties and Characteristics of Cast, Wrought, and Powder Metallurgy (P/M) Processed Cobalt-Chromium-Molybdenum Implant Materials

REFERENCE: Kumar, P., Hickl, A. J., Asphahani, A. I., and Lawley, A., "**Properties and Characteristics of Cast, Wrought, and Powder Metallurgy (P/M) Processed Cobalt-Chromium-Molybdenum Implant Materials,**" *Corrosion and Degradation of Implant Materials: Second Symposium, ASTM STP 859,* A. C. Fraker and C. D. Griffin, Eds., American Society for Testing and Materials, Philadelphia, 1985, pp. 30–56.

ABSTRACT: Cobalt-chromium-molybdenum alloys have been highly successful as implant materials due to their corrosion resistance to body fluids, excellent mechanical properties, and biocompatibility. Limitations associated with the conventional processing of these alloys, such as precision casting or casting followed by hot working, have been overcome recently by the use of powder metallurgy (P/M) processing techniques. This paper presents results from an on-going program of research on cobalt 26 to 28Cr, 5 to 6Mo alloys.

Cast, conventionally hot-rolled, and P/M-processed materials were evaluated. The effect of cooling rates in the range of 10 to $10^{4\circ}$C/s was investigated. The results of tension, localized corrosion, corrosion fatigue, and stress corrosion cracking tests are discussed.

KEY WORDS: implant materials, corrosion, biocompatibility, tension tests, stress corrosion cracking, corrosion fatigue

Cobalt-chromium-molybdenum (Co-Cr-Mo) alloys are generally used in applications that take advantage of their creep-resistance and mechanical strength at elevated temperatures, or their high resistance to wear. In addition, their ambient temperature strength and resistance to aqueous corrosion would suggest that this class of alloys would be suitable for implants and prostheses. In fact, biocompatibility was demonstrated with cast Co-Cr-Mo alloys in the 1930's. [1][3] Major property improvements for these alloys have been achieved since then through heat treatment of the cast materials and through the development

---

[1]Group leader, section manager, and director, respectively, Cabot Corporation, Kokomo, IN 46901.

[2]Professor, Drexel University, Philadelphia, PA 19104.

[3]The italic numbers in brackets refer to the list of references appended to this paper.

of a wrought Co-Cr-Mo alloy [2]. More recently, powder metallurgy (P/M) processing has offered another alternative. Considerable attention has been given to the production of high-performance alloys from rapidly solidified powder. Powder making and powder consolidation techniques are being developed in order to take advantage of refined microstructures and the potential property improvements that such materials possess. The present study is aimed at assessing the magnitude of such improvements and at identifying the ability of current manufacturing technology to produce and retain fine structures.

Cast and Wrought Co-Cr-Mo Alloys

Cast Co-Cr-Mo alloys exhibit extremely high wear resistance and are moderately hard and strong but with limited ductility [3–6]. Cast microstructures exhibit large grains, large carbide particles, severe chemical inhomogeneity, and interdendritic shrinkage porosity [5,6]. The wrought Co-Cr-Mo alloys are harder and stronger than the cast composition and with a similar level of ductility [3,6]. Hot working reduces porosity and promotes a fine microstructure.

Usage of the cast or wrought Co-Cr-Mo alloys for prosthetic devices reflects their high resistance to corrosion and high strength. Several studies have confirmed that the cast Co-Cr-Mo alloy is resistant to crevice corrosion attack [3,7]. Both cast and wrought alloys display excellent resistance to pitting corrosion [8,9]. Resistance to fretting is higher in the cast than the wrought alloys [3,10]. Theoretically, and it would appear true from practical experience, the Co-Cr-Mo alloys are passive under the prevailing conditions regarding metallic implants [11]. Both the cast and the wrought Co-Cr-Mo alloys have been reported to be immune to stress corrosion cracking in magnesium chloride [8] and in Tyrode's solution [12]. However, recent studies showed the Co-Cr-Mo alloy to stress corrosion crack in $MgCl_2$ solution boiling at 154°C [13]. Although Co-Cr-Mo metallic implants rarely fracture in service, some failures are observed. In specific cases, failure of the implant has been shown to be the result of fatigue. Fatigue crack initiation can be initiated by fretting [10] or by surface defects in the cast alloy [14].

Powder Metallurgy Co-Cr-Mo Alloys

The inherent limitations in the microstructure, and hence properties, of the cast Co-Cr-Mo alloy can be avoided by utilizing powder metallurgy processing methods. Powder processing offers several advantages including chemical homogeneity and fine-scale microstructures. Thus, an improved combination of mechanical and chemical properties is expected in implants.

Hirschorn and Reynolds [15] fabricated surgical implants of cobalt-base alloys by a combination of cold isostatic pressing and sintering. Modest strength levels ~105 MN/m^2 (15 ksi) were achieved since the sintered alloy was only ~70% theoretical density. Pilliar et al [16] have examined standard powder metallurgy techniques for forming a porous coating of cobalt-chromium on a solid cast cobalt-alloy implant. The open-pore structure of the surface coating leads to biologically stable long-term bonding in the implant because of bone/tissue ingrowth into the prosthesis.

Bardos [5] has described the fabrication of fully-dense Co-Cr-Mo alloys from argon-atomized powder by means of hot isostatic pressing. The material exhibited a microstructure consisting of fine carbides of $Cr_{23}C_6$ in an ultra-fine grain matrix of the Co-Cr-Mo solid solution. Tensile properties and fatigue strength were superior to the same alloy in cast form. Both the powder-processed and cast alloys displayed identical corrosion behavior, identified in terms of corrosion potential and electrochemical polarization curves. The powder-pressed material was equivalent to the cast form with respect to biocompatibility. Furthermore, fully-dense hip implants of the Co-Cr-Mo alloy, ASTM Specification for Cast Cobalt-Chromium-Molybdenum Alloy for Surgical Implant Applications (F 75–76), have been produced from gas-atomized powder using the fluid die concept [17].

In this paper, cast alloys and commercial wrought alloys used as bioimplant materials (hip prostheses) were compared along with similar alloys produced by gas-atomization of powder and consolidation by hot extrusion. Microstructure, mechanical properties, localized corrosion (electrochemical polarization and pitting temperature), fatigue/corrosion fatigue, and stress corrosion cracking tests were characterized.

Experimental

The initial program utilized bar material produced in commercial and pilot plant facilities. Materials included:

Type 1—Cast bar (1.9 cm diameter by 12.7 cm) made by investment casting.

Type 2—Commercial wrought bar made by vacuum melting, forging, bar rolling, and hot extrusion (1177°C).

Type 3—P/M wrought bar made by encapsulating gas atomized powder in a steel can and hot extrusion (1149°C). Two powder size distributions were used:

a. $^{60}/_{100}$ mesh, or 150 to 250 μm (Type 3a); and

b. 250 mesh, or less than 60 μm (Type 3b).

Bars of Types 1 and 3 were not given any additional heat treatment, bars of Type 2 were given a heat treatment of 871°C for 30 min and slow (furnace) cooled. This heat treatment was given to simulate the post-forging heat treatment of implants.

The chemical compositions of the three materials are listed in Table 1. The P/M wrought bar had higher levels of nickel and iron than the other two types; its iron content was higher than allowed in ASTM F 75–76 (0.75%). The wrought alloy bar had lower levels of carbon and silicon to allow for improved hot workability.

Evaluation of the materials' mechanical properties involved room-temperature tension testing using standard ASTM procedures (ASTM Tension Testing of Metallic Materials (E 8–79)) and rotating beam, fatigue testing in air (ASTM

TABLE 1—*Chemical analysis (percent by weight) of starting material.*

Element	Powder[a]	Vitallium Wrought Bar	Castings
B	<0.007	. . .	0.007
C	0.22	0.06	0.22
Cr	27.8	25.8	27.9
Fe	1.57	0.27	0.54
Mn	0.46	0.6	0.78
Mo	5.83	5.4	5.82
N_2	0.14	0.14	. . .
Ni	2.0	0.3	0.83
O_2	0.03	. . .	. . .
P	<0.005	. . .	0.009
S	0.011	. . .	0.004
Si	0.7	0.46	0.78
W	<0.01	0.1	0.13

[a]Interstitial contents varied as follows with the particle size:

Size	C	O_2	N_2
60/100 mesh	0.22	0.028	0.135
250/D mesh	0.23	0.038	0.114

The extruded bars had the same interstitial contents as the starting powder.

Recommended Practice for Constant Amplitude Axial Fatigue Tests of Metallic Materials (E 466–76)). Environmental resistance testing included electrochemical potentiodynamic polarization curves in Ringer's solution [18] (ASTM Recommended Practice for Standard Reference Method for Making Potentiostatic and Potentiodynamic Anodic Polarization Measurements G 5–78)), and immersion tests for resistance to localized corrosion.

The susceptibility of these materials to stress corrosion cracking was determined using the slow strain-rate technique [19]. Jones et al [12] did not find any effect of Tyrode's solution on alloys of this type so a more severe solution, boiling 30% $MgCl_2$, was used in these tests in addition to tests in Ringer's solution.

Light microscopy was used to determine grain structure and carbide morphology. Scanning electron micrography was used to examine fracture surfaces.

Results

Figures 1 and 2 show the microstructures of Co-Cr-Mo alloy bars produced by the first two processes. The cast material (Type 1) has a very coarse dendritic structure (secondary dendrite arm spacing equal to 60 μm). Etching (HCl + H_2O_2) revealed recrystallized grain boundaries and some indication of residual solidification segregation (coring) in the metal matrix. The wrought materials (Type

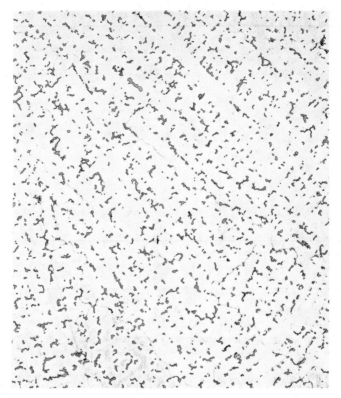

FIG. 1—*Investment cast, Co-Cr-Mo alloy (Type 1) (original magnification ×50).*

2) exhibited an equiaxed grain structure (ASTM 8 to 9) with carbides appearing only as a fine, grain boundary precipitate. Annealing twins were present.

Figure 3 displays the microstructure of both the coarse-powder and fine-powder extruded material (Types 3*a* and 3*b*). Both had similar microstructures with a grain size of ASTM 9 to 10. Micrographs at this magnification (×500) revealed very little volume fraction of carbide. A fine distribution of precipitate within the grains and the irregular, jagged grain boundaries (particularly in Type 3*b* material) indicate the likelihood of a carbide size distribution much finer than that found in the wrought material. Some nonmetallic inclusions were seen as stringers in both Type 3*a* and Type 3*b* material. No attempt was made to quantify the amount or size of these inclusions. It was, however, expected that the size of the inclusions are less in the finer-screened powder (that is, Type 3*b*).

Mechanical Properties

The results of tension tests on these materials are presented in Table 2. The cast material exhibited significantly lower strength and ductility as would be

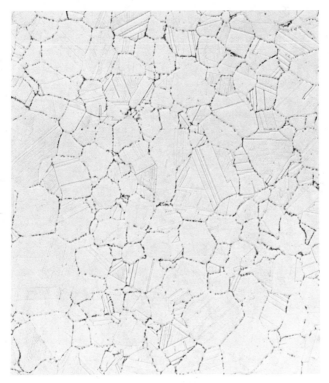

FIG. 2—*Wrought Co-Cr-Mo alloy, extruded plus heat treated (Type 2) (original magnification ×500).*

expected. Type 2 material and Type 3*b* had similar strength and ductility. Type 3*a*, the bar made from coarse powders, showed reduced ductility as compared to the wrought materials made from fine powders (that is, Type 3*b*).

Fractography

Figure 4 shows optical micrographs of the cast sample after tension testing. The scanning electron fractograph of the cast sample, Fig. 5, shows micro-cracks and gross carbide pull out.

Figures 6 through 8 show fractured surfaces of Type 2, 3*a*, and 3*b* materials. In all cases, fracture initiated at or near the surface and the fracture is primarily intergranular. Micro-cracking can also be seen in all three figures with Type 3*a* showing the highest amount and Type 2 showing the least amount of micro-cracking. Bars made from powders (Types 3*a* and 3*b*) also show voids on the fractured surfaces, which are reminiscent of carbide precipitates along grain boundaries. Note that the void size is larger in the bar of coarse powder.

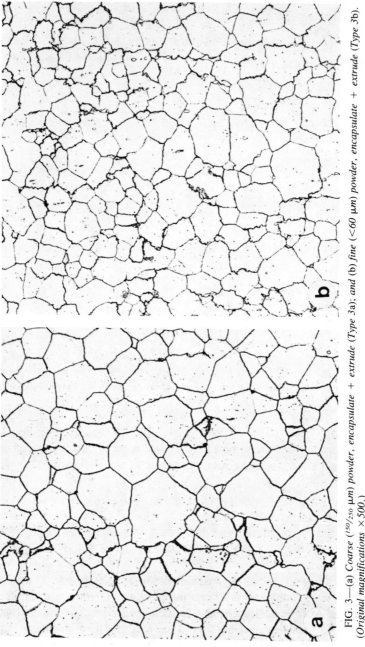

FIG. 3—(a) Coarse ($^{150}/_{250}$ μm) powder, encapsulate + extrude (Type 3a); and (b) fine (<60 μm) powder, encapsulate + extrude (Type 3b). (Original magnifications ×500.)

TABLE 2—*Tensile properties[a] of Co-Cr-Mo alloys.*

Type	Material	Yield Strength,[b] MN/m² (ksi)	Ultimate Tensile Strength, MN/m² (ksi)	Ductility	
				Elongation, %	Reduction in Area, %
1	Cast	475 (69)	620 (90)	8	11
2	Wrought	717 (104)	1200 (174)	38	30
3a	Wrought-P/M (⁶⁰⁄₁₀₀ mesh powder)	730 (106)	1170 (170)	19	18
3b	Wrought-P/M (<250 mesh powder)	744 (108)	1280 (186)	34	28

[a]Reported values are average of four tests.
[b]0.2% offset yield strength is reported.

Localized Corrosion

Electrochemical polarization tests were conducted on the four types of Co-Cr-Mo alloy in Ringer's solution, at 37°C. A relatively slow scan rate of $\simeq 1000$ mV/h was used. All the samples responded in a very similar manner, and the polarization curves for the four types of materials were almost identical in this aerated solution (a typical plot is shown in Fig. 9 for the cast material). No unusual electrochemical features were observed, and none of the samples showed any signs of pitting when examined with a light microscope at $\simeq \times 25$. The experiments were repeated on new samples in Ringer's solutions deaerated by bubbling nitrogen. Again the curves for the four types of the Co-Cr-Mo alloy were identical. However, these curves were slightly different from the ones generated in aerated solutions in that a second passivation "hump" was observed and a loop was formed on the back scan from the noble potentials (Fig. 10). Examination at higher magnifications showed no pitting attack, despite the formation of the back scan loop. A similar experiment conducted on Type 316L stainless steel showed a much more pronounced back scan loop (Fig. 11), and the sample showed signs of localized pitting attack.

Since electrochemical tests were unable to discern any differences between the various types of materials, immersion type tests were conducted in two oxidizing chloride solutions, acidified to a pH $\simeq 2$ with hydrochloric acid or sulfuric acid (4% NaCl + 0.1% Fe$_2$ (SO$_4$)$_3$ plus 0.01 M HCl or plus 0.01 M H$_2$SO$_4$). Various samples were exposed at various temperatures for a 24-h duration then examined for pitting with a light microscope at $\simeq \times 25$. While all samples showed no sign of attack in the sulfuric acid medium up to testing temperatures of 95°C, the hydrochloric acid medium was capable of inducing pitting on all materials. Furthermore, this latter medium clearly distinguished the Type 3a material (that is, ⁶⁰⁄₁₀₀ mesh powder) as inferior material in that it

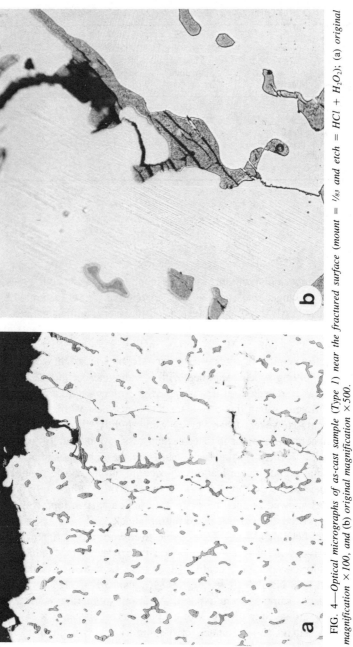

FIG. 4—Optical micrographs of as-cast sample (Type 1) near the fractured surface (mount = 1/83 and etch = HCl + H$_2$O$_2$); (a) original magnification ×100, and (b) original magnification ×500.

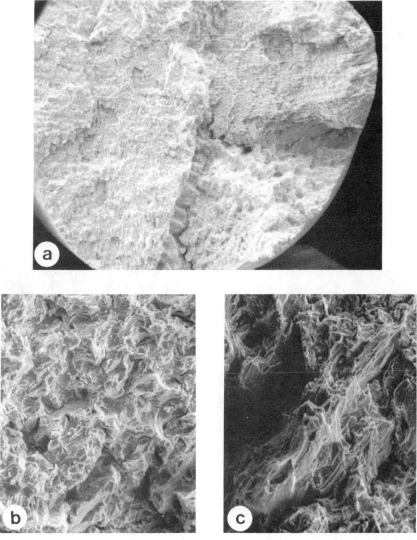

FIG. 5—*Fractographs of as-cast sample (Type 1) after the tension testing; original magnifications* (a) ×16, (b) ×200, *and* (c) ×200.

suffered pitting at temperatures of 40°C and above while the other three types resisted pitting until the solution temperature was raised to 70°C and above. The Type 3b materials (that is, <250 mesh powder) had better pitting resistance than Type 3a (Fig. 12), and all four types of Co-Cr-Mo alloy showed better pitting resistance (that is, higher pitting temperature) than Type 316L stainless steel (Table 3).

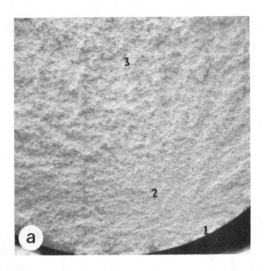

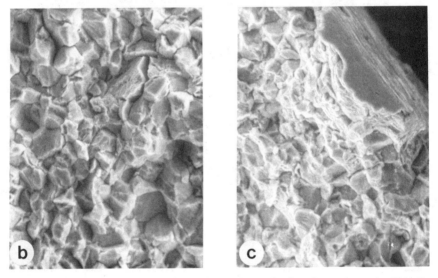

FIG. 6—*Fractographs of extruded plus heat-treated wrought alloy after tension testing (Type 2); original magnifications* (a) ×16, (b) ×500, *and* (c) ×500.

Fatigue/Corrosion Fatigue

The rotating beam fatigue test results are plotted in Fig. 13. The results show the cast material to have a much lower air fatigue limit, 220 MN/m² (32 ksi), than those of the wrought, Type 3a and Type 3b materials (496, 455, and 620 MN/m², respectively). Fractographs of the Type 1 and 3b materials can be compared in Figs. 14 and 15. Cracks were observed along stacking faults with

FIG. 7—*Fractographs of* $^{60}/_{100}$ *mesh as-extruded powder* (*Type 3*a) *after tension testing; original magnifications* (a) ×*16,* (b) ×*500, and* (c) ×*500.*

secondary cracking appearing along annealing twins in Stage I cracks. Fatigue striations, characteristic of Stage II crack growth were also observed. The final failure section exhibited a mixed mode involving intergranular and microvoid failure that is typical of high-strength, precipitation-hardened material.

Upon repeating the fatigue tests on specimens exposed to the Ringer's solution, the fatigue strengths of all the materials were affected. Shorter fatigue lives were observed in the aqueous environment, as illustrated in Fig. 16 for the cast material. Similarly, samples from the wrought and wrought-P/M (Types 3*a* and

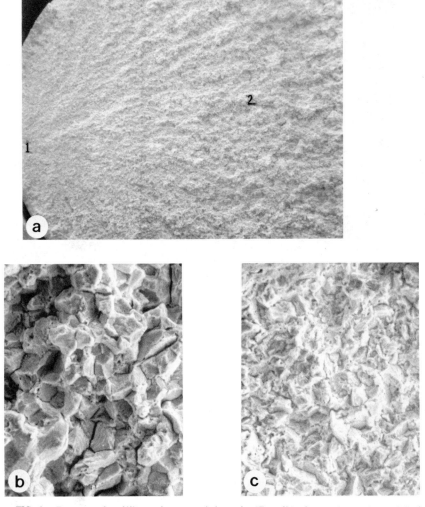

FIG. 8—*Fractographs of $^{250}/_D$ mesh as-extruded powder (Type 3*b) *after tension testing; original magnifications* (a) ×16, (b) ×500, *and* (c) ×500.

3*b*) materials failed in the Ringer's solution at cyclic stresses below the corresponding air fatigue limits (Table 4).

Stress Corrosion Cracking (SCC)

Slow strain rate tests (SSRT) were conducted using specimens with reduced section of 0.191 cm diameter by 1.905 cm length. A slow crosshead speed of 0.0178 cm/cm/h (average strain rate of $\approx 2.6 \times 10^{-6}$ s^{-1}) was adopted for the

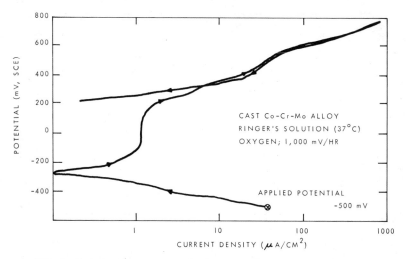

FIG. 9—*Typical polarization curve for Co-Cr-Mo alloys in an aerated solution.*

test. After fracture, the specimens were mounted in epoxy and polished to delineate a longitudinal cross section.

The tests were run in boiling 30% $MgCl_2$ and in the Ringer's solution at 37°C. The measurable parameters (percent elongation, ultimate strength, and test duration) were compared to those in air for the four types of the Co-Cr-Mo alloy (Table 5). As discussed earlier [20], these parameters were not very practical in indicating whether SCC occurred or not. The metallographic examination revealed that in $MgCl_2$ solution, all specimens suffered SCC except the cast product. The short test duration (9 h for cast versus 17 h for wrought) gave the

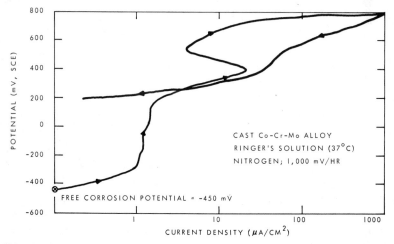

FIG. 10—*Typical polarization curve for Co-Cr-Mo alloys in a deaerated solution (by bubbling N_2 through solution).*

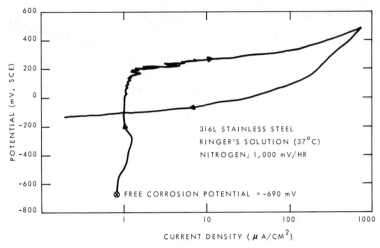

FIG. 11—*Typical polarization curve for wrought Type 316L stainless steel in a deaerated solution.*

impression that the cast material was immune to SCC, when compared to the wrought specimen (Fig. 17). The test in boiling $MgCl_2$ solution was repeated for the cast material using a slower crosshead speed (average strain rate of $\simeq 1.1 \times 10^{-6} \, s^{-1}$) that allowed for a longer test duration of about 14 h. Under these conditions, SCC developed in the cast product (Fig. 18).

FIG. 12—*Comparison of pitting resistance between the two types of Co-Cr-Mo wrought-P/M alloys (4% NaCl + 0.01 M HCl + 0.1% $Fe_2(SO_4)_3$; 40°C; 24 h) (original magnification ×25). (a) Type 3a material ($^{60}/_{100}$ mesh powder), and (b) Type 3b material (<250 mesh powder).*

TABLE 3—*Immersion corrosion tests for pitting resistance in chloride solutions.*

Material	Pitting Temperatures[a] (in °C), 24-h exposure	
	H$_2$SO$_4$ Solution[b]	HCl Solution[b]
316L	30	20
Cast	>95	70
Wrought	>95	70
Wrought-P/M ($^{60}/_{100}$ mesh powder)	>95	40
Wrought-P/M (<250 mesh powder)	>95	70

[a]Temperature at (and above) which pitting is observed; temperature increment = 5°C.
[b]4% NaCl + 0.1% Fe$_2$(SO$_4$)$_3$ + 0.01 M (acid).

No SCC was detected in the Ringer's solution (Table 5). As to the SCC in the 30% MgCl$_2$ medium, the cracking mode was somewhat unexpected. It was predominantly intergranular (Fig. 19) for the wrought material. However, upon repeating the test for the wrought-P/M material in boiling 45% MgCl$_2$, the typical transgranular SCC of the Co-Cr-Mo alloy [*13*] was observed (Fig. 20). Another interesting observation was the presence of "mechanically" induced voids in the wrought-P/M ($^{60}/_{100}$ mesh powder) specimen after SCC testing (Fig. 21) and none in the wrought-P/M (<250 mesh powder) specimens.

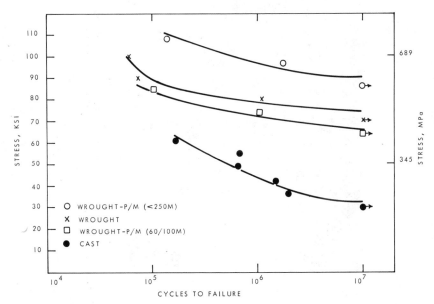

FIG. 13—*Rotating beam fatigue test (ASTM E 466-76) results comparing the cast, wrought, and P/M wrought materials.*

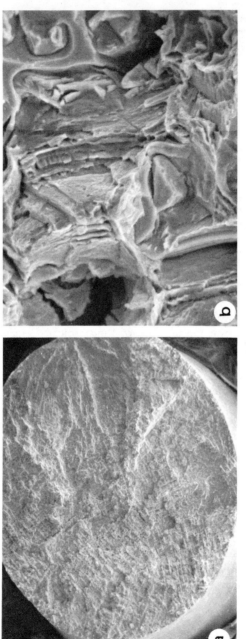

FIG. 14—(a) Overall view of the air-fracture surface in an as-cast material (Type 1) fatigued by rotating beam (original magnification ×13). The dentritic structure can be identified on the fracture surface. (b) As-cast material (Type 1) (original magnification ×1000). Striations can be seen in the dentritic structure also cracking around carbide particles.

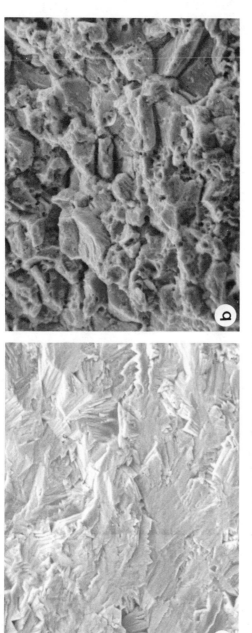

FIG. 15—(a) Air fatigue crack initiation area in Type 3b material (P/M) (original magnification ×500). The striations appear to be crystallographic, typical of Stage I fatigue in low, stacking fault alloys. (b) Air fatigue in Type 3b material (P/M) near the final area of fracture (original magnification ×500). Note the fracture mode is intergranular similar to overload tensile fracture. A few striations are seen. The voids are probably due to carbide particles.

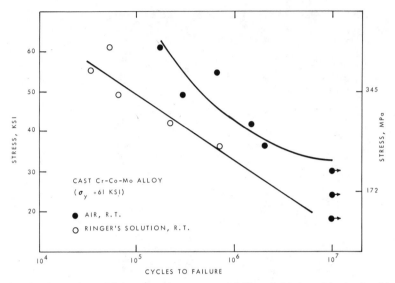

FIG. 16—*Comparison of fatigue lives for cast material (Type 1) in air and in simulated body solution.*

Discussion

Preliminary evaluations of cast, wrought, and wrought P/M versions of Co-Cr-Mo bioimplant alloys (ASTM F 75-76 type materials) have shown significant advantages for the wrought-processed materials. The coarse, cast structure severely limits ductility as well as strength. By reducing the carbon content of this alloy, it is possible to produce this alloy as a hot-worked product with marked improvements in strength and ductility. Processing by P/M techniques does not require the reduction of carbon level while providing similar properties to the cast and wrought product. The considerably higher solidification rate of the gas-atomized powder over the cast ingot allows for a very fine carbide size in the wrought product even though the higher carbon level is present.

The effect of particle size on mechanical properties of the two P/M materials (Types 3a and 3b) is striking. Two differences in these materials were noted. First, the coarser powders have lower cooling rates than the finer powder [20]. This can result in coarser primary carbides formed during solidification. Secondly, the coarser powder fraction had larger average size of nonmetallic inclusions due to the screening process. Since nonmetallic inclusions were observed in both Type 3a and 3b materials, the larger size of the inclusions in Type 3a is a likely cause for mechanical property limitations. This difference in inclusion size appears to be the major cause of the lower ductility in Type 3a, especially when "mechanically" induced voids were only seen in Type 3a specimens (Fig. 19). Grain size and yield strengths for the two materials are very similar, while the effects of inclusion size on ductility are well known [21].

TABLE 4—*Results of fatigue testing in air and in simulated body fluid.*

| Applied Stress, MN/m² (ksi) | Number of Cycles to Failure | |
	Air	Ringer's Solution
	CAST	
420 (61)	0.17×10^6	0.05×10^6
379 (55)	0.69×10^6	0.03×10^6
338 (49)	0.28×10^6	0.06×10^6
296 (43)	1.5×10^6	0.22×10^6
248 (36)	69×10^6	0.61×10^6
206 (30)	$>10 \times 10^6$	
165 (24)	$>10 \times 10^6$	3.9×10^6
124 (18)	$>10 \times 10^6$	$>10 \times 10^6$
	WROUGHT-P/M ($^{60}/_{100}$)	
586 (85)	0.11×10^6	
510 (74)	1.0×10^6	
441 (64)	$>10 \times 10^6$	1.4×10^6
365 (53)	$>10 \times 10^6$	2.1×10^6
290 (42)	$>10 \times 10^6$	$>10 \times 10^6$
	WROUGHT-P/M (<250)	
741 (108)	0.2×10^6	0.1×10^6
668 (97)	1.8×10^6	0.2×10^6
593 (86)	$>10 \times 10^6$	0.7×10^6
524 (76)	$>10 \times 10^6$	0.7×10^6
448 (65)	$>10 \times 10^6$	0.2×10^6
372 (54)	$>10 \times 10^6$	$>10 \times 10^6$
	WROUGHT	
689 (100)	0.06×10^6	
620 (90)	0.7×10^6	
551 (80)	1.1×10^6	
482 (70)	$>10 \times 10^6$	

TABLE 5—*Slow strain rate (average strain rate $\approx 2.6 \times 10^{-6}/s$) SCC test results for the Co-Cr-Mo alloy.*

Alloy	Environment	Elongation, %	Ultimate Tensile Strength, MN/m² (ksi)	Test Duration, h	Observation
Cast	air	5	545 (79)	9	. . .
	boiling 30% MgCl₂	3	538 (78)	7	No SCC
	boiling 30% MgCl₂[a]	1	607 (88)	14	SCC
	ringer, 37°C	6	799 (116)	16	No SCC
Wrought	air	24	1180 (171)	35	. . .
	boiling 30% MgCl₂	10	785 (114)	17	SCC
	ringer, 37°C	39	1253 (182)	50	No SCC
Wrought-P/M	air	12	1157 (168)	27	. . .
($^{60}/_{100}$ mesh	boiling 30% MgCl₂	3	661 (96)	10	SCC
powder)	ringer, 37°C	19	1116 (162)	30	No SCC
Wrought-P/M	air	29	1212 (176)	43	. . .
(<250 mesh)	boiling 30% MgCl₂	5	744 (108)	14	SCC
	ringer, 37°C	36	1267 (184)	48	No SCC

[a]Average strain rate $1.1 \times 10^{-6}/s$.

FIG. 17—*Cross section of SSRT specimens after testing in boiling 30% MgCl$_2$; (a) cast Type 1 and (b) wrought Type 2.*

The results of the fatigue testing also show the inferiority of the coarser powder product. While the yield strengths of Type 3a and 3b materials were similar, the fatigue limit of Type 3a was significantly less than that of Type 3b (that is, about 175 MN/m^2 (25 ksi) lower). Once again, the effect of inclusions is likely to be the dominant one [22,23]. The fine-powder product has the excellent fatigue behavior of the extensively hot-worked-wrought alloy (Type 2) indicating that under these test conditions little effect of inclusions is observed.

The limited corrosion fatigue data show a reduction of the fatigue strengths in the Ringer's solution when compared to those established in air. Since none of the materials showed any obvious signs of uniform or localized corrosion in the Ringer's solution, such reduction of the fatigue strength is attributed to a surface wetting effect [24]. During the corrosion fatigue testing, adsorbed species (water or chlorides) induce plastic deformation and enhance slip-step formation, thus causing premature failures at lower cyclic stress levels.

The electrochemical polarization tests fail to reveal any differences between

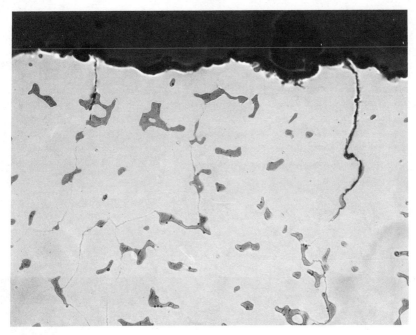

FIG. 18—*Micrograph of cast Co-Cr-Mo alloy after stress corrosion test showing the cracking (original magnification ×200).*

the four types of the Co-Cr-Mo alloy. The polarization curves in the Ringer's solution (aerated or deaerated) all look about the same. In such highly passive materials, no major variation is expected to alter the various electrochemical parameters (passive current density, breakdown potential), even though a relatively slow scan rate of 1000 mV/h was used.

The immersion pitting temperature test in a slightly aggressive chloride environment appears more able to distinguish between the various localized corrosion resistances of those four types of materials (Table 3). The Type 3*a* material (processed from coarse powders) shows a much lower resistance to pitting attack than do the other three types. Furthermore, this immersion-type test is also more selective because it shows that Type 3*b* material (processed from fine powders) has a corrosion rate lower than that of the cast material. Still, the relevancy of testing at temperatures above ≃40°C is questionable for human body implants, and a more adequate immersion-type test may be necessary.

The apparent insensitivity of the cast material to the SSRT in boiling 30% $MgCl_2$ is definitely related to the strain rate and to the test duration. Since the cast specimens show very low ductility, the short test duration and the limited plastic deformation make the cast material appear resistant to stress cracking unless slower strain rates are used (Table 5). It has been known that susceptibility to SCC is time dependent and that plastic deformation is often a major factor

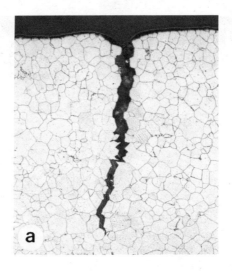

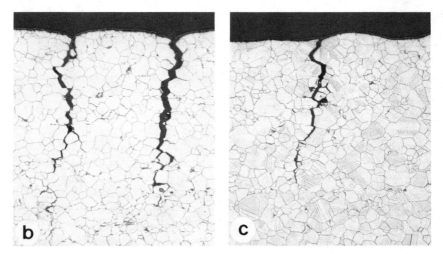

FIG. 19—*Micrographs of wrought Co-Cr-Mo alloys after stress corrosion tests in boiling MgCl₂ showing intergranular stress corrosion cracks (original magnification × 200); (a) Type 3a, (b) Type 3b, and (c) Type 2.*

inducing stress cracking. Hence, the more ductile wrought and P/M-processed materials appear in this SSRT to be less resistant to SCC than the mechanically inferior cast material. This observation supports the belief that one should be careful on how to use the SSRT [*11*] and on how to interpret the results.

Finally, it is of interest to note that the boiling 30% $MgCl_2$ solution induces intergranular SCC. Since cobalt-base, low-nickel/low-molybdenum alloys suffer excessive dissolution in reducing chloride environments, the observed intergranular cracks may rather be the result of a stress assisted intergranular corrosion

FIG. 20—*Micrographs of Type 3a material after stress corrosion cracking tests in two concentrations of MgCl₂ showing stress corrosion cracking mode change (original magnification ×200); (a) 30% MgCl₂, boiling, and (b) 45% MgCl₂, boiling.*

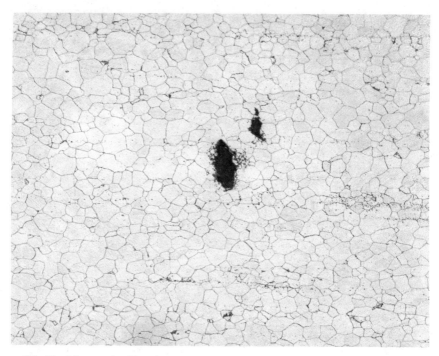

FIG. 21—*Micrograph of Type 3a material after slow strain rate, stress corrosion cracking test. Mechanically induced voids are present (original magnification ×200).*

rather than SCC. The 30% $MgCl_2$ may not be sufficiently concentrated to cause typical transgranular SCC (for example, like 45% $MgCl_2$ solution does), and the samples only corrode in this less concentrated solution. The imposition of a dynamic strain thus may help to open up the most active corrosion path, that is, the grain boundary, and lead to the observed intergranular cracks.

Conclusions

1. Preliminary tests have been conducted that have characterized Co-Cr-Mo alloys intended for bio-implant applications requiring high strength, fatigue resistance, and environmental resistance. These tests will be useful in evaluating new or improved materials for these applications.

2. Direct extrusion of gas atomized powders of fine size (<60 μm) results in material equivalent to the commercially available low-carbon alloy. Similar processing of coarser fractions of this powder (150 to 250 μm) yields material with generally degraded properties.

3. Electrochemical polarization tests are not successful in delineating fine differences between the various types of the Co-Cr-Mo alloy. Immersion-pitting temperature tests are capable of distinguishing between materials processed from fine or coarse powders.

4. Susceptibility to stress corrosion cracking of these materials has been demonstrated in boiling 30% $MgCl_2$. The slow strain rate test method suggests that the more ductile wrought and P/M-processed materials are more susceptible to SCC than the low ductility, cast product. In the Ringer's solution at 37°C, no SCC is detected on all materials tested.

References

[1] R. T. Rylee, Wright Manufacturing Company, Memphis, TN.

[2] Hodge, F. G. and Lee, T. S., III, *Corrosion III*, Vol. 31, March 1975, p. 3.

[3] Mears, D. C., *International Metals Reviews*, Review No. 218, June 1977, p. 119.

[4] Crimmins, D. C., *Journal of Metals*, Jan. 1969, p. 38.

[5] Bardos, D. I. in *Current Concepts of Internal Fixation of Fractures*, H. K. Unthoff, Ed., Springer-Verlay, New York, 1980, p. 111.

[6] Fraker, A. C. and Ruff, A. W., *Journal of Metals*, May 1977, p. 22.

[7] Syrett, B. C., *Corrosion/76*, Paper No. 116, National Association of Corrosion Engineers, 1976.

[8] Thomas, C. R. and Robinson, F. P. A., *Journal South African Institute of Mining and Metallurgy*, Nov. 1976, p. 93.

[9] Brettle, J. and Hughes, A. N., *English Medicine*, Vol. 7, No. 3, 1978, p. 142.

[10] Lisagor, W. B., *ASTM Standardization News*, May 1975, p. 20.

[11] Williams, D. F., *Metals and Materials*, Sept. 1972, p. 388.

[12] Jones, R. L., Wing, S. S., and Syrett, B. C., *Corrosion*, Vol. 34, No. 7, 1978, p. 226.

[13] Uhlig, H. H. and Asphahani, A. I., *Materials Performance*, Vol. 18, No. 11, 1979, p. 9.

[14] White, W. E. and LeMay, I., *Microstructural Science*, Vol. 3, 1975, p. 9.

[15] Hirschhorn, J. S. and Reynolds, J. T., *Research in Dental and Medical Materials*, Plenum Press, New York, 1969, p. 137.

[16] Pilliar, R. M., MacGregor, D. C., MacNab, I., and Cameron, H. U. in *Modern Developments in Powder Metallurgy*, H. H. Hansner and P. W. Tansenslat, Eds., Vol. 11, Metal Powder Industries Federation, Princeton, NJ, 1977, p. 263.

[17] Kelto, C., Kelsey Hayes Corporation; private communication.

[18] Cahoon, J. R. and Cheung, C. T. F., *Canadian Metallurgical Quarterly*, Vol. 21, No. 3, 1982, pp. 289–292.

[19] Asphahani, A. I. in *Stress Corrosion Cracking—The Slow Strain-Rate Technique, ASTM STP 665*, Ugiansky and Payer, Eds., American Society for Testing and Materials, Philadelphia, 1979, p. 279.

[20] Smugeresky, J. E., "Characterization of a Rapidly Solidified Iron-Base Superalloy," SAND81-8865, Sandia National Laboratories, Albuquerque, NM, 1981.

[21] Lankford, J., "Effect of Oxide Inclusions on Fatigue Failure," *International Metal Reviews*, Sept. 1977.

[22] Murakami, Y. and Endo, M., *Engineering Fracture Mechanics*, Vol. 17, No. 1, 1983, pp. 1–15.

[23] Hyzak, J. M., "The Effect of Defects on the Fatigue Crack Initiation Process in Two P/M Superalloys," Ph.D. thesis, Carnegie-Mellon University, Pittsburgh, PA, 1980.

[24] Asphahani, A. I. and Sridhar, N., *Corrosion*, Vol. 38, No. 11, 1982, p. 587.

DISCUSSION

N. D. Greene[1] *(written discussion)*—What cyclic frequency was used for the corrosion fatigue tests? It is important to recognize that corrosion fatigue re-

[1] Department of Metallurgy, University of Connecticut, Storrs, CT 06268.

sistance is very dependent on cyclic frequency. Fatigue performance decreases with decreasing cyclic frequency.

P. Kumar, A. J. Hickl, A. I. Asphahani, and A. Lawley (authors' closure)— The corrosion fatigue testing was done at 1000 cycles/minute.

*Henry R. Piehler[2] (written discussion)—*Did you do any fatigue testing of HIPed powder compacts?

P. Kumar, A. J. Hickl, A. I. Asphahani, and A. Lawley (authors' closure)— No, we did not test HIPed powder compacts.

*L. Gustavson[3] (written discussion)—*Were the fatigue and tensile data presented for wrought Co-Cr-Mo alloy obtained from bar stock?

P. Kumar, A. J. Hickl, A. I. Asphahani, and A. Lawley (authors' closure)— Yes, commercial wrought bar made by vacuum melting, forging, and bar rolling were used. These bars were extruded and heat treated prior to testing.

*A. Craig Hood[4] (written discussion)—*Your data showed the improved fatigue life of the wrought-P/M <250 to be greater than either wrought or cast materials. What was the relative fatigue life of the various process histories in the *in vitro* corrosion metal?

P. Kumar, A. J. Hickl, A. I. Asphahani, and A. Lawley (authors' closure)— Table 4 gives fatigue lives of various materials in Ringer's solution. It is apparent that the P/M-processed material of finer particle (<250 mesh) size had longer life than any other material.

*Manfred Semlitsch[5] (written discussion)—*Is the increase in carbon (0.06 to 0.22) in P/M wrought Co-Cr-Mo alloy responsible for the increase in wear resistance of this newly produced alloy? Can this alloy composition be used, for example, for total hip balls, and is a metal/metal pairing for joint end prosthesis possible?

P. Kumar, A. J. Hickl, A. I. Asphahani, and A. Lawley (authors' closure)— We do not have any information about the effect of carbon content on the corrosion resistance of Co-Cr-Mo alloy. One would expect the wear resistance to improve with the increase in carbon.

This alloy is already in use for the total hip prosthesis. The ball of the prosthesis is made of this alloy. However, the cup in which the ball rotates is made of plastic. We do not have any information on metal/metal pairing of this alloy.

[2]Department of Metallurgical Engineering and Materials Science, Carnegie-Mellon University, Pittsburgh, PA 15213.
[3]Howmedica, Inc., Rutherford, NJ 07070.
[4]Advanced Metallurgical Products Division, SPS Technologies, Benson East, Jenkintown, PA 19046.
[5]Sulzer Bros., Ltd., Dept. R & D, CH-8401 Winterthur, Switzerland.

John P. Sheehan,[1] *Charles R. Morin,*[1] *and Kenneth F. Packer*[1]

Study of Stress Corrosion Cracking Susceptibility of Type 316L Stainless Steel *In Vitro*

REFERENCE: Sheehan, J. P., Morin, C. R., and Packer, K. F., **"Study of Stress Corrosion Cracking Susceptibility of Type 316L Stainless Steel *In Vitro*,"** *Corrosion and Degradation of Implant Materials: Second Symposium, ASTM STP 859*, A. C. Fraker and C. D. Griffin, Eds., American Society for Testing and Materials, Philadelphia, 1985, pp. 57–72.

ABSTRACT: Post-failure analyses of Type 316L stainless steel implants have frequently implicated stress corrosion cracking as the mode of failure. If stress corrosion cracking of this stainless steel is likely *in vivo*, this material would have questionable utility for such applications. However, this material is widely used by manufacturers and approved by government regulating agencies. This study was designed to critically evaluate this issue. No susceptibility to stress corrosion cracking of 316L was found in manufactured implants or in specimens from implant quality material *in vitro* by both static loading and slow strain rate testing techniques. An adjunct *in vitro* fatigue study disclosed multiple fatigue cracks with branching tendencies. It is concluded that crack branching and secondary cracking in 316L implants are not adequate indicia of stress corrosion cracking of implants.

KEY WORDS: implant materials, fatigue (materials), stainless steels, corrosion, cracking, stress, strain rate

The failure of stainless steel orthopedic implants *in vivo* is a recognized occurrence, and has been the subject of failure analyses by a number of investigators of varied technical and medical backgrounds [*1–6*][2] The results of the investigations frequently implicate environmentally-affected mechanisms of crack initiation and growth—that is, stress corrosion cracking and corrosion fatigue. The rationale for this is that the implant is certainly subjected to both steady and cyclic stresses, both residual and applied, while in the relatively hostile body environment, and that the susceptibility of austenitic stainless steels to such failure in solutions containing chloride ion is "well known."

The stress corrosion cracking of austenitic stainless steels in chloride solutions was first recognized in the late 1940s. It was also recognized shortly after this

[1]Metallurgist, vice president, and president, respectively, Packer Engineering Associates, Inc., Naperville, IL 60566.

[2]The italic numbers in brackets refer to the list of references appended to this paper.

period that the main problem appears to arise during exposure of stressed stainless steels to high-temperature solutions containing a high concentration of chloride, particularly where the pH due to either hydrolysis or to deliberate or inadvertent acid addition is below neutral. A well-known magnesium chloride test came into being to determine susceptibility of different austenitic stainless steels to stress corrosion cracking because of this tendency. In this test, the specimen of steel to be tested is stressed in a solution of magnesium chloride, approximately 45% by weight, which is boiling at 154°C. Rapid failures of susceptible materials ensue in this solution (Fig. 1). The advent of the nuclear reactor with its exceedingly high pressures of water, and the likely use of Type 304 stainless steel in critical components, led to the testing of Type 304 and other candidate stainless steels in water at temperatures up to 315.6°C (600°F), with traces of chloride and oxygen. In this research, it was clear that as the temperature increases substantially, the threshold level of chloride to initiate and propagate stress corrosion cracking decreases dramatically, and that with traces of oxygen at around the 1 ppm level, less than 10 ppm of chloride will allow stress corrosion cracking to occur.

At the other extreme, Acello and Greene [7] showed that by using strong sulfuric acid solutions with different amounts of chloride added, the austenitic stainless steels can be made to stress corrosion crack at room temperature. Bianchi

FIG. 1—*Stress corrosion cracks in tension surface of 316L stainless steel intramedullary nail tested in three-point static bending near yield strength in boiling MgCl₂. Etchant: 10% oxalic acid— electrolytic (original magnification, ×50).*

and coworkers [8] have expanded this and have shown that stress corrosion cracking can be made to occur in refrigerated hydrochloric acid at near 0°C. Note, however, that the pH in these situations is very low.

To briefly summarize this information, it appears that the austenitic stainless steels containing 10 to 20% of nickel are very susceptible to stress corrosion cracking at elevated temperatures in the presence of relatively small quantities of chloride. As the temperatures decrease, either the chloride concentration must be increased or the pH must be decreased in order to maintain susceptibility to stress corrosion. At room temperature, exceedingly acid solutions under specific conditions must be utilized in order for stress corrosion to occur. It would thus appear on this basis that the stress corrosion cracking of Type 316L stainless steel in physiological environments would be very unlikely. However, because of apparent fractographic and morphological similarities between implant failures observed *in vivo* and those shown in stress corrosion cracking investigations of stainless steels, the stress corrosion cracking failure mode has frequently been implicated in the *in vivo* failure of orthopedic implants [1–6].

Relatively few critical studies have been published with the aim being a definition of the conditions necessary to cause stress corrosion cracking versus corrosion fatigue in the body environment. Gilbertson [9] recently presented some work that showed that even at 66°C and the pH adjusted to 2 to 3 with hydrochloric acid, Type 316 low carbon, vacuum melt (LVM) four-point bend specimens did not fail in 15 000 h of testing. Similarly, Jones, Wing, and Syrett [10] showed that under the conditions that they utilized, Type 316L cold-worked bars would not fail by stress corrosion cracking in a physiological saline solution.

Objectives

The purpose of the experiments to be described in this paper is to determine whether or not Type 316L stainless steel implants and implant-quality material are susceptible to stress corrosion cracking *in vitro* under conditions that may exist *in vivo*. Furthermore, one facet of the program included both cold-worked and annealed material, inasmuch as these conditions are encompassed in ASTM Specification for Stainless Steel Bar and Wire for Surgical Implants (F 55-76) and ASTM Specification for Stainless Steel Bars and Wire for Surgical Implants (Special Quality) (F 138-76). Also, the crack morphology of specimens tested in fatigue in the simulated human environment was of interest in order to compare these features with those typical of stress corrosion cracking.

Materials

1. Type 316L stainless steel of commercial quality in the annealed condition and cold rolled 80%.
2. Type 316 LVM[3] stainless steel of implant quality in 0.79 cm[4] ($^5/_{16}$-in.) round rod for slow strain rate tension testing.

---

[3]Low carbon, vacuum melt of ASTM F 138-76 quality.
[4]Original measurements made in English units.

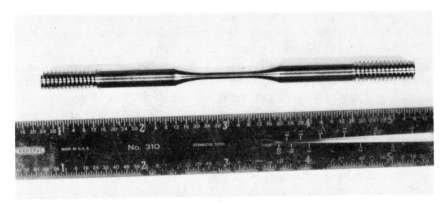

FIG. 2—*Slow strain rate tension specimen.*

3. Type 316 LVM stainless steel Schneider intramedullary implant nails for three-point static bend testing.

4. Type 316 LVM stainless steel implant quality bars of 4.37-cm (1.72-in.) diameter for notched fatigue testing and static cantilever bending tests.

Testing Procedures and Results

Slow Strain Rate Tension Tests

Threaded end tensile specimens were machined to 0.64-cm (0.25-in.) diameter from 0.79-cm (5/16-in.) diameter implant-quality rod. A 1.6-cm (5/8-in.) long gage section was reduced to 0.25-cm (0.1-in.) diameter (Fig. 2). Prior to testing, the specimens were electropolished by a manufacturer of implants. The procedure employed was identical with that used for surgical implants provided to the medical community.

Three specimens were pulled to failure in lactated Ringer's[5] solution (aerated) at 37 ± 1°C (98 ± 2°F) at strain rates of 10^{-5}, 10^{-6}, and 10^{-7} cm/cm/s (in./in./s). Two additional specimens were tested in the same aerated solution adjusted to a pH of 2 with hydrochloric acid (HCl) or with ferric chloride (FeCl$_3$) at a strain rate of 10^{-6} cm/cm/s (in./in./s). The results are shown in Table 1.

There is no significant change in strength or ductility as a function of strain rate or pH level. If the material was susceptible to stress corrosion cracking in the test environment for the range of conditions investigated, a decrease in strength or ductility or both would be found with decreasing strain rate or at the lower pH values.[6] This has not occurred. Furthermore, photomicrographs of longitudinal sections through the fracture ends of the test specimens show ductile cup-cone fractures and no evidence of secondary cracking (Fig. 3). In assessing

[5]A physiological solution containing mixtures of sodium, potassium, and calcium chlorides and bicarbonates that approximate the electrolyte and corrosive environment of the human body; pH = 6.5.

[6]Mon et al [11] have shown that Type 316L stainless steel suffers a 50% decrease in fracture strain at 6.4 × 10^{-6} cm/cm/s (in./in./s) strain rate when tested in 123°C (253°F) MgCl$_2$.

TABLE 1—*Results of slow strain rate tension tests of implant-quality 316L stainless steel.*

Specimen	Strain Rate, cm/cm/s (in./in./s)	Ultimate Tensile Strength, MPa	Ultimate Tensile Strength, psi	Elongation, %	Reduction in Area, %	Time to Failure, h
3A	10^{-5}	956	138 000	27	43	4.7
3B	10^{-6}	947	137 000	31	49	54
3C	10^{-7}	930	135 000	24	47	432
3D[a]	10^{-6}	934	136 000	28	43	46.5
3E[b]	10^{-6}	947	137 000	27	46	47.6

[a]HCl acidified to pH of 2.
[b]FeCl₃ acidified to pH of 2.

these data, it should be noted that the times to fracture range from about 4½ h to 432 h, depending upon the strain rate selected.

Static Bend Tests

Schneider intramedullary nails, 0.79-cm ($^5/_{16}$-in.) diameter and certified as 316 LVM stainless steel, were procured from a manufacturer,[7] and a 10.16-cm (4-in.) long specimen was cut from the central section of each nail. Four such specimens were strain-gaged and loaded in three-point bending fixtures, constructed wholly of 316L stainless steel, to obtain a 0.1% yield strain. The strain gages were removed and the entire assemblies placed in aerated, circulated Ringer's solution controlled to 38 ± 1°C (100 ± 2°F) and pH of 6. The Ringer's solution was changed biweekly, and the static bend tests were continued for a period of 4320 h (six months). At this time, microscopic examination (×7 to ×40) and dye penetrant checks were made on the tension surfaces of each specimen with no evidence of cracking found. Photographs illustrating the tension surfaces are provided in Fig. 4.

Additional static bend tests utilizing square, notched specimens similar to the standard Izod impact test specimen were prepared from the 4.37-cm (1.72-in.) diameter bars and tested in cantilever bending.

Based upon the mechanical properties of the material determined as follows,

Ultimate Tensile Strength	0.2% Yield Strength	Elongation	Reduction in Area
888 MPa (129 000 psi)	765 MPa (111 000 psi)	20%	77%

the specimens were stressed to 0.2% yield strength with the stress at the root of the notch being considerably higher. Again, these tests were conducted in aerated, circulated Ringer's solution controlled to 38 ± 1°C (100 ± 2°F) and at pH levels of 6, 4, and 1, the latter two achieved by additions of HCl. A schematic diagram

[7]Zimmer, Inc., Warsaw, IN 46580.

FIG. 3—*Polished and etched section through fracture of Specimen 3A in Table 1. Etchant: 10% oxalic acid—electrolytic (original magnification, ×50).*

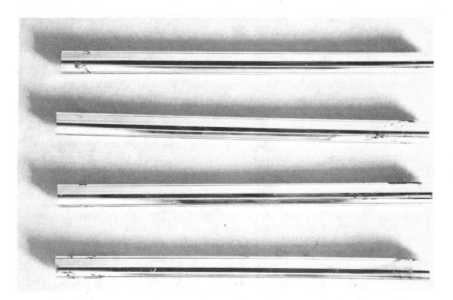

FIG. 4—*Photograph of specimens from intramedullary nails after six months in Ringer's solution 38°C (100°F) under static bending stress of 0.1% yield.*

of the test cell and specimen configuration is provided in Fig. 5. Figure 6 shows the test stand and loading arrangement. The static bend tests utilized the test setup shown in Fig. 6 without the motor, eccentric drive, and gear reducer. The latter equipment was used for the fatigue tests to be described later. These static bend tests were continued for eight months (approximately 5760 h) with the Ringer's solution being changed weekly. When the tests were discontinued, there was no evidence of cracking found by means of microscopic examination and dye penetrant testing in the highly stressed notches of the specimens.

Fatigue Tests

With the same type of notched specimen and loading arrangement shown in Figs. 5 and 6, several fatigue tests of the 316 LVM material from the 4.37-cm (1.72-in.) diameter bars were conducted in Ringer's solution. Stresses were selected to provide low- and high-cycle fatigue cracking. One specimen that failed in 15 000 cycles and another that required over one million cycles were sectioned and prepared for metallographic examination. The photomicrographs in Figs. 7 and 8 illustrate the morphology of the fatigue cracks. It is noted that secondary fatigue cracks are present, and many of these exhibit some crack

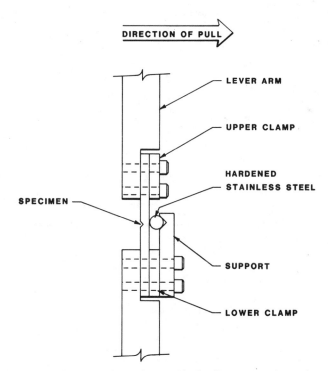

FIG. 5—*Specimen and loading device that provides bending stress at the notch root radius.*

branching characteristics similar to that associated with stress corrosion cracking in other environments.

Similar fatigue testing of a Schneider intramedullary nail in Ringer's solution (Figs. 9–11) also produced secondary cracks of identical branching characteristics on the tension side of the nail (Fig. 12). These branching characteristics are attributed to a high ΔK associated with a rapidly advancing fatigue crack [12].

Effects of Cold Work

Other investigators have reported a deleterious effect of cold work on the times to failure of 316 stainless steel subjected to boiling 42% $MgCl_2$ [13]. Increasing amounts of localized residual stresses produced by the cold working were reported to be responsible for this behavior. Whether or not such a condition

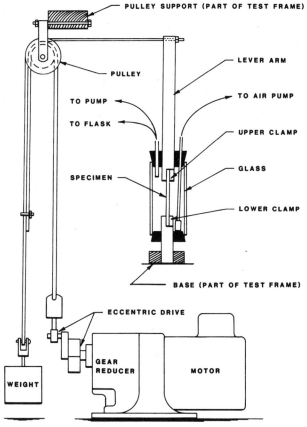

FIG. 6—*Mechanical testing apparatus that provides cyclic loading to the specimen during the fatigue tests and static loading during the stress corrosion tests. The details of the specimen arrangement are shown in Fig. 5.*

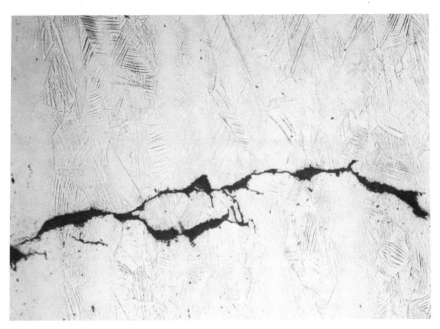

FIG. 7—*Tip of fatigue crack in partially cracked specimen after 15 000 cycles of stress. Note branching characteristics. Etchant: 10% oxalic acid—electrolytic (original magnification, ×200).*

FIG. 8—*Subsidiary cracks branching off main fatigue crack in specimen tested for one-million cycles. Etchant: 10% oxalic acid—electrolytic (original magnification, ×400).*

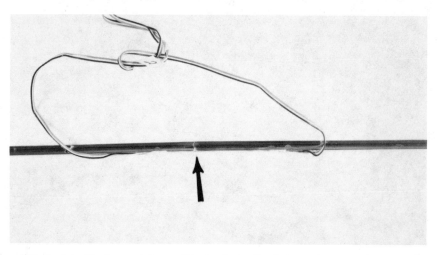

FIG. 9—*Schneider intramedullary nail broken by bending fatigue in Ringer's solution at 38°C (100°F). Arrow shows fatigue failure.*

FIG. 10—*Fatigue fracture surface of Schneider intramedullary nail.*

FIG. 11—*Scanning electron micrograph of fatigue fracture surface showing numerous secondary cracks branching off main fracture (original magnification, ×200).*

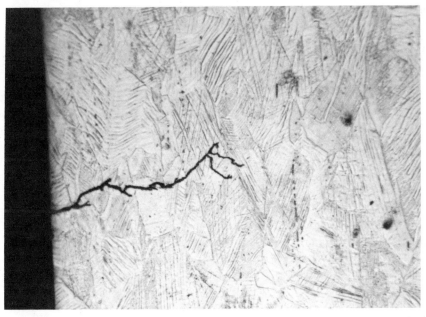

FIG. 12—*Secondary fatigue crack on tension surface of intramedullary nail, adjacent to main fatigue fracture. Etchant: 10% oxalic acid—electrolytic (original magnification, ×200).*

FIG. 13—*Dog-bone tension specimen used to evaluate effects of cold work on 316L stainless steel.*

might influence the behavior of implant material in the human environment has been questioned and evaluated in the present program.

Annealed commercial-quality 316L stainless steel was cold rolled 80% to sheet stock and flat dog-bone type tension specimens (Fig. 13) were machined from the annealed and cold-rolled material. The test specimens were ground and polished to 600 grit finish and degreased in acetone prior to testing. Slow strain rate tension testing was carried out at 10^{-4}, 10^{-5}, and 10^{-6} cm/cm/s (in./in./s) in either Ringer's solution or mineral oil, as shown in Table 2.

These results show no adverse effects of testing in Ringer's solution as compared to mineral oil at the same strain rate (Tests A and B, D and E), or of varying strain rates in Ringer's solution (Tests B, C and D, F and G). Figure 14 is a representative photomicrograph of longitudinal sections through the fractures that showed no evidence of secondary cracks or branching.

Summary and Discussion

The test results from this investigation contribute to and support existing data [9,10] that stress corrosion cracking is not a mode of failure of 316L stainless steel implants *in vitro* and provide no reason to suggest that it might exist *in vivo*. Even with high levels of residual stress from severe cold working of the material, or with acidified Ringer's solution as the environment, there was no evidence of stress corrosion cracks with either the slow strain rate tension tests or the static bending tests conducted beyond the yield strength of the material.

In view of the branching characteristics of the cracks found in the fatigue-tested specimens, it is understandable that some investigators might misinterpret similar cracks found in post-failure analyses of implants as stress corrosion cracks. The authors have also noted these types of cracks in numerous broken implants or prostheses examined in the laboratory where the general characteristics of the fracture surface have been indicative of fatigue.

TABLE 2—Results of slow strain rate tension tests of commercial-quality 316L stainless steel in the severely cold-worked and annealed conditions.

Specimen	Cold Work, %	Strain Rate, cm/cm/s (in./in./s)	Test Medium	Ultimate Tensile Strength, MPa	Ultimate Tensile Strength, psi	Elongation, %	Time to Failure, h
A	80	10^{-4}	mineral oil	1140	165 000	9.4	0.16
B	80	10^{-4}	Ringer's	1120	162 000	9.8	0.17
C	80	10^{-5}	Ringer's	1060	154 000	14	2.4
D	80	10^{-6}	Ringer's	1200	174 000	11.3	19.6
E	80	10^{-6}	mineral oil	1150	167 000	11.9	20.7
F	annealed	10^{-5}	Ringer's	517	75 000	62	10.8
G	annealed	10^{-6}	Ringer's	510	74 000	48	83.3

FIG. 14—*Polished and etched section through fracture of Specimen G in Table 2. Etchant: 10% oxalic acid—electrolytic (original magnification, ×50).*

References

[1] Zapffe, C. A., *Metal Progress*, July 1955, p. 95.

[2] White, W. E. and LeMay, I., *Microstructural Science*, Vol. 3, American Elsevier, 1975, p. 911.

[3] Gray, R. J., *Journal of Biomedical Materials Research Symposium*, Vol. 5, 1974, pp. 27–38.

[4] White, W. E., Postlewhaite, J., and LeMay, I., *Microstructural Science*, Vol. 4, American Elsevier, 1976, p. 145.

[5] Bombara, G. and Cavallini, M., *Corrosion Science*, Vol. 17, Pergamon Press, London, 1977, p. 77–85.

[6] Brettle, J., "Stress Corrosion Cracking of Surgical Implant Materials," DHSS Project A129 (Progress Report), United Kingdom Atomic Energy Authority, AWRE Metallurgy Division, Feb. 1972.

[7] Acello, S. J. and Greene, N. D., *Corrosion*, Vol. 18, Aug. 1962.

[8] Bianchi, G., Cerquetti, A., Mazza, F., and Torchio, S., *Corrosion Science*, Vol. 12, 1972.

[9] Gilbertson, L. N., "Stress Corrosion Cracking of 316 LVM," 4th Annual Meeting of the Society for Biomaterials.

[10] Jones, R. L., Wing, S. S., and Syrett, B. C., "Stress Corrosion Cracking and Corrosion Fatigue of Some Surgical Implant Materials in a Physiological Saline Environment," *Corrosion*, Vol. 34, No. 7, July 1978.

[11] Mom, A. J., Dencher, R. T., Wekken, C. J., and Schultze, W. A. in *Stress Corrosion Cracking: The Slow Strain Rate Technique*, ASTM STP 665, American Society for Testing and Materials, Philadelphia, 1979, pp. 305–319.

[12] Gilbertson, L. N., "The Relationship of Microstructure to Fracture Topography in Orthopedic Alloys," *Microstructural Science*, Vol. 4, 1976.

[13] Kraft, G., Anacker, J., Saxer, R., and Myers, J., "Threshold Stress and Incubation Period in Stress Corrosion of Types 302 and 316 Wire in Boiling MgCl₂," *Corrosion*, Vol. 21, No. 6, June 1965.

DISCUSSION

N. D. Greene[1] *(written discussion)*—Were the branching cracks observed during the corrosion fatigue tests inter- or transgranular in nature?

J. P. Sheehan, C. R. Morin, and K. F. Packer (authors' closure)—The main corrosion fatigue cracks were transgranular as were the secondary cracks (see Figs. 7, 8, 11, and 12).

Henry R. Piehler[2] *(written discussion)*—Were you able to see fatigue striations on the secondary cracks you observed? (If they were present, it would settle the issue as to whether fatigue as opposed to stress corrosion causes these secondary cracks to form.)

J. P. Sheehan, C. R. Morin, and K. F. Packer (authors' closure)—The secondary cracks observed in the corrosion fatigue specimens were not opened for examination in the SEM to search for striations. No clearly defined striations were observed on the main crack faces so the likelihood of finding them on the secondary cracks was deemed to be low. We do not consider the absence of striations to be a reliable criterion for determining failure mode. However, the conditions necessary to cause striation formation on the fracture surfaces of corrosion fatigue cracks requires further research.

Jonathan Black[3] *(written discussion)*—Since many of these 316L parts are used in fracture fixation and thus undergo repeated autoclaving performance, how would you expect your results to be affected if you had performed one to two dozen autoclave cycles first?

J. P. Sheehan, C. R. Morin, and K. F. Packer (authors' closure)—Autoclaving implants for sterilization purposes prior to insertion is not likely to affect performance in the body or in our tests. The passive film on the implant may be modified somewhat by exposure to high-temperature steam, but unless the steam is contaminated with significant levels of chloride or other species known to adversely affect passivity, the autoclaving cycles are judged to be harmless. It must be remembered that for cracking to occur, the passive film will be disrupted either by wear, fretting, or pitting and the cyclic plasticity of the fatigue inducing stresses. These effects probably greatly overshadow the effect of any passive film modification during routine autoclaving.

H. Ravi Shetty[4] *(written discussion)*—What is the purpose of using mineral oil in slow strain rate stress corrosion testing of 316L stainless steel? Don't you

[1]Department of Metallurgy, University of Connecticut, Storrs, CT 06268.
[2]Department of Metallurgical Engineering and Materials Science, Carnegie-Mellon University, Pittsburgh, PA 15213.
[3]School of Medicine, University of Pennsylvania, Philadelphia, PA 19104.
[4]Zimmer, Inc., Warsaw, IN 46580.

think that it is more realistic to compare the stress corrosion susceptibility data of the alloy in Ringer's solution with tests in air?

J. P. Sheehan, C. R. Morin, and K. F. Packer (authors' closure)—Mineral oil was used as an easily reproducible inert environment in which being liquid, the temperature could be controlled as was done in the Ringer's solution. Oil[5] or glycerine [11] are commonly used for such purposes.

[5]Parkins, R. N. in *Stress Corrosion Cracking—The Slow Strain Rate Technique, ASTM STP 665,* American Society for Testing and Materials, Philadelphia, 1979, pp. 5–25.

Kirk J. Bundy[1] and Vimal H. Desai[2]

Studies of Stress-Corrosion Cracking Behavior of Surgical Implant Materials Using a Fracture Mechanics Approach

REFERENCE: Bundy, K. J. and Desai, V. H., **"Studies of Stress-Corrosion Cracking Behavior of Surgical Implant Materials Using a Fracture Mechanics Approach,"** *Corrosion and Degradation of Implant Materials: Second Symposium, ASTM STP 859,* A. C. Fraker and C. D. Griffin, Eds., American Society for Testing and Materials, Philadelphia, 1985, pp. 73–90.

ABSTRACT: Results of stress-corrosion cracking (SCC) studies of surgical implant materials that have employed a fracture mechanics type of specimen are presented. The behavior of Type 316L stainless steel and Ti-6Al-4V ELI in a number of different environments has been examined. Tests have been performed in boiling 44% magnesium chloride solution (154°C), 5% hydrochloric acid (37°C), and a physiological saline solution (37°C). Crack propagation velocity versus stress intensity curves have been obtained. Plateau velocities and threshold stress intensities have been determined. Some specimens were tested under open circuit conditions. Others were potentiostatically polarized with applied potentials similar in magnitude to those that can be generated by bioelectric effects *in vivo.* Propagation of stress corrosion cracks in Type 316L in physiological saline solution at 37°C was observed in situations where there was a high stress intensity and a passive film that had been disrupted by polarization.

KEY WORDS: implant materials, surgical implants, corrosion, fracture mechanics, stress-corrosion cracking, Type 316L stainless steel, Ti-6Al-4V ELI alloy, polarization, passive film, biomaterials, biological degradation, fatigue (materials)

A number of reports have tentatively identified stress-corrosion cracking (SCC) in medical devices or components that have seen *in vivo* service [1–10].[3] Other researchers have disputed these findings and claim that the environmental conditions (chloride ion, Cl$^-$, concentration; temperature; pH; and oxygen partial pressure, pO_2) to which implants are subjected are not severe enough to cause SCC [11].

[1]Assistant professor, Materials Science and Engineering Department (School of Engineering) and Biomedical Engineering Department (School of Medicine), The Johns Hopkins University, Baltimore, MD 21218; presently, associate professor, Biomedical Engineering Department, Tulane University, New Orleans, LA 70118.

[2]Ph. D. candidate, Materials Science and Engineering Department, School of Engineering, The Johns Hopkins University, Baltimore, MD 21218.

[3]The italic numbers in brackets refer to the list of references appended to this paper.

Extrapolating the results of SCC tests in saline solutions at 37°C to the prediction of performance *in vivo*, however, presupposes that the presence of organic materials and the electrical and electrochemical activity of the body associated with physiological processes (nerve action potentials, tissue piezoelectric effects, variable d-c potential shifts induced by a changing chemical environment, etc.) play no part in influencing the corrosion behavior of surgical implant materials. There have been several studies that suggest or demonstrate that these effects actually are important in influencing corrosion behavior *in vivo* [12–15].

Also, it should be noted that it is difficult, from retrospective studies of failed orthopedic devices, to firmly establish the circumstances for which SCC susceptibility *in vivo* exists because the service conditions for the failed devices are not definitely known and because fractographic evidence can sometimes be subject to more than one interpretation.

There is thus a need for laboratory studies that provide carefully controlled conditions under which only SCC can occur (in susceptible alloys) and that simulate as closely as possible the relevant physiological conditions of *in vivo* service. Previous laboratory research concerning SCC of surgical implant materials, however, gives virtually no quantitative data with respect to the identification of damaging stress levels or rates of crack propagation.

In the work described here, SCC studies using fracture mechanics specimens have been performed on surgical implant alloys in a variety of environments. Tests have been performed in boiling magnesium chloride ($MgCl_2$) to determine SCC susceptibility in Cl^- environments. Other experiments have been conducted in hydrochloric acid (HCl) to examine whether SCC can occur in these alloys at 37°C. Finally, tests have been conducted in a physiological saline solution. In some of the experiments, the specimens were subjected to applied polarization to determine whether bioelectrical effects might be important.

Materials and Methods

The specimens used in this investigation were made from Type 316L stainless steel and Ti-6Al-4V ELI that met the compositional and mechanical property limits of the applicable ASTM Standards: ASTM Specification for Stainless Steel Bar and Wire for Surgical Implants (F 55–76), and ASTM Specification for Titanium 6Al-4V ELI Alloy for Surgical Implant Applications (F 136–79). The geometry of the specimen is given in Fig. 1. The specimens were not pre-cracked prior to SCC testing. The specimen was stressed by very carefully driving a wedge into the gap between the two sides of the specimen until a pre-selected displacement, δ, between the two sides was reached (see Fig. 1). The wedge was made of the same material as the sample to avoid galvanic coupling effects.

A measurement of δ and of the length, *a*, allowed the stress intensity present at the crack tip after crack growth occurs to be calculated according to the compliance calibration technique of Irwin and Kies [16]. This analysis shows

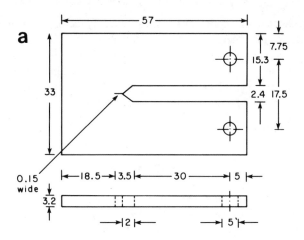

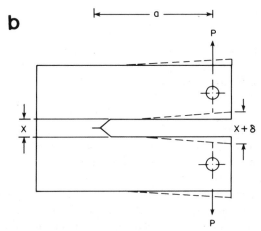

FIG. 1—*Fracture mechanics specimen (dimensions in millimetres; figures are not drawn to scale);* (a) *specimen geometry and* (b) *geometry of stressed specimen.*

that the strain energy release rate, G, (also known as the crack extension force) is given by

$$G = (P^2/2b) \cdot \frac{d}{da}(\delta/P) = (P^2/2b)\frac{dC}{da} \qquad (1)$$

where P is the load applied, b is the specimen thickness, a is the crack length (or more precisely the distance between the point of load application and the crack tip), and C is the specimen compliance (that is, δ/P).

The G is related to the more familiar stress intensity factor, K, by the expression [17] $K = (EG/\alpha)^{1/2}$ where $\alpha = 1 - \nu^2$ for plane strain and $\alpha = 1$ for plane stress. The E is Young's modulus of the material and ν is its Poisson's ratio. As is described in the ASTM Test Method for Plane-Strain Fracture Toughness of Metallic Materials (E 399–78a), the value of α to be used depends upon sample thickness, yield stress, and the stress intensity level.

It has often been found [18] that for specimen geometries similar to that used here

$$C = Aa^n \tag{2}$$

where A and n are constants dependent upon the material and the specimen geometry. This type of relationship was observed as well for the specimen used here. By substitution of Eq 2 into 1, it can be shown that the stress intensity is given by

$$K = \left[\frac{En\delta^2}{2Aba^{n+1}\alpha} \right]^{1/2} \tag{3}$$

In Eq 3, P does not appear explicitly. To determine K, one must only measure δ and a once the C versus a relationship (the compliance calibration curve) has been determined. To obtain the compliance calibration curve, specimens were loaded with a tension testing machine, and δ and a were measured with a traveling telemicroscope (Gaertner Model M101AT).

After the initial compliance measurement, the specimen was immersed in an environment that produced SCC crack growth. When sufficient time had passed so that a new value of crack length was reached, the specimen was removed from the environment, placed in the tension testing machine, and reloaded to obtain a further C, a data point. The sample was then reimmersed in the environment, and the procedure just described was repeated. The testing continued in this manner until a sufficient number of data points were obtained for the compliance calibration curve. The compliance calibration curves are shown in Fig. 2. For Type 316L, three specimens were used to obtain the compliance calibration curve. The data is well approximated by the relationship $C = 1.48 \times 10^{-6} a^{4.15}$ (when a is given in centimetres). The C is in units of centimetres/kilogram. For the titanium alloy, four specimens were used to obtain the C versus a curve. The relationship obtained via regression was $C = 1.07 \times 10^{-7} a^{7.14}$.

During compliance calibration testing, δ was measured along the center line connecting the pin holes through which the load was applied. The distance, a, was measured from this center line to the crack tip. For the initial compliance measurement, the distance between the center line and the base of the machined groove, a_0, was used. For the wedge loaded specimens used in the SCC tests the same coordinate system for measuring δ and a was employed. The δ was

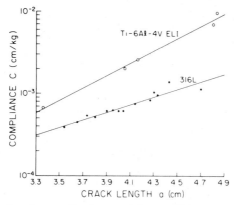

FIG. 2—*Compliance calibration curves for Type 316L and Ti-6Al-4V ELI.*

measured along the center line connecting the holes, and a was measured from this line to the crack tip (or, initially, to the base of the machined groove).

The surfaces of the specimens used in SCC tests were prepared by finishing with successive abrasive papers down to No. 600 grit. The specimens were ultrasonically cleaned in acetone and then in distilled water.

The apparatus used to conduct SCC tests of unpolarized specimens in boiling 44% $MgCl_2$ solution (154 ± 1.5°C) was that described in the ASTM Recommended Practice for Performing Stress-Corrosion Cracking Tests in a Boiling Magnesium Chloride Solution (G 36–73) (1979). For the polarized specimens, the corrosion cell was a reaction kettle as shown in Fig. 3. Polarization was applied with an EG & G PAR Corrosion Measurement System and Accessories. This apparatus was also used for the other polarization tests performed in this work. The salt bridge used was made from a thistle tube with an attached Vycor frit filled with a 33% $MgCl_2$ solution and containing a Fiber Frax wick (Carborundum Corp.) that ran the length of the tube.

A simple experimental arrangement consisting of a 37°C constant temperature bath was employed for the tests in the HCl solution. Each SCC specimen was contained in its own battery jar immersed in the bath. The specimen was suspended by a polymer thread, and totally immersed in an environment of 5% HCl by weight (pH 0.5 to 1.0).

For the tests in physiological saline solution (8.6 g/L NaCl, 0.33 g/L $CaCl_2$, 0.3 g/L KCl), the temperature was also maintained at 37°C. The solution was exposed to the ambient atmosphere. By additions of NaOH or HCl, the pH was maintained in the 7.0 to 7.5 range. The concentrations of oxygen (O_2) and carbon dioxide (CO_2) were measured to be 4.5 to 6.5 ppm and 1 to $1.4 \times 10^{-4}M$, respectively. As with the $MgCl_2$ tests, both polarized and unpolarized specimens were tested in physiological saline.

In all of the SCC tests, the crack propagation velocity, V, was determined by measuring the crack length and remeasuring it after a time interval, Δt, sufficient

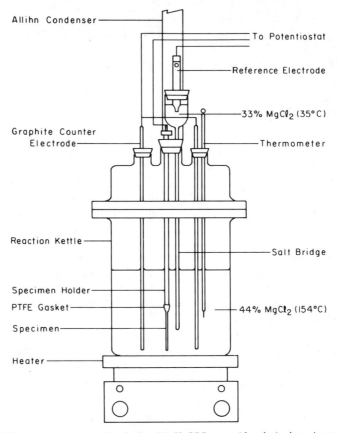

FIG. 3—*Apparatus used for boiling MgCl₂ SCC tests with polarized specimens.*

for significant crack growth, Δa, to occur. The V is thus $\Delta a/\Delta t$. This procedure was repeated until the crack either ceased growing (when the threshold stress intensity, K_{ISCC}, was reached) or reached the end of the specimen. The stress intensity corresponding to a velocity point can be determined by Eq 3 since δ is constant and crack length, a, is measured. The stress intensity declines as the crack propagates for this type of constant δ test. For both alloys, the stress-corrosion cracks in the environments investigated propagated in a fairly straight fashion and without crack branching except at the threshold stress intensity. Crack length measurements were thus easily made.

Strictly speaking, the stress intensity that appears in Eq 3 applies only when there is a sharp crack tip, that is, after the stress-corrosion crack is growing, and not for a machined groove. For convenience in describing the initial loading conditions before crack growth has occurred, a factor, $K°$, is used

$$K° = \left[\frac{En\delta^2}{2Aba_0^{n+1}\alpha} \right]^{1/2} \tag{4}$$

where a_0 is the initial (and minimum) value of a, and the other quantities have the same meaning as in Eq 3. The $K°$ increases as δ increases and is in some sense a relative measure of the severity of the initial loading conditions. It is emphasized, however, that $K°$ is not a stress intensity factor as defined by linear elastic fracture mechanics. After crack growth occurs, though, the stress intensity factor, K, that appears in Eq 3 is valid.

Some potentiodynamic polarization tests were also conducted in boiling $MgCl_2$ and physiological saline. For these tests, the surfaces of the specimens were prepared in the same manner as the fracture mechanics specimens and then given a final polish with 15 μm alumina. Potentiodynamic polarization curves were measured at a scan rate of 0.1 mV/s. The procedures used in these polarization tests were in accordance with those outlined in the ASTM Recommended Practice for Standard Reference Method for Making Potentiostatic and Potentiodynamic Anodic Polarization Measurements (G 5–78). For the $MgCl_2$ tests, the cell and salt bridge/reference electrode arrangement was as previously described. For the tests in physiological saline, the PAR K47 Cell was employed.

Results

$MgCl_2$ Tests

The velocity versus stress intensity curve (the V versus K curve) obtained from measurements with six specimens is shown in Fig. 4 for Type 316L in boiling $MgCl_2$. Although significant scatter was observed, the data clearly shows the existence of the classical pattern, often observed in SCC experiments using fracture mechanics specimens, of a plateau (Region II) where the crack propagation velocity V_{II} is independent of K, a region of declining V with decreasing K, and a K_{ISCC} below which growth of sub-critical cracks does not occur. Data

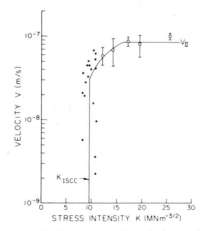

FIG. 4—*Crack propagation velocity versus stress intensity for unpolarized Type 316L in boiling 44% $MgCl_2$.*

points below $K \cong 10$ MNm$^{-3/2}$ are shown individually to illustrate that the threshold (about 9.5 MNm$^{-3/2}$) can clearly be delineated.

This value is very similar to that recently measured by other investigators who have tested other austenitic stainless steels in this environment (or one closely similar to it) [19–22]. The plateau velocity for crack propagation, V_{II}, in Type 316L was shown to be 8.5×10^{-8} m/s. The SCC cracks propagated in a transgranular fashion as can be seen in Fig. 5 which shows several crack branches present at the tip of the arrested crack.

Cracks in Ti-6Al-4V ELI propagated so rapidly in MgCl$_2$ that full determination of the V versus K curve was not possible using the methods just described. However, K_{ISCC} values could be determined since the crack was arrested before reaching the end of the specimen. Tests with nine specimens showed that $K_{ISCC} = 22.5 \pm 1.6$ MNm$^{-3/2}$. An approximate average plateau velocity was obtained by dividing total crack length by the immersion time in solution. This was found to be 8.0×10^{-5} m/s.

HCl Tests

The results of the V versus K measurements in HCl at 37°C are shown in Fig. 6. Although the data points are much sparser than for the MgCl$_2$ tests, the classical

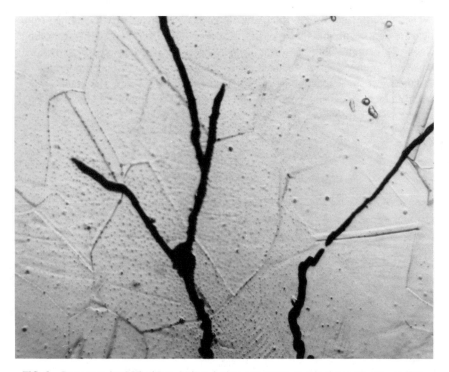

FIG. 5—*Transgranular SCC of Type 316L in boiling 44% MgCl$_2$ (original magnification $\times 520$).*

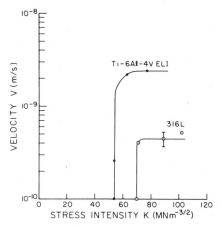

FIG. 6—*Crack propagation velocity versus stress intensity for Type 316L and Ti-6Al-4V ELI in 5% HCl at 37°C.*

pattern for the V versus K curve is nevertheless apparent. As would be expected, the V_{II} velocities for both materials are much lower and the K_{ISCC} values are much higher in HCl at 37°C than in $MgCl_2$ at 154°C. Table 1 compares the measured crack propagation parameters of Type 316L and Ti-6Al-4V ELI in the $MgCl_2$, HCl, and physiological saline environments. The table includes data from both polarized and unpolarized specimens. The available data indicates that Type 316L stainless steel is more resistant to propagation of stress-corrosion cracks in HCl at 37°C than is Ti-6Al-4V ELI because it has a higher K_{ISCC} (70 $MNm^{-3/2}$ as compared to 53.9 $MNm^{-3/2}$ for Ti-6Al-4V ELI) and a lower plateau velocity (4.5 × 10^{-10} m/s as compared to 2.4 × 10^{-9} m/s for Ti-6Al-4V ELI).

Influence of Potentiodynamic Polarization on SCC Behavior

Figure 7 shows the results of SCC tests in boiling $MgCl_2$ in which Type 316L specimens otherwise identical to those described previously were potentiostatically polarized. Three samples were anodically polarized with a 50 mV overpotential; two samples were polarized in the cathodic direction with overpotentials of 50 and 150 mV, respectively. The results for unpolarized specimens (given in Fig. 4) are shown here as the broken line.

The effect of anodic polarization on the V versus K curve is to raise the threshold stress intensity modestly (from 9.5 to 13.0 $MNm^{-3.2}$) and to leave V_{II} essentially unchanged. The points near the threshold shown as shaded circles are discrete data points. The open circles are average values. Cathodic polarization seems to promote inhibition of SCC in Type 316L in $MgCl_2$. As shown in Fig. 7, K_{ISCC} is substantially higher than that of the unpolarized state—about 19 $MNm^{-3/2}$ (for the 150-mV overpotential specimen) as compared to 9.5 $MNm^{-3/2}$. Another specimen cathodically polarized by 50 mV at $K° = 13.8$ $MNm^{-3/2}$ was below K_{ISCC} and showed no cracking in 105.5 h.

TABLE 1—Summary of stress-corrosion crack propagation parameters for surgical implant alloys in different environments.

Material	Environment	T, °C	Polarization Conditions	Average V_{II}, m/s	Average K_{ISCC}, MNm$^{-3/2}$	Number of Specimens
Type 316L	44% $MgCl_2$	154	none	8.5×10^{-8}	9.5	6
Type 316L	44% $MgCl_2$	154	50 mV anodic to E_{corr}[a]	9.0×10^{-8}	13	3
Type 316L	44% $MgCl_2$	154	150 mV cathodic to E_{corr}	$\geq 6.3 \times 10^{-8}$	19	1
Type 316L	5% HCl	37	none	4.5×10^{-10}	70	5
Type 316L	physiological saline	37	none	no cracks observed after 12 weeks of immersion		2
Type 316L	physiological saline	37	100 to 375 mV anodic to E_{corr}	2.4×10^{-10}	...	2
Ti-6Al-4V ELI	44% $MgCl_2$	154	none	8.0×10^{-5}	22.6	3(for V_{II}) 9(for K_{ISCC})
Ti-6Al-4V ELI	5% HCl	37	none	2.4×10^{-9}	53.9	1

[a]Free corrosion potential.

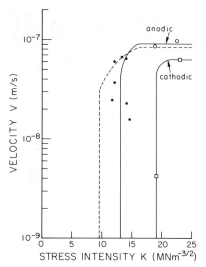

FIG. 7—*Comparison of crack propagation velocity versus stress intensity curves for polarized (——) and unpolarized (– – – –) Type 316L in boiling 44% MgCl$_2$.*

Physiological Saline Solution Tests

Several Type 316L specimens have been tested in physiological saline solution. So far, two control specimens with no applied potential have been immersed in this environment for twelve weeks at $K°$ values of 80 and 106 MNm$^{-3/2}$ without any crack propagation. This does not mean, however, that they are immune to SCC; these specimens could still be within the incubation period. The specimen with the lower stress intensity was pre-cracked in MgCl$_2$.

Two additional Type 316L specimens were loaded to $K°$ values of 60 and 79 MNm$^{-3/2}$ and were pre-cracked in MgCl$_2$ to create initiation sites. The pre-cracks are the dark, relatively wide areas in Fig. 8. These specimens were anodically polarized to disrupt the passive film and then were held at overpotentials between 100 and 375 mV anodic to the free corrosion potential E_{corr}. The cracks propagated by extending from the tip to the initiation sites at a velocity of about 2.4×10^{-10} m/s, a value that is somewhat below that for unpolarized Type 316L specimens in HCl. The crack growth that occurred in physiological saline solution is indicated by arrows in Fig. 8. It has not yet been possible to determine K_{ISCC} for these specimens because the cracks have not stopped growing.

Discussion

The most important question related to this work is whether or not SCC can occur in surgical implant alloys under *in vivo* conditions. There have been two controlled studies using SCC specimens *in vivo*. One of them showed evidence for *in vivo* SCC [13], the other did not [23]. A number of observations of retrieved implants, however, indicate that *in vivo* SCC is possible [2–10]. It

FIG. 8—*Stress-corrosion cracking of Type 316L in physiological saline solution at 37°C (original magnification ×40, arrows indicate crack growth from pre-cracked initiation sites, initial stress intensity = 79 MNm$^{-3/2}$).*

should also be pointed out that SCC susceptibility could possibly lead to failure of orthopedic devices by other mechanisms that are more often observed in failure analysis of retrieved devices. For example, in some alloy systems SCC has been known to initiate fatigue cracking. This is the main reason for concern with SCC in airframe components [24].

With respect to the Type 316L alloy, the SCC literature contains many allusions to the immunity of austenitic stainless steels in chloride environments below a certain temperature that has variously been stated as 60°C [24,25] or 80°C [26]. These observations are based upon many years of practical engineering experience [25,27]. This engineering experience has presumably been acquired for structures that have been safely designed and not overloaded. In severely overstressed conditions, the situation could be somewhat different. One of the references [24] quotes the 60°C temperature but says that "one must be prepared for ugly and usually unexplained surprises." One researcher [22] using fracture mechanics type specimens finds SCC in Type 304 stainless steel in 22% NaCl at 50°C and, for sensitized grades, at 20°C.

It appears from the tests so far in saline that SCC Type 316L in physiological fluids at 37°C is possible at a slow rate under certain conditions. The important question then becomes what is the likelihood that *in vivo* conditions will become severe enough to propagate SCC cracks. One essential feature for crack propagation appears to be a disrupted passive film. In our tests, this was accomplished by anodic polarization, but there are other possibilities under *in vivo* conditions. It has been suggested that SCC in implant alloys could initiate at sites of fretting corrosion [28]. Fretting would be a mechanical means whereby breakdown was affected.

Polarization tests performed here and those in other work [13] show that the corrosion potential of Type 316L in Ringer's solution is only about 100 mV below the breakdown potential for pitting. Furthermore, these tests show that high stress levels lower the breakdown potential. Fluctuation in applied potential due to physiological causes even of intermittent or transient character, could possibly disrupt the passive film by momentarily raising the potential above the threshold for pitting, which has been shown to be the critical potential for SCC in austenitic stainless steels in Cl^- environments [29]. There are a number of physiological effects that might be causative [13].

Another essential factor for crack propagation is a very high value of the stress intensity factor. In normal practice, such high values of the stress intensity factor will not be present (because the nominal stress levels within implants produced by applied loads will not be sufficient to cause them), and so SCC *in vivo* will be a rare event as clinical observations show. However, there are several situations in which the stress level can exceed the yield stress thus damaging the implant by causing plastic deformation and introduction of residual stress [2,5,30,31]. Thus, in situations where improper fabrication has produced a low yield stress material or where the stress level due to the applied loads and residual stresses are exceptionally high, SCC of surgical implants *in vivo* could be a danger.

Although the data so far obtained in physiological saline solution is limited, there has been a more extensive series of tests in the more aggressive HCl and boiling $MgCl_2$ environments that gives some insight into the behavior in saline because all three environments tested have certain features in common. The pH in a crack in Type 316L stainless steel in boiling $MgCl_2$ has been estimated to be $\cong 0$ [20]. We have measured the pH of the HCl solution to be in the 0.5 to 1.0 range. Low values of pH (≤ 1.0) have often been reported in cracks including cracks in neutral NaCl solutions [26]. The crack, like a pit, represents a type of occluded cell where the solution chemistry is much different than in the bulk solution. The data that we have obtained support the idea that SCC *in vivo* is certainly a rare event, yet one that is conceivable under extreme conditions where high stresses and occluded cell conditions that disrupt passive films are present in the physiological environment.

All three environments, boiling $MgCl_2$, HCl, and the physiological saline solution (and thus by extension, the *in vivo* solution) can be considered as situations where there are H^+ and Cl^- ions in the vicinity of the crack tip that presumably strongly influence the SCC behavior. Qualitatively, then, similarities exist with respect to the conditions at the crack tip that promote SCC for all three environments, and the same basic mechanisms can be expected to proceed in each case. Quantitatively, the rates of the processes controlling initiation, growth, and arrest of SCC cracks will be quite different in each solution since they will depend strongly on temperature, overpotential, pH, and Cl^- concentration. Under *in vivo* conditions, organic constituents in the electrolyte and polarization of areas on the implant surface due to bioelectric effects are additional effects that might be important. Further research should be pursued to determine the roles of organic materials and bioelectric effects in influencing *in vivo* SCC behavior.

As would be expected for thermally activated processes, the plateau velocities are much lower at 37°C as compared to 154°C. The V_{II} at body temperature is, at most, less than 0.5% as fast as at 154°C. Since pH and Cl^- concentrations of the two solutions differ, it is not possible to determine reliably an activation energy in general. It has been shown in one study [22], however, that the V_{II} versus temperature data for annealed Type 304, sensitized Type 304, and annealed Type 304L obtained in 42% $MgCl_2$ and 22% NaCl over a wide range of temperature can be fitted to the same functional relationship. If the Type 316L alloy behaved in the same manner, the apparent activation energy determined for V_{II} from the data in Table 1 is 11.74 kcal/mole. This value is consistent with the range of measurements of other investigators of SCC behavior of austenitic stainless steels in Cl^- containing solutions. There is thus at least a future possibility that $MgCl_2$ could be used as an accelerated test environment and that the results could be extrapolated to *in vivo* conditions.

Conclusion

The SCC experiments with fracture mechanics specimens were conducted using Type 316L stainless steel and Ti-6Al-4V ELI as test samples. Three environments were employed for these measurements: boiling $MgCl_2$ (154°C), HCl (37°C), and physiological saline solution (37°C). The most interesting observation was that crack propagation occurred in pre-cracked specimens of Type 316L in the physiological saline solution when they were subjected to high levels of stress intensity while polarized in a manner that disrupted the passive film.

When reliable data for *V* versus *K* and incubation time are obtained for saline solution conditions, this should allow prediction of the lifetime before failure due to SCC of implants for which a reliable stress analysis is available. However, it should be pointed out that SCC resistance should only be one factor to consider in the design of implants. Biocompatibility, mechanical properties, and other types of electrochemical and mechanical interactions must be considered, and probably with greater priority.

Acknowledgments

We are grateful for the financial support of the NACE Research Committee through a Seed Grant for Corrosion Research that partially funded this work. Special thanks are due to the Grant monitors, Drs. E. H. Phelps and W. E. Berry for their advice and assistance during the course of this investigation. We also thank Dr. L. N. Gilbertson of Zimmer U.S.A. for providing the Type 316L material and Dr. H. B. Bomberger of RMI Company for providing the Ti-6Al-4V ELI material that was used in this project. We wish to express our gratitude to R. G. Kelly and J. G. Fleming for their assistance with the experimental work.

References

[1] Zapffe, C. A., *Metal Progress*, Vol. 67, 1955, pp. 95–98.
[2] Bechtol, C. P., Ferguson, A. B., and Laing, P. G., *Metals and Engineering in Bone and Joint Surgery*, Williams and Wilkins Co., Baltimore, 1959.
[3] Rose, R. M., Schiller, A. L., and Radin, E. L., *Journal of Bone and Joint Surgery*, Vol. 54A, 1972, pp. 854–862.
[4] Gray, R. J., *Journal of Biomedical Materials Research Symposium*, Vol. 5, 1974, pp. 22–38.
[5] Lisagor, W. B., *ASTM Standardization News*, Vol. 3, 1975, pp. 20–24.
[6] White, W. E., Postlethwaite, J., and LeMay, I., *Microstructural Science*, Vol. 4, 1976, pp. 145–158.
[7] Bombara, W. and Cavallini, M., *Corrosion Science*, Vol. 17, 1977, pp. 77–85.
[8] Rostoker, W. and Galante, J. O., *Journal of Biomechanical Engineering*, Vol. 101, 1979, pp. 2–14.
[9] Bandyopadhyay, S. and Brockhurst, P., *Journal of Materials Science*, Vol. 14, 1979, pp. 3002–3003.
[10] Kossowsky, R., Dujouny, H., and Kossovsky, N., "Failure Analysis of an Aneurism Surgical Clip," paper presented at AIME Annual Meeting, Dallas, TX, American Institute of Mining, Metallurgical, and Petroleum Engineers, 15–18, Feb. 1982.
[11] Taussig, L. M. in *Implant Retrieval: Material and Biological Analysis*, NBS Special Publication 601, A. Weinstein et al, Eds., National Bureau of Standards, 1981, pp. 201–222.
[12] Revie, R. and Greene, N. D., *Journal of Biomedical Materials Research*, Vol. 3, 1969, pp. 465–470.
[13] Bundy, K. J., Marek, M., and Hochman, R., *Journal of Biomedical Materials Research*, Vol. 17, 1983, pp. 467–487.
[14] Brown, S. A. and Merritt, K., *Journal of Biomedical Materials Research*, Vol. 15, 1981, pp. 479–488.
[15] Black, J., *Biological Performance of Materials*, Marcel Dekker, Inc., New York, 1981.
[16] Irwin, G. R. and Kies, J. A., *Welding Journal Research Supplement*, Vol. 33, 1954, pp. 193–198.
[17] Knott, J. F., *Fundamentals of Fracture Mechanics*, Butterworths, London, 1973, p. 108.
[18] Williams, D. P. and Nelson, H. G., *Metallurgical Transactions*, Vol. 3, 1972, pp. 2107–2113.

[*19*] Speidel, M., *Corrosion*, Vol. 33, 1977, p. 199.
[*20*] Russell, A. J. and Tromans, D., *Metallurgical Transactions*, Vol. 10A, 1979, pp. 1229–1238.
[*21*] Dickson, J. I., Russell, A. J., and Tromans, D., *Canadian Metallurgical Quarterly*, Vol. 19, 1980, pp. 161–167.
[*22*] Speidel, M., *Metallurgical Transactions*, Vol. 12A, 1981, pp. 779–789.
[*23*] Galante, J. and Rostoker, W., *Clinical Orthopedics and Related Research*, Vol. 86, 1972, pp. 237–244.
[*24*] Brown, B. F., *Stress Corrosion Cracking Control Measures*, NBS Monograph 156, National Bureau of Standards, June 1977.
[*25*] Hoxie, E. C., "Some Corrosion Considerations in the Selection of Stainless Steels for Pressure Vessels and Piping," *Pressure Vessels and Piping: A Decade of Progress*, Vol. 3, American Society of Mechanical Engineers, New York, 1977.
[*26*] Dean, S. W. in *Stress Corrosion—New Approaches, ASTM STP 610*, H. L. Craig, Ed., American Society for Testing and Materials, 1976, pp. 308–337.
[*27*] Truman, J. E., *Methods Available for Avoiding SCC of Austenitic Stainless Steel in Potentially Dangerous Environments*, ISI Publication 117, The Iron and Steel Institute, London, 1969, p. 101.
[*28*] Jones, R. L., Wing, S. S., and Syrett, B. C., *Corrosion*, Vol. 34, 1978, pp. 226–236.
[*29*] Uhlig, H. H. and Lincoln, J., *Journal of the Electrochemical Society*, Vol. 105, 1958, pp. 325–332.
[*30*] Hughes, A. N. and Jordon, B. A., *Journal of Biomedical Materials Research*, Vol. 6, 1972, pp. 33–48.
[*31*] Hochman, R. F. and Taussig, L. M., "Improved Properties of 316L Stainless Steel Implants by Low Temperature Stress Relief," paper presented at the Joint ASM-ASTM Symposium on Orthopedic Materials, American Society for Metals and American Society for Testing and Materials, 19 Oct. 1965.

DISCUSSION

N. D. Greene[1] (*written discussion*)—How did you determine that the cracks formed in magnesium chloride continued to grow in Ringer's solution?

Type 316 stainless steel can exhibit stress-corrosion cracking at ambient temperature (for example, 20°C) in acidic, high chloride environments. This material does not normally exhibit this behavior in near-neutral, dilute sodium chloride solutions. I suspect that your observations are the result of magnesium chloride retained within the cracks or the acidic conditions caused by anodic polarization, or both.

K. J. Bundy and V. H. Desai (*authors' closure*)—In response to your first question, prior to immersion in the physiological saline solution, photomicrographs were taken of the area of the specimen where pre-cracks were formed in magnesium chloride. After immersion in the physiological saline solution, photomicrographs were again made of this area. By comparison of the photomicrographs taken before and after exposure to physiological saline solution, the crack growth that occurred in this environment was readily identified.

In response to your second comment, a number of interesting points are raised.

[1]Department of Metallurgy, University of Connecticut, Storrs, CT 06268.

The cleaning procedures used (which are further described in answer to Dr. Lemons' question), make retention of significant amounts of residual magnesium chloride only a very remote possibility. Quantitatively, the values of stress intensity level, potential, temperature, pH, and chloride concentration above or below which SCC of austenitic stainless steels will or will not occur have not been precisely established. This is especially true with regard to how these values might depend upon each other.

In near-neutral dilute NaCl solutions, SCC of austenitic stainless steels has been observed at elevated temperatures. The important question here is what is the lowest temperature at which this can occur. It has been established that the threshold temperature for SCC decreases as pH decreases and as chloride concentration increases, and, as you point out, Type 316 stainless steel can exhibit SCC in acidic, high chloride environments at ambient temperature. We suspect that, everything else being equal, an increasing stress intensity level will also decrease the threshold temperature.

It is certainly true that SCC of Types 316 or 316L stainless steel is not normally exhibited in near-neutral dilute NaCl solutions at ambient temperature or body temperature either at stress levels employed in conservatively designed engineering structures or in laboratory specimens for the exposure times usually used in SCC tests. We are not totally certain that SCC in this environment would not occur in extreme circumstances. To our knowledge, conclusive experiments with very highly loaded samples exposed for extremely long periods under open-circuit conditions have not been performed. Whether SCC will ultimately be observed in our unpolarized specimens is uncertain.

With regard to applied anodic polarization, it does accelerate the SCC process in physiological saline solution at 37°C and this could be due to crack tip solution acidification. Acidic conditions in contact with metal surfaces can occur without externally applied polarization, though, even in neutral (bulk) solutions under occluded cell conditions as you and Dr. Fontana have pointed out. Also, in implants, applied anodic polarization could in effect occur under conditions where there is improper galvanic coupling. Conceivably, *in vivo* bioelectric effects of various types might also act as sources of anodic overpotential, although it should be pointed out that much further research would need to be done to established the merit, if any, of this hypothesis.

J. Lemons[2] (*written discussion*)—The cleaning of MgCl₂ from cracks on these type specimens is quite difficult. How specifically did you clean the crack region and confirm and absence of residue?

K. J. Bundy and V. H. Desai (*authors' closure*)—After removal from the MgCl₂ solution, the pre-cracked specimens were washed for about 4 min in running water. This was followed by ultrasonic cleaning for about 2 min in

[2]University of Alabama, Birmingham, AL 35294.

acetone and then about 3 min of ultrasonic cleaning in distilled water. After this procedure, the pre-cracked specimen was again subjected to the usual surface preparation procedure used in this investigation that consists of polishing through 600 grit abrasive paper, ultrasonic cleaning in acetone for 2 min, ultrasonic cleaning in distilled water for 2 min, and finally drying with hot air. Visual and microscopic observation confirmed the absence of residue.

Metallic Materials: *In Vitro*

Henry R. Piehler,[1] Marc A. Portnoff,[2] Lewis E. Sloter,[3] E. J. Vegdahl,[1] Jeremy L. Gilbert,[1] and Mary Jo Weber[4]

Corrosion-Fatigue Performance of Hip Nails: The Influence of Materials Selection and Design

REFERENCE: Piehler, H. R., Portnoff, M. A., Sloter, L. E., Vegdahl, E. J., Gilbert, J. L., and Weber, M. J., **"Corrosion-Fatigue Performance of Hip Nails: The Influence of Materials Selection and Design,"** *Corrosion and Degradation of Implant Materials: Second Symposium, ASTM STP 859,* A. C. Fraker and C. D. Griffin, Eds., American Society for Testing and Materials, Philadelphia, 1985, pp. 93–104.

ABSTRACT: Corrosion-fatigue performance is evaluated for several Jewett-type hip nails of different design tested in Ringer's solution. Other devices of similar design fabricated from Type 316L stainless steel and Ti-6Al-4V are tested as well. Microstructural and fractographic analyses of all devices are also performed. Present results confirm initial observations that the corrosion-fatigue performance of large-plate Jewetts exceeds that of small-plate devices. Devices of similar design except for the placement of the proximal screw hole in a centered or offset position are evaluated; the latter devices improved corrosion-fatigue performance. Comparison of Jewetts of similar design reveals that the corrosion-fatigue performance of the Ti-6Al-4V devices markedly exceeds that of the Type 316L, even in the presence of substantial fretting and wear.

KEY WORDS: implant materials, stainless steels, titanium alloys, orthopedic implants, corrosion fatigue, fretting corrosion, fretting fatigue, hip nails, performance tests, fatigue (materials), fatigue-crack initiation, medical devices

It has long been recognized that the mechanical properties of orthopedic implants [1,2],[5] particularly their corrosion fatigue performance [3–6], are essential characteristics that must be controlled in an effort to increase the *in vivo* performance of these devices. While existing standards for materials (for example, ASTM Specification for Titanium 6Al-4V ELI Alloy for Surgical Implant Applications (F 136–79)) and devices (for example, ASTM Specification for Hip Nail—Jewett Type (F 369–73) and ASTM Recommended Practice for Static Bend Testing of Nail Plates (F 384–73)) prescribe only material, configurational,

[1]Professor, former graduate student, and graduate student, respectively, Department of Metallurgical Engineering and Materials Science, Carnegie-Mellon University, Pittsburgh, PA 15213.
[2]Associate scientist, Mellon Institute, Pittsburgh, PA 15213.
[3]Senior scientist, Vought Corporation, Dallas, TX 75265.
[4]Program engineer, General Electric Company, San Jose, CA 95112.
[5]The italic numbers in brackets refer to the list of references appended to this paper.

and monotonic mechanical property requirements, a number of previous investigators have recognized and documented the pivotal importance of the corrosion-fatigue performance of orthopedic implants. Examples of these efforts are the studies of Semlitsch et al [3,4], Brunner and Simpson [5], Imam et al [6], and previous studies at Carnegie-Mellon University [7].

Semlitsch et al have employed tests using both rotating beam bending and simulation-loosened femoral prostheses corrosion-fatigue tests [3,4]. These corrosion-fatigue tests, coupled with the results of clinical experience, have led Semlitsch et al to conclude that the risk of stem breakage in instances of stem loosening is minimized (perhaps eliminated) in hot-forged high-strength cobalt and titanium-based alloys if:

1. the fatigue strength of the material lies in the range of 400 to 900 N/mm^2 as determined in rotating bending, and
2. the pulsating fatigue strength of simulation-loosened and corrosion-stressed femoral prostheses exceeded 2800 N (3.5 times body weight).

The efforts of Brunner and Simpson [5] were directed at determining the corrosion-fatigue performance of bone plates and employed potential measurements to determine the onset of fatigue-crack initiation. Among their results were data that show the strength of titanium alloy bone plates to have decreased as a result of bending the plates prior to testing. Imam et al [6] have concentrated their efforts on investigating the effects of microstructure on the corrosion-fatigue performance of titanium alloys. Their results indicate that the corrosion-fatigue behavior of titanium alloys can be improved by alterations in heat treatment and the resulting microstructure. Semlitsch et al did not show a substantial difference in corrosion-fatigue performance in their simulation-loosened and corrosion-stressed femoral prostheses tests on titanium prostheses with different heat treatments, a discrepancy that might be related to the fact that Semlitsch's prostheses were shot peened prior to testing. The previous efforts at Carnegie-Mellon University [7,8], like the present, have been directed at documenting and interpreting the corrosion-fatigue performance of Jewett-type hip nails. The principal conclusions for stainless steel devices from these previous studies are:

1. The *in vitro* performance test employed reasonably simulated *in vivo* performance and failure.
2. The corrosive environment decreased the fatigue lifetime of the devices tested.
3. Crevice corrosion, although it may have clinical consequences, is of secondary importance to fretting corrosion in the mechanical failure of multicomponent devices.
4. Fretting corrosion, while present in the countersinks of all devices, initiated fatigue failure only when it occurred in highly stressed locations.

The objectives of this study are threefold:

1. to compare the corrosion-fatigue performance of devices of identical design fabricated from Type 316L stainless steel and Ti-6AI-4V,
2. to document the role of proximal screw-hole placement on the corrosion fatigue performance of Type 316L devices of otherwise identical design, and
3. to report additional results that compare the corrosion-fatigue performance of the previously investigated Type 316L large-plate and small-plate hip nails.

In addition, observations will be reported concerning the effects of torsional loading (introduced as a result of inadvertent trifin asymmetry) on corrosion-fatigue performance of the straight-plate devices with either offset or centered proximal screw-holes.

Experimental Procedure

The experimental techniques employed in this study are similar to those reported previously [7]. The effects of a nonunion are simulated by applying a bending moment to the plate via an oscillatory (tension/release) load to the end of the nail portion of the device at a frequency of 3 Hz. The plate is screwed into an assembly that simulates the mechanical characteristics of the femur, and the fatigue-critical areas are bathed in aerated Ringer's solution, which is kept at a pH of 6.5 and a temperature of 37°C.

The devices for comparing the corrosion-fatigue performance of Type 316L stainless steel and Ti-6AI-4V were of the large-plate design studied previously [7]. The effects of proximal screw-hole placement (centered or offset) were studied using the straight-plate design. The large-plate design was comprised of a 18.9-mm (5-in.) nail and plate length, an 135° angle, and five screw holes with the proximal screw hole offset. The straight-plate design was comprised of a 18.9-mm (5-in.) nail length, a 22.7 mm (6-in.) plate length, a 138° angle, and six screw holes. As reported previously [7], the commercial straight-plate design employed centered proximal screw holes. The straight-plate devices tested in this study were custom devices fabricated from the same lot of Type 316L stainless steel but with proximal screw holes in either the centered or offset position.

Recently obtained corrosion-fatigue data comparing the performance of the small- and large-plate designs, along with the initial results reported previously [7], are included as well.

Scanning electron microscopy was also employed to characterize the fatigue process, especially the initiation step. Particular attention was paid to the presence of fretting corrosion and its causal relationship, if any, to the fatigue-crack initiation process.

Experimental Results and Discussion

Microstructures

Typical transverse microstructures of the Type 316L stainless steel and Ti-6Al-4V Jewett hip nail proximal plates are shown in Figs. 1 and 2. As shown in Fig. 1, the Type 316L structure is uniform and fine grained (approximately ASTM No. 7). The a-β structure shown in Fig. 2 is again uniform and even finer structure than that in Fig. 1, similar to the structures depicted previously [3].

Performance of Large-Plate and Small-Plate Stainless Steel Devices

New corrosion-fatigue data, as well as those reported previously, for the small-plate and large-plate designs are shown in Fig. 3. These newly tested devices failed at the same locations as those observed previously, that is, the small-plate devices failed at the driver hole and the large plate at the proximal screw hole. As reported previously, the corrosion-fatigue performance of the large-plate design is superior to that of the small plate, especially in the high stress, low cycles to failure regime. This difference in fatigue lifetimes is even greater than the load data shown in Fig. 3, since the nail length of the small plate is 20% shorter than that of the large.

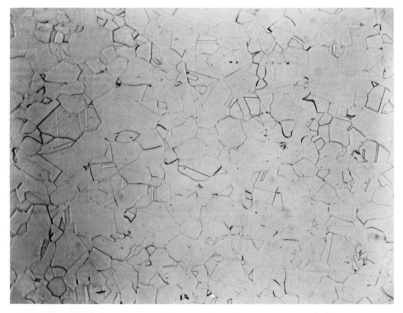

FIG. 1—*Optical micrograph of typical Type 316L stainless steel microstructure (transverse section). Glyceregia etchant (28% HNO₃, 28% acetic acid, 42% HCl, and 2% glycerol). (Original magnification ×250.)*

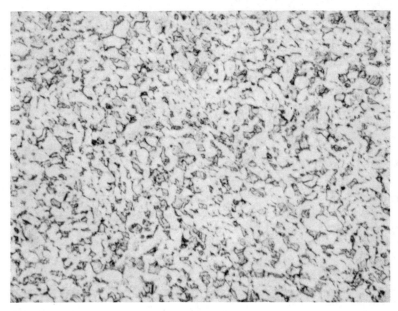

FIG. 2—*Optical micrograph of typical Ti-6Al-4V microstructure (transverse section). Kroll's reagent etchant (2% HF, 5% HNO₃, balance water). (Original magnification ×800.)*

Performance of Centered and Offset Proximal Screw-Hole Stainless Steel Devices

A summary of the results obtained from all of the straight-plate devices with centered or offset proximal screw holes is shown in Table I. The specimen identification, initial applied load, number of cycles to failure, location of failure,

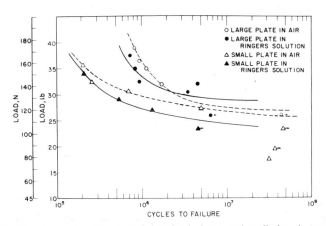

FIG. 3—*Load versus number of cycles to failure for the large- and small-plate devices. Note that the moment arm for the small-plate devices is 20% shorter than that for the large.*

TABLE 1—*Summary of results for centered and offset straight-plate designs.*

Specimen Number	Initial Load, N	Number of Cycles to Failure	Location of Failure	Mechanism of Initiation
		TYPE 316L CENTERED DESIGNS, STRAIGHT PLATE		
PC1	172.6	228 000	DH[b]	FSI
PC2	143.4	480 000	DH	FSI
PC3[a]	154.6	537 000	PSH[c]	fretting
PC4	141.1	562 000	DH	FSI
PC5[a]	125.4	651 000	PSH	fretting
PC6[a]	112.0	1 015 000	PSH	fretting
PC7	123.2	1 255 000	PSH	fretting
PC8[a]	103.0	3 350 000	PSH	fretting
PC9[a]	89.6	8 260 000	NF[d]	NA[f]
PC10[a]	89.6	15 333 300	PSH	fretting
		TYPE 316L OFFSET DESIGN, STRAIGHT PLATE		
PO1	147.8	740 000	PSH	fretting
PO2	132.2	1 830 000	DH	pitting
PO3	136.6	2 400 000	DH	FSI
PO4	127.7	656 000	PSH	fretting
PO5	118.7	3 956 000	NF	NA

[a]Results from previous study.
[b]DH = driver hole.
[c]PSH = proximal screw hole.
[d]NF = no failure.
[e]FSI = free surface initiation.
[f]NA = not available.

and the mechanism of fatigue-crack initiation are all contained in Table I. The location of failure in these devices varied as it does *in vivo*. Failure occurred in one of two locations: either at the proximal screw-hole countersink or at the nail-plate junction (driver hole). The mechanism of initiation refers to whether or not fretting facilitated the initiation process or whether a free surface mechanism was operative. It should be noted that a majority of these stainless-steel devices tested failed at the proximal screw hole for both centered and offset designs (8 out of 14). The remaining devices failed at the driver hole or did not fail at all. Of the five offset devices tested, two failed at the proximal screw hole, two failed at the driver hole, and one did not fail. One of the driver-hole failures was initiated by what appeared to be corrosion pits in the driver hole vicinity. Of the ten centered devices tested, six failed at the proximal screw hole, three failed at the driver hole, and one did not fail at all. All six proximal screw hole failures were fretting assisted fatigue failures and all of the driver hole failures were initiated by a free surface mechanism.

The results of the corrosion-fatigue tests on stainless-steel centered and offset proximal screw designs are summarized in the *S-N* curve of Fig. 4. It is apparent that the results for the centered proximal screw hole devices tested previously are consistent with the present results and that the offset proximal design has a superior fatigue performance in the high cycle to failure regime. This suggests

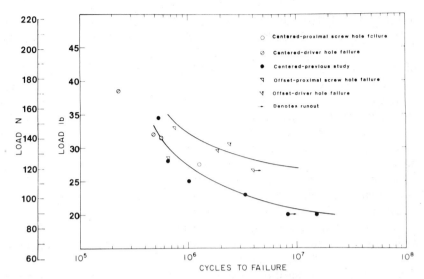

FIG. 4—*Load versus number of cycles to failure for straight-plate Type 316L stainless-steel Jewetts with centered or offset proximal screw holes tested in Ringer's solution. Failure locations are indicated. (All failures from previous study occurred at proximal screw hole.)*

that the difference in design affects the initiation stage of the fatigue process, which is dominant in high cycle fatigue [9].

This superiority in high-cycle fatigue performance of an offset proximal screw-hole device would have been expected from previous results [7], which indicated that fretting, while present in the countersinks of all devices, initiated fatigue only in those of centered proximal screw-hole design. This superiority in high-cycle fatigue performance was indeed observed here, even though the present results were complicated by the superposition of torsional loads of varying magnitude and orientations. Additional differences in fatigue mechanisms and failure sites were also noted in the present study compared to those observed previously [7]. All previously tested offset devices failed at the proximal screw hole by free surface initiation away from areas of fretting. In the present study, driver-hole failures were obtained as were fretting-initiated proximal screw-hole failures. Yet, even in the presence of these complications, the high-cycle fatigue performance of the offset proximal screw-hole devices was found to be superior to that of the centered.

It is surprising to note that, for lifetimes under about one million cycles, there appears to be very little scatter among devices of different design, failure locations, and torsional loading stemming from nail trifin asymmetry. The scatter in the results for all the centered proximal screw-hole devices is strikingly small compared to typical fatigue data, even though three devices failed at the driver hole rather than the proximal screw hole. These three proximal screw-hole failures all occurred in the high-stress, low cycle to failure regime. However, a device previously tested in this same regime (PC3) failed at the proximal screw hole, which might reflect differences in surface finish between the two different lots of centered proximal screw-hole devices. It should also be noted that two of the present devices (PC2 and PC4) were subjected to torsional loading from trifin asymmetry but in the opposite sense. While the differences in torsional loading caused differences in the relative amounts of staining in the symmetrically located driver-hole initiation sites, there resulted no substantial differences in fatigue performance. The PC1 device, which contained a symmetrical trifin and hence no torsional loading, was stained equally in the two symmetric driver-hole initiation sites and also fell on the same trend line for all the centered proximal screw-hole devices. Finally, it should be noted that the superposition of a torque on the PC7 device produced no change in fatigue performance and failure location when compared to the previously tested PC6 device.

The scatter in the fatigue response of the offset proximal screw-hole devices, while larger than that for the centered, is still not large when compared to typical S-N data. The trend in the failure location with cycles to failure is opposite to that noted for the centered proximal screw-hole devices in that the two driver hole failures occurred in the high cycle rather than the low cycle to failure regime. All of the offset devices were subjected to an additional torsion from trifin asymmetry, but no systematic variation in fatigue performance could be associated with changes in either the magnitude or the orientation of this superimposed torsion.

Performance of Identically Designed Type 316L and TI-6AL-4V Devices

A total of six Type 316L and five Ti-6Al-4V large-plate devices were tested in Ringer's solution. The results of these tests are contained in Table 2 and shown in Fig. 5. All of the stainless-steel large-plate devices failed at the proximal screw hole via a free surface fatigue-crack initiation mechanism. The titanium devices, however, failed either at the proximal screw hole or at the second screw hole after the proximal screw head had failed or loosened. Two of the three titanium devices that failed (Ti2 and Ti3) experienced proximal screw-head loosening or fracture prior to device failure. This indicates that it may well be the screws, not the fixation device itself, which is the design limiting component of the bone implant system for these titanium devices. The screws failed by fatigue initiated by fretting of the screw head on the countersink. The fracture surfaces of the screw head and Ti2 device are shown in Fig. 6a. The screw head failed by fatigue prior to the device fatigue failure. Failure of the screw head appears to have initiated at the site of maximum contact stress between screw and countersink. Figure 6b is a scanning electron microscope (SEM) fractograph of the cross section of Ti2 after removal from the mount. Fretting corrosion/wear is clearly evident on the right side of the countersink where the fatigue crack initiated. Figure 6c is a higher magnification SEM fractograph of the fatigue-crack initiation site for Ti2. The fretting/wear damage seen on the left

TABLE 2—*Summary of results for stainless steel and Ti-6A1-4V large-plate devices.*

Specimen Number	Initial Load, N	Number of Cycles to Failure	Location of Failure	Mechanism of Initiation
		TYPE 316L LARGE PLATE		
SS[a]1	143.4	4 580 000	PSH[c]	FSI[e]
SS2	156.8	885 000	PSH	FSI
SS3	136.6	3 560 000	PSH	FSI
SS4	116.5	6 570 000	NF[d]	NA[f]
SS5	168.0	757 000	PSH	FSI
SS6	143.4	973 000	PSH	FSI
		TI-6A1-4V LARGE PLATE		
Ti1	327.4	74 000	PSH	FSI at scratch in bore
Ti2[b]	280.0	1 440 000	PSH	fretting
Ti3[b]	262.1	1 690 000	2nd SH	wear or fretting or both
Ti4	174.7	7 000 000	NF	NA
Ti5	134.4	10 000 000	NF	NA

[a]SS = stainless steel.
[b]Proximal screw failure.
[c]PSH = proximal screw hole.
[d]NF = no failure.
[e]FSI = free surface initiation.
[f]NA = not available.

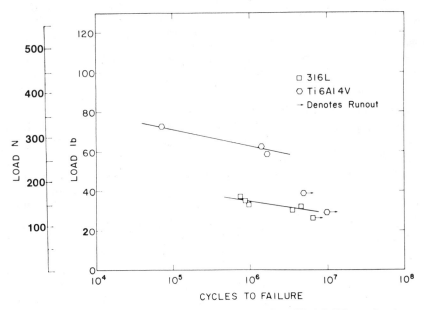

FIG. 5—*Load versus number of cycles to failure for Type 316L and Ti-6Al-4V large-plate Jewetts tested in Ringer's solution.*

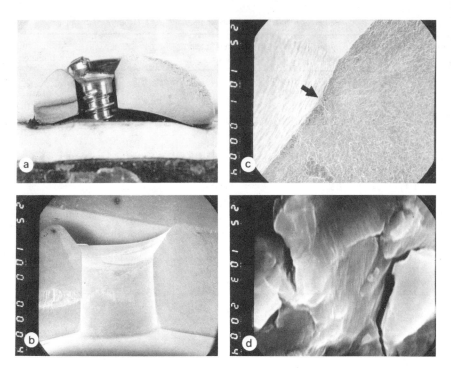

FIG. 6—*Ti-6Al-4V fatigue failures.* (a) *Distal fracture surface of Ti2 device and fractured proximal screw head (original magnification ×5).* (b) *Scanning electron micrograph fracture surface, bore and countersink of Ti2 showing extensive wear/fretting in the countersink from contact with screw (original magnification ×10).* (c) *Higher magnification scanning electron micrograph of the area of wear/fretting shown in (b). Arrow indicates fatigue-crack initiation site (original magnification ×100).* (d) *Fatigue striations and crack branching in region of Stage II fatigue-crack propagation of Ti1 device (original magnification ×10 000).*

side of this fractograph and the initiation site is at the edge of the area of fretting/ wear damage farthest from the neutral axis of bending. Figure 6*d* is an SEM fractograph of Ti1 in the region of Stage II fatigue-crack propagation [*10*] showing both striations and extensive crack branching.

The deflections encountered by the titanium devices loaded at the increased load levels necessary to cause failure appear to be well beyond those that are anatomically feasible. This appears to indicate that the *in vivo* loading of these devices would not be nearly as high as that imposed in this test. Consequently, the possibility of fatigue failure of those titanium devices *in vivo* would be significantly reduced, perhaps eliminated.

The *S-N* curve shown in Fig. 5 clearly shows a marked difference in the corrosion-fatigue life between the titanium and stainless steel. For example, at 10^6 cycles, the fatigue load sustainable by the Ti-6Al-4V device is approximately twice that of stainless steel. Hence, the use of Ti-6Al-4V in orthopedic implants is strongly indicated on the basis of the superior corrosion-fatigue performance of the titanium large-plate devices compared to the Type 316L. This choice is further supported by the favorable biocompatibility response of titanium [*11*]. In addition, the use of titanium is indicated by the fact that its low modulus (approximately one half that of stainless steel) is closer to that of bone than stainless-steel or cobalt-chromium alloys.

Perhaps the most important mechanistic observation to be drawn from this study of Ti-6AI-4V and Type 316L Jewetts is that fretting, while it clearly occurs extensively in the titanium devices, does not compromise the corrosion-fatigue performance compared to that of stainless steel. Hence, previous findings that fretting will cause substantial reductions in corrosion-fatigue lifetimes of titanium alloys [*12,13*] do not prevail for these devices.

Conclusions

The corrosion-fatigue performance tests on Jewett nails tested in this and the previous study indicated that:

1. Corrosion-fatigue performance is affected by implant design. Superior corrosion-fatigue performance was noted for the large-plate Jewett compared to the small plate in the high-stress, low cycle to failure regime. The straight-plate Jewett devices with offset proximal screw holes resulted in a superior corrosion-fatigue performance compared to the centered screw-hole devices, but in this case only in the low-stress, high cycle to failure regime.

2. For the straight-plate devices with both centered and offset proximal screw holes, the observed differences in loading mode, failure sites, and mechanisms of fatigue-crack initiation appeared to have little or no influence on corrosion-fatigue performance.

3. The corrosion-fatigue performance of Ti-6AI-4V large-plate Jewetts was markedly superior to that of Type 316L stainless-steel Jewetts of identical design. This superiority in the corrosion-fatigue performance of the titanium devices was superior even in light of the fact that a substantial amount of fretting/wear occurred in the countersinks of all these devices.

References

[*1*] Bechtol, C. H., Ferguson, A. B., and Laing, P. G., *Metals and Engineering in Bone and Joint Surgery,* Balliere, Tindall & Cox, London, 1959, p. 8.
[*2*] Mears, D. C., *Orthopedic Survey,* Vol. 1, No. 1, July, Aug. 1977, p. 64.
[*3*] Semlitsch, M. and Panic, B., *Biomedizinische Techniln,* Band 28, Heft 4, 1983, pp. 66–78.
[*4*] Semlitsch, M. F., Panic, B., Weber, H., and Schoen, R. in *Titanium Alloys in Surgical Implants, ASTM STP 796,* H. A. Luckey and F. Kubli, Jr., Eds., American Society for Testing and Materials, Philadelphia, 1983, pp. 120–135.
[*5*] Brunner, H. and Simpson, J. P., *Injury, the British Journal of Accident Surgery,* Vol. 11, No. 3, 1979, pp. 203–207.
[*6*] Imam, M. A., Fraker, A. C., Harris, J. S., and Gilmore, C. M. in *Titanium Alloys in Surgical Implants, ASTM STP 796,* H. A. Luckey and F. Kubli, Jr., Ed., American Society for Testing and Materials, Philadelphia, 1983, pp. 105–119.

[7] Sloter, L. E. and Piehler, H. R. in *Corrosion and Degradation of Implant Materials, ASTM STP 684,* B. C. Syrett and A. Acharya, Eds., American Society for Testing and Materials, Philadelphia, 1978, pp. 173–195.

[8] Sloter, L. E., "The Performance of Surgical Implants: An Engineering and Policy Analysis," Ph.D. thesis, Carnegie-Mellon University, Pittsburgh, PA, 1979.

[9] Rolfe, S. T. and Barsom, J. M., *Fracture and Fatigue Control in Structures: Applications of Fracture Mechanics,* Prentice-Hall, Englewood Cliffs, NJ, 1977, p. 23.

[10] Forsythe, P. J. E., *The Physical Basis of Metals Fatigue,* American Elsevier, NY, 1969, p. 64.

[11] Laing, P. G., Ferguson, A. B., Jr., and Hodge, E. S., *Journal of Biomedical Materials Research,* Vol. 1, 1967, pp. 135–149.

[12] Liu, H. W., Corten, H. J., and Sinclair, G. M., *Proceedings,* American Society for Testing and Materials, Philadelphia, Vol. 57, 1957, pp. 623–641.

[13] Lutynski, C., Simansky, G., and McEvily, A. J., *Materials Evaluation under Fretting Conditions, ASTM STP 780,* American Society for Testing and Materials, Philadelphia, 1982, pp. 150–164.

Stanley A. Brown[1] and Katharine Merritt[1]

Fretting Corrosion of Plates and Screws: An *In Vitro* Test Method

REFERENCE: Brown, S. A. and Merritt, K., **"Fretting Corrosion of Plates and Screws: An *In Vitro* Test Method,"** *Corrosion and Degradation of Implant Materials: Second Symposium, ASTM STP 859*, A. C. Fraker and C. D. Griffin, Eds., American Society for Testing and Materials, Philadelphia, 1985, pp. 105–116.

ABSTRACT: A laboratory fretting corrosion simulator was constructed for studying the fretting corrosion of the contact areas between screw heads and plate hole countersinks of surgical implants used for fixation of broken bones. The system was constructed so that the tests could be done under sterile conditions, thus permitting the use of proteinaceous as well as conventional electrolyte solutions. Studies were done to investigate the effects of contact stress as controlled by screw torque, the effects of blood serum proteins, the effects of proteins at different solution pH values, and the effects of mixing metals. The alloys used were Type 316L stainless steel, MP35N, titanium, and titanium-6Al-4V. The results demonstrated that statistically significant data can be obtained with a small number of specimens. The rate of fretting corrosion was independent of load over a limited range. The presence of proteins significantly reduced the fretting corrosion of stainless steel and MP35N, had no effect of Ti-6Al-4V, but increased the fretting corrosion of titanium. In solution with pH values acidic to its isoelectric point, the protein albumin had a negligible effect; whereas at solution pH values basic to its isoelectric point, fretting corrosion was reduced. Mixing of metals had an effect on the fretting corrosion of MP35N and titanium, but no effect on stainless steel.

KEY WORDS: fretting corrosion, stainless steel, titanium, cobalt-nickel-chromium-molybdenum alloys, proteins, albumin, serum, implant materials

There have been a number of reports in recent years concerning the effects of proteinaceous fluids used for corrosion testing of implant materials. The interest has been fostered by concern about the systemic effects of metal ions and corrosion products [1,2][2] and by interest in the immune response to metal ions complexed with proteins [3], as is discussed in detail in another paper in this volume [4]. Corrosion studies *in vivo* have shown that corrosion rates change with time [5,6], that protein binding is not necessarily in proportion to the alloy composition [7], and that corrosion rates *in vivo* are less than those measured *in vitro* [8]. All these effects presumably are due to the interactions of proteins

---

[1]Associate professors of Biomedical Engineering, Department of Biomedical Engineering, Case Western Reserve University, Cleveland, OH 44106.
[2]The italic numbers in brackets refer to the list of references appended to this paper.

with corrosion processes. Laboratory studies of stainless steel corrosion in proteins have shown that the corrosion rate in blood and serum proteins is greater than that in saline [9], that the addition of serum to saline results in a 10-fold increase in electrochemical corrosion rate [10], and that it is difficult to get reproducible polarization curves because of specimen pitting [11]. Speck and Fraker [12] showed that the amino acid, cysteine, caused an increased current density with Ti-6Al-4V, and lowered the breakdown potential of the TiNi "memory alloy," perhaps due to protein interactions with the nickel. Clark and Williams [13] reported that serum proteins had no effect on the corrosion rate of titanium but caused a slight increase in chromium and nickel corrosion and significantly increased the corrosion rate of cobalt. While there is no total agreement between these reports, it is clear that the serum proteins have a significant effect on the electrochemical corrosion rates of surgical alloys.

Our interest in fretting corrosion and the effects of proteins came from examination of retrieved implants that demonstrated evidence of fretting corrosion [14,15], and the need to generate corrosion products in serum for metal sensitivity studies, in a mode analagous to *in vivo* production. In a previous publication, it was shown that the metal weight loss and nickel release due to fretting corrosion in a 10% solution of calf serum is less than 10% of that in saline [16–18]. Starkebaum [19] reported that the addition of serum to distilled water had minimal effect on the wear (fretting corrosion) rate of stainless steel but resulted in a significant increase in the wear of titanium. The studies to be reported here were designed to identify some of the variables involved in the fretting corrosion system, and to determine the effects of proteins and mixed metals on fretting corrosion rates.

Methods

Fretting corrosion was studied using a simulator as previously reported [16,18]. Screws were placed through the holes of two-hole plates, and contact pressure between the head and the countersink areas was maintained by compressed elastomeric spacers. The system generated a rocking motion of the screw head against the plate. The plates and screws used were commercially available implants for internal fixation of fractures. Two-hole plates were cut from the two ends of longer plates, and were used as matched pairs for matched experiments. Screws with spherical heads were passed through the plate holes, through silicone rubber spacers, and into threaded holes in plastic posts. One post was held fixed while the other mated with a horizontal plunger that was driven with an oscillating motion by a rotating cam. The cam motion produced 2° of relative motion between the two posts. This resulted in a relative motion between screw head and plate countersink of between 30 μm and 120 μm, which is within the range studied by Sherwin et al [20].

To conduct an experiment, the plates and screws were cleaned, dried, weighed and then secured to the posts. A torque screw driver was used to ensure that the torque and thus axial load was uniform throughout the series. The plate and post assemblies were placed in a beaker, covered with a rubber membrane with two holes through which the post tops were inserted, and steam sterilized. The sterile test solution (20 cm³) then was added using a sterile syringe and needle. The simulator consisted of eight test chambers, as is shown in Fig. 1, set up as four matched pairs; one on each side of the central shaft and cam. The horizontal shaft oscillated at one cycle per second for 16 h a day. Experiments were run for 14 days (806 000 cycles). At the completion of an experiment, the screws and plates were removed from the posts, the implants ultrasonically cleaned in the test solution to remove any loose corrosion products, and then ultrasonically cleaned in detergent, dried, and weighed. The test solutions were stored for subsequent chemical analysis. The amount of fretting corrosion was assessed as the weight loss of each plate, the weight loss of the two screws, the "total" weight loss of the plate and two screws weighed together, and as the concentration of corrosion products in the test solution determined by chemical analysis of the solutions for nickel (ASTM Test for Nickel in Water (D 1886-77)) or for all elements by atomic absorption spectroscopy.

The solutions used were 0.9% NaCl (USP for injection), 10% solutions of new born calf serum[3] in saline, or 0.5% albumin in saline (0.5% albumin is the concentration of albumin in 10% serum). The screws used were all spherical head made by the same manufacturer as the plates; the plates used were round

FIG. 1—*The fretting corrosion simulator.*

[3]Gibco Laboratories, Grand Island, NY 14072.

hole stainless steel[4] (ASTM Specification for Stainless Steel Bars and Wire for Surgical Implants (Special Quality) (F 138-76)) or DCP-Type[5] of stainless steel, MP35N[6] (ASTM Specification for Wrought Cobalt-Nickel-Chromium-Molybdenum Alloy for Surgical Implant Applications (F 562-78)), commercially pure titanium (ASTM Specification for Unalloyed Titanium for Surgical Implant Applications (F 67-77)) or Ti-6Al-4V (ASTM Specification for Titanium 6Al-4V ELI Alloy for Surgical Implant Applications (F 136-79)).

Results

Some initial studies were conducted to compare the amount of weight loss of implants in saline solution with that in air. Those in air did not lose any measurable amount of weight, and thus it is concluded that the test is a measure of fretting corrosion and not simply fretting and wear.

The amount of weight loss and nickel in solution as a function of different screw torques is shown in Table 1. These were done with round hole plates and screws[7] in 0.9% saline for 14 days with four sets per group. The results are expressed here and in all subsequent tables and figures as the mean and standard deviation in parentheses ( ), for the plates, the two screws, and the "total" of the three weighed together. There were no differences between the low loads of 260 N and 560 N. The weight loss and nickel concentration in solution at 860 N were all less than at 560 N. The level of significance as determined by using the Student's t-test are also shown.

The results of matched fretting corrosion experiments DCP plates and screws[8] in saline and 10% serum are given in Table 2. The level of significance of the differences (P values from Student's t-test) are given below each pair. These results demonstrate that the presence of serum proteins in saline resulted in a significant decrease in the fretting corrosion of stainless steel and MP35N, an increase in fretting corrosion of titanium, and no effect on Ti-6Al-4V. As a result

TABLE 1—*Weight loss (mg) of plates, screws, total, and nickel concentrations ($\mu g/mL$) in solution expressed as mean (standard deviation), at different screw torques and axial loads, and statistical comparison between 560 N and 860 N.*

Plate	Screws	Total	Nickel	Torque, load	Number of Samples
2.19 (0.14)	2.15 (0.43)	4.35 (0.53)	18.1 (3.5)	0.5 Nm, 260 N	4
2.24 (0.31)	1.97 (0.36)	4.26 (0.52)	18.0 (3.2)	1.0 Nm, 560 N	4
1.20 (0.61)	1.14 (0.51)	2.34 (1.0)	12.9 (2.2)	1.5 Nm, 860 N	4
$p < 0.02$	$p < 0.05$	$p < 0.01$	$p < 0.05$	560 N versus 860 N	. . .

[4]Synthes, Institut Straumann, Waldenburg, Switzerland, and Zimmer, Inc., Warsaw, IN 46580.
[5]Synthes, Institut Straumann, Waldenburg, Switzerland.
[6]MP35N is the trade name for the cobalt-nickel-chromium-molybdenum alloy.
[7]Zimmer, Inc., Warsaw, IN 46580.
[8]Synthes, Institut Straumann, Waldenburg, Switzerland.

TABLE 2—*Weight loss of metals (mg) and nickel concentrate (μg/mL) expressed as mean (standard deviation) in 0.9% saline and in 10% serum in saline, and level of significance of the differences, using Student's t-test.*

Metal	Plates	Screws	Total	Nickel	Solution	Number of Samples
316L	1.23 (0.37)	1.21 (0.30)	2.22 (0.60)	10.9 (3.3)	0.9% saline	5
316L	0.14 (0.07)	0.22 (0.08)	0.32 (0.11)	1.9 (1.4)	10% serum	4
	$p < 0.001$	$p < 0.001$	$p < 0.001$	$p < 0.001$		
MP35N	0.31 (0.23)	0.28 (0.20)	0.66 (0.35)	10.4 (4.2)	0.9% saline	4
MP35N	0.02 (0.03)	0.08 (0.10)	0.14 (0.10)	0.29 (0.2)	10% serum	4
	$p < 0.05$	NS[a]	$p < 0.025$	$p < 0.005$		
Titanium	0.07 (0.03)	0.07 (0.04)	0.14 (0.08)	NA[b]	0.9% saline	6
Titanium	0.13 (0.05)	0.17 (0.07)	0.26 (0.08)	NA	10% serum	6
	$p < 0.05$	$p < 0.01$	$p < 0.025$			
Ti-6Al-4V	+0.02 (0.02)	0.001 (0.06)	0.02 (0.03)	NA	0.9% saline	4
Ti-6Al-4V	+0.03 (0.03)	0.025 (0.05)	0.02 (0.06)	NA	10% serum	4
	(no significant differences)					
Titanium	0.14 (0.11)	0.21 (0.09)	0.29 (0.18)	NA	10% serum	2
Ti-6Al-4V	0.01 (0.01)	0.11 (0.01)	0.09 (0.01)	NA	10% serum	2
	(only two per group, no statistics)					

[a]NS = not significant.
[b]NA = not applicable.

of the somewhat surprising results with titanium and Ti-6Al-4V, two matched experiments of titanium versus Ti-6Al-4V were conducted to verify the difference. The results with stainless steel, MP35N, and titanium are shown graphically in Fig. 2.

In previous studies, we have reported that the effect of serum is due principally to the protein albumin [*18*]. To investigate the mechanism of the inhibition of fretting corrosion of stainless steel due to proteins, we have initiated a series of experiments with proteins in solutions with different pH values relative to their isoelectric points. The preliminary results with albumin are shown in Table 3. These are from 14-day experiments with round hole plates with an albumin concentration equivalent to that in 10% serum. It can be seen that at a solution pH of 3, which is acidic to the isoelectric point (pI) of albumin, the protein had minimal effect on the fretting corrosion rate. At its pI of 4.5 and at the more alkaline solution pH level, the protein had a significant inhibitory effect.

The results of mixing metals on the fretting corrosion in saline are given in Table 4, and are shown graphically in Fig. 3. Alloy designations are abbreviated as MP35N = P, 316L = S, titanium = T, titanium 6Al-4V = 6,4. The experiments with MP35N plates with MP35N screws, P/P, versus MP35N plates with stainless steel screws, P/S, were done as matched pairs. The titanium on titanium, T/T, versus titanium against steel, T/S, was not done as matched experiments. It can be seen that the fretting corrosion of MP35N increased and that of titanium decreased when the plates were fretted against stainless steel screws. The weight loss of the stainless steel screws in the different combinations

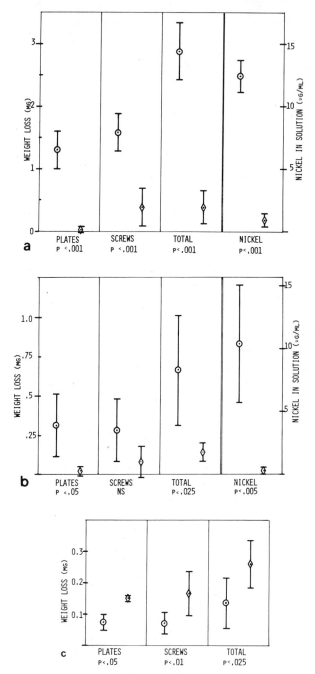

FIG. 2—*Weight loss of plates, screws, and total due to fretting corrosion in 0.9% saline (circle) and 10% serum (diamond) after 14 days for* (a) *stainless steel,* (b) *MP35N, and* (c) *titanium. Data are expressed as mean value with one standard deviation error bars.*

TABLE 3—*Total weight loss (mg) and nickel concentration ($\mu g/mL$) from stainless steel plates and screws in saline and albumin at different pH values.*

	Saline	pH 3	pH 4.5	pH 8
Weight loss, mg	3.5	3.3	2.5	2.1
Nickel concentration, $\mu g/mL$	17	22.5	14	7.2

was not significantly different. Figure 3 compares the results of several of the experiments with fretting in saline.

Discussion

Sherwin et al [20] reported that the fretting corrosion of stainless steel increases with increasing contact stress. The results, in Table 1, would suggest the opposite. However, at the low range of axial screw load in the present experiment, we probably were seeing relatively small changes in the contact stress and thus minimal fretting corrosion rate effects. At the higher stress, it is reasonable to assume that the magnitude of the screw-plate motion was reduced due to the construction of the machine. The simulator was constructed with a positive drive in the outward direction of the plunger by the cam, and an inward return by a small coil spring. The springs probably were not stiff enough to maintain full motion at the higher load, and thus the fretting corrosion rate was reduced. The fact that the rate is not affected by small load changes in the lower range implies that the system reproducibility is not critically affected by this variable.

The results of fretting corrosion of steel DCP plates in saline and 10% serum confirmed our previously reported results for round hole plates [16–18]; the addition of serum to saline resulted in a significant decrease in the fretting corrosion rate of stainless steel. The DCP plates used in the present study have an oval-shaped hole, and thus the contact area between plate and screw head is smaller than with round holes. As a result, the actual amount of fretting corrosion with the DCP plates was less than the round holes, due to the smaller contact

TABLE 4—*Weight loss and metal release from mixed metal experiments.*

Metal	Plates	Screws	Total	Nickel	Cobalt	Number of Samples
P/P	0.37 (0.10)	0.44 (0.09)	0.69 (0.10)	12.2 (4.0)	16.8 (4.9)	6
P/S	0.98 (0.30)	0.77 (0.40)	1.48 (0.40)	17.1 (3.0)	33.0 (16)	6
	$p < 0.001$	NS^a	$p < 0.001$	$p < 0.05$		
T/T	0.07 (0.03)	0.07 (0.04)	0.14 (0.08)	NA^b	NA	6
T/S	0.04 (0.04)	0.71 (0.16)	0.66 (0.20)	4.4 (0.60)	NA	5

^{a}NS = not significant.
^{b}NA = not applicable.

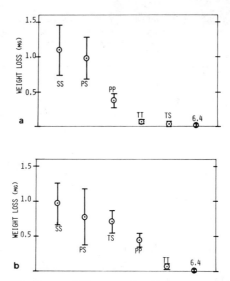

FIG. 3—*Weight loss of* (a) *plates and* (b) *screws after 14 days of fretting in saline. The letters with the bars indicate the alloy combination as plate and screw; SS = stainless steel plate with stainless steel screw, PS = MP35N plate with stainless steel screw, T = titanium, and 6,4 = Ti-6Al-4V.*

areas involved. The similarity between the effects on stainless steel and MP35N are as we would expect, since these are both self-passivating alloys.

The results with titanium are somewhat surprising. The comparisons between saline and serum were conducted as matched pair experiments and demonstrated that the addition of serum resulted in a small but nonetheless statistically significant increase in weight loss of the implants. The differences between commercially pure titanium and Ti-6Al-4V are also difficult to explain. To verify that these two forms of titanium behaved differently, we ran two matched experiments in 10% serum. These confirmed that the metals do not behave the same in serum.

The effect of serum is presumably due to the presence of the serum proteins. In other experiments, we have demonstrated that pH of the solutions does not change significantly during the test period, and that the pH of the saline solution is not significantly different from that of the serum solutions. We also have shown that a 0.5% solution of albumin has similar effects as does 10% serum [18]. In an effort to explain the effects of serum proteins on fretting corrosion of stainless steel, we have initiated a series of experiments at different pH values. At solution pH values basic to the protein's isoelectric point, the molecule carries a strong electronegative charge and, as shown in Table 3, its presence has a significant inhibitory effect on fretting corrosion. In reflecting on possible mechanisms, we have reported [16] that fretting is associated with a negative shift of the rest potential, as a manifestation of the cathodic reaction. The magnitude of the potential shift with fretting corrosion is significantly less in the presence of serum. In saline, the shift from rest is 350 to 450 mV, whereas it is only 100 to 150 mV in 10% serum [16]. In a basic solution, the cathodic reaction is

$$2H_2O + O_2 + 4e^- \rightarrow 4OH^-$$

The presence of highly electronegatively charged protein molecules would tend to block this reaction and cathodically polarize the system. At more acidic levels, which were more acidic than normal pH, they would not be as charged and thus would not compete with the cathodic reaction. Since we are dealing with small sites of the anodic reaction (the contact areas), the reaction is cathodically controlled and thus the reaction could be inhibited by the proteins. If the cathode reaction is blocked by excess negatively charged moieties, such as proteins, then no corrosion current can flow from the anode to the cathode, and thus the anodic reaction is blocked. This would be analogous to fretting in air, which, as we have shown, does not result in implant weight loss.

In contrast to the inhibitory effects of proteins seen in these fretting corrosion experiments, electrochemical dissolution type experiments [9,10,11,13] have demonstrated that the presence of proteins tends to accelerate the corrosion rate. These types of experiments were not done with such small anodic sites, but rather with the possibility of equal-sized anodes and cathodes. Thus, different rate limiting phenomena may apply. In these tests, the anodic reaction may be rate-limiting due to metal ion diffusion, or accumulation of metal corrosion products. Thus, the presence of the negatively charged proteins that would result in metal ion binding could result in acceleration of the anodic reaction. In our own study [10], we observed significant pitting corrosion without a proportional increase in corrosion current that suggested a localized reaction. The localized formation of metal protein complexes and thus reduction metal ion concentration at the anode could be part of the reason for this phenomena. This may explain why proteins inhibit fretting corrosion, yet enhance corrosion in other types of corrosion test.

The comparisons between MP35N and stainless steel in the mixed metal experiments demonstrated several interesting facts. By itself MP35N is more resistant to corrosion than is stainless steel [21]. In the present study, MP35N was more resistant to fretting corrosion to such an extent that the release of nickel was not significantly different than that from stainless steel in spite of MP35N's higher nickel concentration: 35% versus 12%. However, the fretting corrosion rate of MP35N was increased when coupled with steel screws, such that the nickel release was higher than steel/steel. Of the metals studied, stainless steel is the most susceptible to fretting and electrochemical corrosion, and therefore it is not surprising that in the mixed metal combinations the weight loss of the steel screws was not significantly different from steel against steel.

The increase in weight loss of titanium plates fretted with titanium screws as compared with steel screws is easier to explain. Titanium is a comparatively soft metal and tends to gall in wear situations. The harder steel screws would tend to minimize this effect. The results presented here are consistant with reported differences in tissue concentrations of titanium around plates with steel versus titanium screws [15,22].

Conclusion

The method presented here was designed to permit the study of fretting corrosion in saline and proteinaceous solutions, under sterile conditions to avoid bacterial growth. The results have demonstrated that the system can generate statistically significant data with comparatively small sample sizes. It has been shown that the addition of serum proteins has a significant effect on the fretting corrosion of surgical alloys. However, the addition of proteins has not changed the relative ranking of the alloys, and thus valid comparisons could be made with this type of system without the additional complication of sterile conditions and the use of proteins. In addition, it has been shown that mixing metals in a fretting corrosion situation can have a significant effect on the corrosion behavior of these alloys.

Acknowledgments

This work was supported in part by grants from the U.S. Public Health Service, AM-20271 and 29684, and the International A.O. (Arbeitsgemeinschaft für Osteosynthesefragen). Donation of implants for these studies by Zimmer, Inc., and Institute Straumann is gratefully acknowledged. The authors wish to thank Mr. Neil Sharkey for his time spent on these studies.

References

[1] Black, J., *Biological Performance of Materials,* Marcel Dekker, New York, 1981.
[2] Greene, N. D., Onkelinx, C., Richelle, L. J., and Ward, P. A., in *Biomaterials,* NBS SP 415, Horowitz and Torgesen, Eds., U. S. Dept. of Commerce, Washington, DC 1975, pp. 45–54.
[3] Merritt, K. and Brown, S. A., *International Journal of Dermatology,* Vol. 20, 1981, pp. 89–94.
[4] Merritt, K. and Brown, S. A., "Biological Effects of Corrosion Products from Metals," in this volume.
[5] Colangelo, V. J., Greene, N. D., Kettlelkamp, D. B., Alexander, H., and Campbell, C. J., *Journal of Biomedical Materials Research,* Vol. 1, 1967, pp. 405–414.
[6] Steinemann, S. in *Evaluation of Biomaterials,* Winter, Leray, and de Groot, Eds., Wiley, Chichester, 1980, pp. 1–34.
[7] Woodman, J. L., Black, J., and Jimenez, S. in *Transactions,* Society for Biomaterials, Vol. 4, 1981, p. 18.
[8] Syrett, B. C. and Davis, E. E., in *Corrosion and Degradation of Implant Materials, ASTM STP 684,* B. C. Syrett and A. Acharya, Eds., American Society for Testing and Materials, Philadelphia, 1979, pp. 229–244.
[9] Samitz, M. H. and Katz, S. A., *British Journal of Dermatology,* Vol. 92, 1975, pp. 287–290.
[10] Brown, S. A. and Merritt, K., *Journal of Biomedical Materials Research,* Vol. 14, 1980, pp. 173–175.
[11] Danko, G. A., Howell, B. F., and Fraker, A. C. in *Transactions,* Society for Biomaterials, 1981, Vol. 4, p. 15.
[12] Speck, K. M. and Fraker, A. C., *Journal of Dental Research,* Vol. 59, 1980, pp. 1590–1595.

[13] Clark, G. C. F. and Williams, D. F., *Journal of Biomedical Materials Research*, 1982, Vol. 16, pp. 125–134.

[14] Brown, S. A. and Simpson, J. P., *Journal of Biomedical Materials Research*, 1981, Vol. 15, pp. 867–878.

[15] Simpson, J. P., Geret, V., Brown, S. A., and Merritt, K. in *Implant Retrieval: Material and Biological Analysis*, Weinstein, Gibbons, Brown, and Ruff, Eds., NBS SP-106, National Bureau of Standards, Washington, DC, 1981, pp. 395–422.

[16] Brown, S. A. and Merritt, K., *Journal of Biomedical Materials Research*, Vol. 15, 1981, pp. 479–488.

[17] Brown, S. A. and Merritt, K., *Biomaterials, Medical Devices, and Artificial Organs*, Vol. 9, 1981, pp. 57–63.

[18] Brown, S. A. and Merritt, K. in *Clinical Applications of Biomaterials*, Lee, Albrektsson, and Branemark, Eds., Wiley, New York, 1982, pp. 195–202.

[19] Starkebaum, W., "Abrasion resistance of Ti-6A1-4V," presented at ASTM F-4 Symposium on Titanium Alloys in Surgical Implants, Phoenix, AZ, 1981.

[20] Sherwin, M. P., Taylor, D. E., and Waterhouse, R. B., *Corrosion Science*, Vol. 11, 1981, pp. 419–429.

[21] Sury, P. and Semlitsch, M., *Journal of Biomedical Materials Research*, Vol. 12, 1978, pp. 723–741.

[22] Ruedi, T. P., Perren, S. M., and Pohler, O. in *Evaluation of Biomaterials*, Winter, Leray, and de Groot, Eds., Wiley, Chichester, 1980, pp. 35–37.

DISCUSSION

N. D. Greene[1] *(written discussion)*—Your results are most interesting. However, it should be noted that applying a 5-V anodic potential to a metal immersed in saline solutions is an extreme test. Under these conditions, the surface pH decreases, and oxygen and chlorine evolution are possible. This environment has little relationship to actual *in vivo* conditions.

The extensive corrosion inhibitor literature shows that proteins and amino compounds inhibit the corrosion of iron and ferrous alloys. Your results with stainless steels seem consistent with this observation.

S. A. Brown and K. Merritt (authors' closure)—We agree with your comments concerning the 5-V experiments and their lack of relationship with the *in vivo* environment. That is why we developed the fretting corrosion simulator. However, there are other reports in the literature that demonstrate that the presence of proteins in "physiologic" solutions enhance the corrosion of some metals and alloys.

H. Ravi Shetty[2] *(written discussion)*—Your results indicate that fretting corrosion resistance of Ti-6Al-4V alloy is better than Type 316L stainless steel in Ringer's solution. It is well known that titanium-based alloys, including the Ti-6Al-4V alloy, has poor galling resistance. Furthermore, Ti-6Al-4V forms TiO_2 film when it is exposed to Ringer's solution. Because of the poor galling re-

[1]Department of Metallurgy, University of Connecticut, Storrs, CT 06268.
[2]Zimmer, Inc., Warsaw, IN 46580.

sistance to Ti-6Al-4V and the abrasive nature of TiO_2 film that forms instantaneously during fretting corrosion, one should expect higher weight loss of Ti-6Al-4V rather than lower weight loss that you have observed in your experiment. How do you explain this phenomenon?

S. A. Brown and K. Merritt (authors' closure)—Your comments concerning the poor galling resistance of titanium are of interest, and may point to some of the differences between wear and fretting corrosion. We have found that running stainless steel plates and screws in air results in no measurable weight loss. Thus, the majority of the weight loss in the electrolytes is due to fretting corrosion. The high corrosion resistance of titanium may account for the differences between titanium and the stainless steel and cobalt alloys.

However, there is one experiment in the series presented in this paper that would support the galling hypothesis. In Table 4, the results of titanium plates fretted against titanium screws are compared with titanium plates against stainless steel screws. The weight loss of the titanium plates was less with the stainless steel screws. This may be due to a reduction in galling.

Gabriel I. Ogundele[1] and William E. White[1]

Polarization Studies on Surgical-Grade Stainless Steels in Hanks' Physiological Solution

REFERENCE: Ogundele, G. I. and White, W. E., "**Polarization Studies on Surgical-Grade Stainless Steels in Hanks' Physiological Solution,**" *Corrosion and Degradation of Implant Materials: Second Symposium, ASTM STP 859,* A. C. Fraker and C. D. Griffin, Eds., American Society for Testing and Materials, Philadelphia, 1985, pp. 117–135

ABSTRACT: The corrosion resistance of a wrought Type 316L stainless steel (WSS316L) and a surgical-grade stainless alloy (SG2SA) were studied. The behavior of the alloys, as influenced by chloride (Cl^-) and bicarbonate (HCO_3^-) ion concentrations, as well as pH and temperature in Hanks' physiological solution, were explored using electrochemical techniques.

The results show that SG2SA exhibited a more noble response and higher passive breakdown potential than WSS316L. Both materials were significantly affected by changes in Cl^- and HCO_3^- ion concentrations. The corrosion and breakdown potentials shifted to more negative values with increasing Cl^- ion concentrations, but to more positive values with increasing HCO_3^- ion concentrations. The alloys displayed most noble characteristics in neutral solutions and at higher temperature conditions.

The alloys were found to be active only in highly acidic solutions at potentials below -300 mV versus saturated calomel electrode (SCE). Furthermore, the materials were found to experience an early passive film breakdown at higher temperatures, indicating that the passive film is more stable at lower temperatures.

Cyclic polarization tests on WSS316L showed that the chance of repassivation of actively growing pits is very limited in environments containing variable Cl^- and HCO_3^- ion concentration as well as at higher temperature.

As both of the test materials were determined, from chemical analyses, to conform to AISI Type 316L stainless steel, the observed differences in corrosion behavior are explained on the basis of different chemical composition and microstructural differences as observed by metallographic analyses.

KEY WORDS: corrosion, pitting, surgical implants, stainless steel, implant materials

The use of surgical metal implants in humans was first recorded in 1562 when a gold prosthesis was used to close a defect in a cleft palate [1].[2] They were primarily developed to function as prosthetic devices (for the replacement of damaged parts that cannot be restored through treatment) or fixation devices (for

[1]Graduate student and adjunct associate professor, respectively, Department of Mechanical Engineering, The University of Calgary, Calgary, Alberta, Canada T2N 1N4.
[2]The italic numbers in brackets refer to the list of references appended to this paper.

holding bone parts in place until healing is established). The short-comings, notably inadequate strength and inertness, exhibited by earlier materials resulted in the adoption of stainless steel AISI Type 316L (SS316L), cobalt-chromium-molybdenum (Co-Cr-Mo) alloy, and titanium alloys [1–4].

Several authors [3–7] have studied and commented on the successes and incessant failure of these alloys. Cahoon [8] reported that the problem remaining is corrosion, while Sonstegard et al [9] stated that the body puts severe demands on metals. Furthermore, implant failures are usually environmentally assisted owing to the combination of complex loading conditions and the hostile environment that is as corrosive as sea-water [10]. In spite of the use of "tissue compatibility" as a criterion for materials' selection, several traumatic side effects, such as hypersensitivity, have been discovered [4,11]. Scales [12], however, pointed out that " . . . the magnitude of the problem has not been investigated." Also "the variability of the chemical and electrochemical reaction . . . is not well identified or understood"[10].

With internal fixation, the disturbance of the blood supply to the bones is often accompanied by severe pathological changes that may affect healing and variation in the equilibrium state electrochemically [13,14]. The ionic species also perform numerous functions that include maintenance of body pH and participation in the oxidation-reduction (e^- transfer) reactions [15]. Normal imbalance occurs in the fluid compartment and different transport of ions and nonuniform changes normally accompany disease states. For example, during intensive care, after accidents or surgical operations, the fluid compartments are often disturbed [15]. From an electrochemical viewpoint, the acceleration of corrosion can be due to the differential conditions existing along the implant surface. These conditions may be responsible for the formation of electrochemical cells accompanied by active metal dissolution at favored localized spots at the implant-body fluid interface.

The corrosion of metal implants has been studied extensively in the literature. Revie and Greene [16,17] published a two-part paper that considered the effects of sterilization [16] and surface preparation [17] in isotonic saline solution (0.9% NaCl) at near body temperature. They have shown that increasing time or temperature of sterilization results in a decrease in corrosion rate. In the other case, electropolished surfaces corroded more slowly than the sandblasted finishes. Mueller and Greener [18] performed polarization studies in Ringer's solution. They found that the pitting tendency of stainless steels was dependent on oxygenation of the electrolyte and microstructure of the materials. Over 50% of several stainless steels and vitallium alloys sampled by Cahoon and Paxton [19] were found to suffer from metallurgical defects. Sandrik and Wragg [20] have shown that artifacts resulting from implant sterilization may produce microanodes that induce local tissue reactions. Bandy and Cahoon [21] performed potentiodynamic cyclic polarization tests on Type 316L stainless steel in Ringer's solution.

Previous electrochemical studies [2,4,12,18,21,22], including those just described, have considered the effects of chloride ion, temperature, and pH as well as stress and aeration tests in isotonic saline Ringer's, and Tyrodes solutions, on implant materials. In this investigation, a wrought Type 316L stainless steel (WSS316L) and a surgical-grade stainless alloy (SG2SA) were subjected to simulated environments by varying some of the ionic species in Hanks' physiological solution. It is intended here to compare and contrast the effects of Cl^- and HCO_3^- ion, as well as pH and temperature on the corrosion behaviors of the stainless alloys.

Experimental Procedure

Materials

The materials selected for this investigation were a wrought stainless alloy (WSS316L) taken from a 12.7-mm ($\frac{1}{2}$-in.) diameter hot-rolled rod and a surgical-grade stainless alloy (SG2SA) taken from a Zimmer total-hip replacement prosthesis. The nominal chemical compositions, in Table 1, are all within the recommended limits for standard surgical-grade stainless steels given in ASTM Specification for Stainless Steel Sheet and Strip for Surgical Implants (F 56-78) [23]. The one exception is the silicon content being quite high for the WSS316L. The minor differences in the chemistry of the alloys may result in different polarization characteristics. However, by superposing the chemical compositions on an iron-chromium-nickel (Fe-Cr-Ni) ternary phase diagram available in the literature [24], the alloys are found to be austenitic single-phase alloys.

Examination of the micro-structures of the alloys is important as variations in grain-size, the degree of prior-cold deformation, the presence of nonmetallic inclusions and other features inherent in the microstructures may affect the surface reactivity and corrosion-resistance properties. Some samples were examined using a Carl Zeiss, ICM 405 Research Metallograph.

From the microstructures shown in Figs. 1 and 2, both materials may be seen to contain minor amounts of oxides, sulfides, and silicates. The grain structures observed are typical for face-centered-cubic (fcc) materials containing numerous annealing twins. Also, the microstructures indicate that the materials are single-phase, austenitic, and relatively free from carbides and delta-ferrite. However, the grain sizes may be seen to be significantly different. The grain size of WSS316L corresponded to ASTM No. 3, (mean grain diameter of approximately

TABLE 1—*Nominal chemical composition of the alloys used for corrosion studies (percent by weight).*

Material/ Components	Carbon	Chromium	Nickel	Molybdenum	Magnesium	Silicon	Cobalt	Iron
WSS316L	0.022	19.600	10.800	2.500	2.000	1.100	. . .	balance
SG2SA	0.02	17.30	14.00	2.61	1.84	0.27	0.12	balance

FIG. 1—*Microstructure of WSS316L material, 10% oxalic acid etch, original magnification* × *100.*

FIG. 2—*Microstructure of SG2SA material, 10% oxalic acid etch, original magnification* × *500.*

0.144 mm) and that of SG2SA to ASTM No. 7, (mean grain diameter of approximately 0.036 mm) [25]. Additionally, numerous strain lines may be observed in the microstructure of WSS316L, whereas the SG2SA alloy is seen to be free from prior cold-straining.

The effects of plastic deformation (cold work) on metal corrosion has been researched extensively [26,27]. Generally, cold rolling introduces a number of dislocations that are arrayed in their slip planes. When a metal is severely cold worked, arrays of dislocations on several intersecting sets of slip planes are produced. Such deformation is known to increase the dislocation density for annealed iron from perhaps 10^8 to 10^{12} dislocations/cm^2 [28]. Greene and Saltzman [27] pointed out that prior cold work increases the rate of the anodic dissolution reaction, Fe $\rightarrow$ Fe^{2+} + $2e^-$, and has a lesser effect on the rate of hydrogen evolution. Greene and Saltzman have also suggested that plastic deformation will cause an increase in the number of active surface sites (emerging dislocations) for activation of dissolution processes.

Hence, any observed differences in polarization characteristics can be related to the preceding chemical and microstructural differences.

Electrolyte

The choice of an appropriate solution to duplicate the physiological conditions in the human body has been considered inadequate in several studies for *in vitro* assessment of the corrosion characteristics of implant alloys [29]. The solutions that can be used for physiological and clinical studies have been listed by Guyton [15]. Hanks' solution [4] (NaCl, 8; CaCl$_2$, 0.14; KCl, 0.4; NaHCO$_3$, 0.35; Glucose, 1; MgCl$_2 \cdot$ 6H$_2$O, 0.1; Na$_2$HPO$_4 \cdot$ 2H$_2$O, 0.06; KH$_2$PO$_4$, 0.06; MgSO$_4 \cdot$ 7H$_2$O, 0.06; in gram/litre) was used in this study because it contains as much of the anion and cation as are found in the body fluid.

Polarization Measurements

The equipment and technique used in this study for potentiodynamic polarization measurements conform with the ASTM Practice for Standard Reference Method for Making Potentiostatic and Potentiodynamic Anodic Polarization Measurements (G 5-78) [30], and detailed experimental procedure are described elsewhere [31]. All the polarization measurements were generated at a scan rate of 0.6 V/h. The specimen surface was polished with 600-grit silicon carbide paper, rinsed in distilled water, and air dried. The solutions were prepared by dissolving reagent-grade chemicals in distilled water. Nitrogen was bubbled through the solution to maintain deaeration. This was done in order to generate adequate active dissolution of the electrode. Moreover, the pitting characteristics of the steel can be studied.

The results presented here were obtained with stationary, vertical, and planar electrodes and partly quiescent solutions. Any form of stirring was due to the purging gas that was maintained at the same flow rate during all the runs. With the exception of the study of the influence of temperature, the test electrolytes were maintained at 37°C. The temperature was controlled to ± 0.5°C using a constant temperature water bath.

Results and Discussion

Corrosion Behavior in Hanks' Solution

The alloys displayed similar trends in the variation of their corrosion potential with time; a change of which was noticed in the noble direction within the first 2 min of immersion followed by a gradual change of about 2 to 5 mV over a 5-min period. Similar observations were made by Mueller and Greener [18] in which a 4-mV change over a 5-min period was recorded within 2 to 3 h. This action, perhaps, might imply the nucleation and growth of a stable oxide film at sites that favor maximum or multilayer absorption of an oxidizing agent [32]. Figure 3 presents the experimental anodic polarization of the alloys in deaerated Hanks' physiological solution maintained at 37°C. Two sections are evident as observed in the present study, the passive region and a "distinct" potential at which passivity locally breaks down with a concomitant increase in the anodic current density. The surgical-grade stainless alloy (SG2SA) displayed more noble characteristics than the wrought stainless steel (WSS316L). In addition, the former possesses more resistance to pitting attack due to the higher breakdown potential, E_b, of +300-mV saturated calomel electrode (SCE) compared with +200-mV SCE WSS316L. The superior corrosion performance of the SG2SA may be attributed to higher molybdenum content and additional cobalt in the

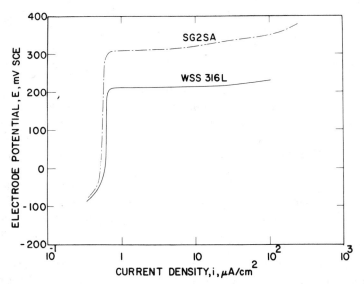

FIG. 3—*Anodic polarization characteristics of the test alloys in deaerated Hanks' physiological solution maintained at 37°C.*

alloy. The molybdenum enhances the resistance to pitting and crevice corrosion in chloride-containing environments and nonoxidizing acids [*33,34*].

Also, it is suggested that the prior cold deformation observed for WSS316L in its microstructure may have contributed to the corrosion potential being more negative than that for SG2SA. The differences in grain sizes too may have contributed in part to the observed electrochemical behavior differences. The lower breakdown potentials noted for WSS316L in the anodic polarization curves may be attributed to a combination of the coarse grain structure and the adsorption of anions (for example, Cl^-) at the active dislocation sites. Finally, significant metal dissolution will be thermodynamically favored dislocation sites due to the lower bonding energies at such sites.

It is sufficient at this time to state simply that even small dissimilarities in the microstructures of materials subjected to hostile service environments may be enough to promote significant increases in chemical or electrochemical reaction rates. Basically, these plots in Fig. 3 serve as reference curves and show mainly the corrosion behavior of the alloys in Hanks' physiological solution. The deviations from these curves due to environmental variation are explained in subsequent sections.

Effect of Cl^- ion

The influence of Cl^- ion was examined by varying the sodium chloride (NaCl) content of the Hanks' solution. It can be seen in Fig. 4 that the corrosion potential shifts to more negative values as the Cl^- ion concentration increases, and that the SG2SA remains more noble than WSS316L. The shapes of the polarization

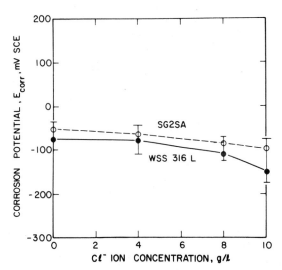

FIG. 4—*Influence of Cl^- ion (NaCl variation) in Hanks' physiological solution on the corrosion potential of the test alloys at 37°C.*

curves were similar to those shown in Fig. 3. However, further anodic polarization might have favored dissolution of metal ions if the passive film were soluble in the surrounding electrolyte or if aggressive anions (for example, Cl^-) adsorbed to the metal surface in place of the oxidizing agent from the structure promoting a decrease in the repairing action of the passive film and inhibiting spontaneous repassivation [32,35,36].

The critical passive breakdown potentials, for both alloys, moved towards less noble values with increasing Cl^- ion concentration. The mechanisms responsible for such action by Cl^- have been explained in the literature [35,36]. During anodic polarization, however, concentration of Cl^- ion at the metal surface may increase, eventually reaching a value that allows Cl^- to displace any adsorbed species. The amount of Cl^- ion adsorbed is controlled by the concentration of Cl^- in solution. Since the maximum amount of Cl^- ion that can be adsorbed on favored sites is not stated anywhere in the literature (that is, the Cl^- adsorption saturation limit is unknown), it is sufficient to say that the adsorption rate will be greater with more Cl^- ion in solution; a situation existing in the body fluid. Hence, enough accumulation of Cl^- responsible for passivation breakdown is obtained at lower electrode potentials during extended anodic polarization. This suggests that below this critical potential, so long as the passive film remains unimpaired, Cl^- cannot displace the adsorbed species; hence pitting or metal dissolution is obviated. As shown in Fig. 5, the breakdown potential for SG2SA is higher than that for WSS316L. This shows that the characteristic passive film on the former alloy is more stable at higher potentials than for the latter material.

Effect of HCO_3^- ion

The effect of HCO_3^- ion was studied by altering the $NaHCO_3$ content of the Hanks' solution. The variation of HCO_3^- ion caused a change in the pH of the solution due to its buffering nature; the pH measured at room temperature varied

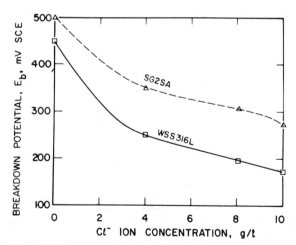

FIG. 5—*Effect of Cl^- ion concentration on the breakdown potential of the test alloys.*

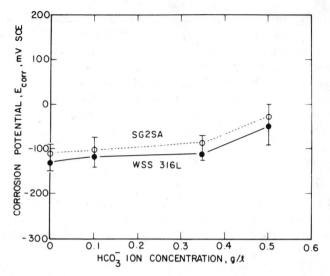

FIG. 6—*Influence of HCO₃⁻ ion (NaHCO₃ variation) in Hanks' physiological solution on the corrosion potential of the test alloys.*

from 6.52 with 0 g/L NaHCO₃ to 7.52 with 10 g/L NaHCO₃. No attempt was made to buffer the pH to that of Hanks' solution of 7.50. The corrosion potential, as shown in Fig. 6, became more positive with increasing HCO₃⁻ ion. Although the shapes of the polarization curves were similar to that shown in Fig. 3, the decrease in HCO₃⁻ ion concentration caused a shift in the passive breakdown potential in the active direction (Fig. 7). This effect is probably due to the reduction in the oxidizing power of the environment, resulting in the formation

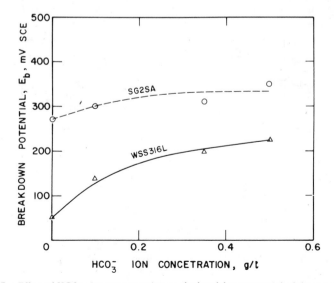

FIG. 7—*Effect of HCO₃⁻ ion concentration on the breakdown potential of the test alloys.*

of a weak oxide-filmed surface. From the polarization measurements, the magnitude of the passive current density (i_{pass}) for WSS316L was determined to vary between 0.45 $\mu A/cm^2$ and 0.70 $\mu A/cm^2$ while it was relatively constant, at approximately 0.5 $\mu A/cm^2$, for the SG2SA alloy. The i_{pass} of 0.5 $\mu A/cm^2$ for these alloys under normal conditions is in agreement with the finding of Mueller and Greener in Ringer's solution [18]. This low passive current is indicative of the general corrosion resistance of the alloys that is attributed to the tough passive film on the metal surface.

Effect of pH

The influence of pH was studied by reducing the NaCl content of Hanks' solution and introducing HCl or NaOH as appropriate. For a pH ≃ 2, 300 mL of 0.2 M HCl was added to Hanks' solution, the NaCl being reduced to 6 g/L. On open-circuit, the WSS316L in solution of pH ≃ 2, while initially active, showed an initial rise in potential followed by a rise and fall that indicates "sequential formation of pits, repassivation, and recurrent local breakdown of passivity" [4]. Similar performance was observed by White et al [37].

The variation of pH as it affected the corrosion potentials of the alloys is illustrated in Fig. 8. These changes in corrosion potentials will promote the formation of local action cells in which microanodes and cathodes on discrete areas are developed. The eventual effect of pH shows that the alloys are more noble in regions where the solution is relatively neutral. This corresponds with situations around the normal state of the body fluid.

The polarization curves for variable pH electrolytes using WSS316L and SG2SA test specimens are shown in Figs. 9 and 10, respectively. The results generally show no active corrosion regions at very low potential except for the solution acidified to pH = 2. Similar observations were made by Cahoon [8];

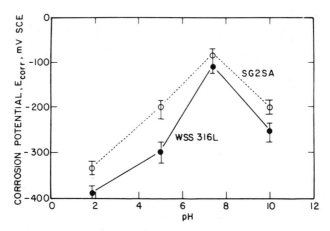

FIG. 8—*Influence of pH of Hanks' physiological solution on the corrosion potential of the test alloys.*

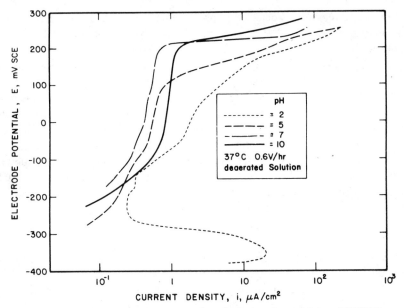

FIG. 9—*Effect of variation in pH on anodic polarization characteristics of WSS316L.*

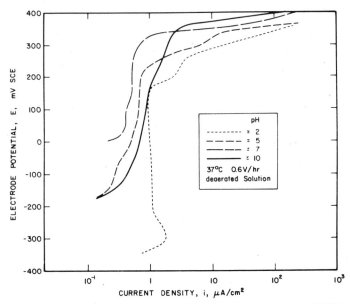

FIG. 10—*Effect of variation in pH on anodic polarization behavior of SG2SA.*

that no active region was detected in physiological solutions for stainless steels at potentials as low as -300 mV versus SCE. An exception to this is that under highly acidic conditions, as active dissolution was observed below -300 mV SCE. The alloys exhibited, at low pH, active passive transition; with the critical current density (i_{crit}) needed for passivation found to be 1.8 $\mu A/cm^2$ for SG2SA alloy and 22 $\mu A/cm^2$ for WSS316L. This shows that the former is more easily passivated in such environments. The wider passive region on SG2SA is an indication of a more stable oxide film on its surface than WSS316L. Hence, crevice areas, where a local drop in pH below 2 has been noted [6,21], are undesirable for stainless steels. The variation of pH as it affected the breakdown potential, E_b, is illustrated in Fig. 11. It may be seen that the breakdown potential increased with increasing pH.

Effect of Temperature

Tests over a small range of temperatures (25 to 60°C) were undertaken that brackets the normal body temperature of 37°C. As seen in Fig. 12, the corrosion potential moved slightly to more positive values with increasing temperature. The polarization curves were observed to be of similar shape to that shown in Fig. 3. The passive current densities (average 0.5 $\mu A/cm^2$ for both alloys) did not change much with temperature. However, the breakdown potentials (see Fig. 13) dropped with increasing temperature leading consequently to a decrease in pitting resistances [38]. This finding agrees with the work of White et al [37]. Also, a similar response was observed by Sury [7] for which the pit repassivation

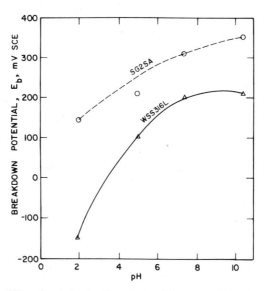

FIG. 11—*Effect of variation in pH on the breakdown potential of the test alloys.*

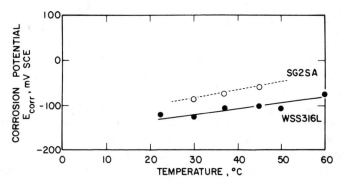

FIG. 12—*Effect of temperature on the corrosion potential of the test alloys in Hanks' physiological solution.*

potentials in Ringer's solution at a current density of 10 $\mu A/cm^2$ dropped sharply with temperature.

Cyclic Polarization Behavior of WSS316L

The alloys have been shown to possess some similarities in chemical composition. In this respect, cyclic polarization is expected to provide similar trends in the polarization characteristics. This appears to be the case in the results presented earlier on. Hence, cyclic polarization was performed on the WSS316L alone to study the chances of repassivation of growing pits under variable environmental conditions.

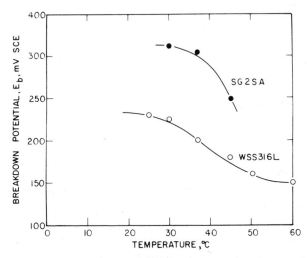

FIG. 13—*Influence of temperature on the breakdown potential of the test alloys in Hanks' physiological solution.*

Effect of Cl⁻ Ion

By using similar NaCl concentrations adopted previously, cyclic polarization curves were generated at a scan rate of 0.6 V/h. The test scans were extended up to a current density of $\simeq 100$ μA/cm^2 and then reversed. The polarization curves are shown in Fig. 14. It can be seen that at constant current density reversal, the potential at which this reversal occurred shifted to more negative values with increasing chloride ion concentration that is consistent with the observed lowering of breakdown potentials with increasing Cl⁻ concentrations noted earlier. Furthermore, a considerable hysteresis loop was observed and is displayed in Fig. 14, the size being dependent on the Cl⁻ ion concentration. The essential parameters from Fig. 14 are given in Table 2. The protection potential is lowered with increasing Cl⁻ ion concentration, but the magnitude of $(E_b - E_p)$ increases somewhat with increasing Cl⁻ ion. This indicates that the chance of repassivation of WSS316L decreases with increasing Cl⁻ ion content. The active value of E_p relative to the E_b will lead to growing pits that have no chance of repassivation [21], a situation highly undesirable in surgical implant performance.

Effect of HCO₃⁻ Ion

The experimental technique for tests to evaluate the influence of HCO$_3^-$ ion concentration was similar to that used for Cl⁻ ion concentration studies. Figure 15 shows the polarization curves with current reversals also made at current

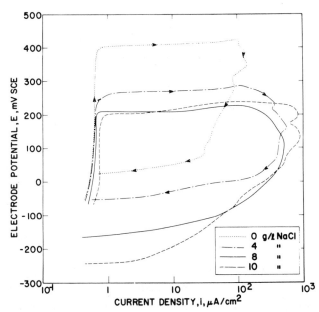

FIG. 14—*Results from potentiokinetic cyclic experiments using WSS316L in Hanks' physiological solution at 37°C with variable concentrations of NaCl showing hysteresis loops characteristic of susceptibility to crevice corrosion or pitting attack or both. Note: the arrows indicate scan directions.*

TABLE 2—*Cyclic polarization data of wrought stainless steel in Hanks' solution.*

Variable	Corrosion Potential, E_{corr}, mV (SCE)	Potential at Reversal, mV (SCE)	Current Density at Reversal, $\mu A/cm^2$	E_b, mV (SCE)	E_p, mV (SCE)	$E_b - E_p$, mV (SCE)
			NaCl			
0 g/litre	− 75	+425	100	+400	+ 20	+380
4 g/litre	− 80	+290	100	+230	− 50	+280
8 g/litre	−110	+230	100	+200	−170	+370
10 g/litre	−150	+240	100	+175	−240	+415
			NaHCO₃			
0 g/litre	−130	+220	100	+150	− 70	+220
0.1 g/litre	−115	+200	100	+170	− 70	+240
0.35 g/litre	−110	+230	100	+200	−170	+370
0.50 g/litre	− 50	+300	100	+210	−175	+385
			Temperature			
25°C	−125	+300	500	+220	+ 90	+130
37°C	−110	+300	3 500	+200	−280	+480
50°C	−110	+300	2 300	+170	−290	+460
60°C	− 80	+300	50 000	+140	−350	+490

densities of 100 $\mu A/cm^2$. The results show some erratic behavior that might be attributed to pitting or occurrence of a different reaction at the electrode surface. The relevant parameters are given in Table 2. The protection potentials, E_p, are not significantly far apart for concentrations of NaHCO₃ of 0 and 0.1 g/L NaHCO₃ and the concentrations of 0.35 and 0.50 g/L HaHCO₃; but the differ-

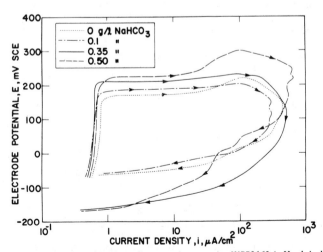

FIG. 15—*Results from potentiokinetic cycling experiments using WSS316L in Hanks' physiological solution at 37°C with variable concentrations of NaHCO₃ showing hysteresis loops characteristic of crevice corrosion or pitting attack or both. Note: the arrows indicate scan directions.*

ences in $(E_b - E_p)$ values tend to increase with increasing HCO_3^- ion concentration. This finding also indicates that the variation in HCO_3^- ion limits the chances of repassivation of growing pits in stainless steels. Although increasing the bicarbonate ion concentration caused an increase in breakdown potential (which is beneficial), it may be seen that the bicarbonate ion influences detrimentally the values for repassivation potential, these latter potentials being lowered with increasing HCO_3^- ion concentrations.

Effect of Temperature

The cyclic polarization scans for all temperatures examined were reversed at values of electrode potential of $\simeq +300$ mV SCE. This differs from the procedure adopted for the studies in variable Cl^- and HCO_3^- ion concentration. It was found in trail tests that at higher temperatures it was difficult to reverse the scans at exactly 100 $\mu A/cm^2$ because of the acceleration in dissolution rate. Moreover, it was intended to provide the effect of extended polarization on the repassivation of growing pits in corroding electrodes. The results from the polarization scans are shown in Fig. 16, and the extracted data are given in Table 2. At the potential of reversal, it can be seen that the current density increased with increasing temperature. The magnitude $(E_b - E_p)$ was higher than that observed for variable Cl^- and HCO_3^- ion tests at 37°C. This behavior demonstates the effect of polarization on the protection potential [21] that moves to more active values with increasing temperature. The large hysteresis loops are the result of extended polarization that increases the amount of localized attack within minute pits [21]. This is due to the fact that more time has been made available for the reaction involving hydrolysis of corrosion products and chloride ion buildup within pit cavities [21]. Comparing the case at 37°C that corresponds

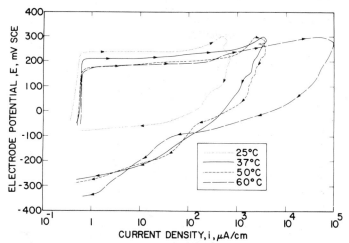

FIG. 16—*Results from potentiokinetic cyclic tests using WSS316L in Hanks' physiological solution at the temperatures indicated. Note: the arrows indicate scan directions.*

to the environment with 8 g/L NaCl and 0.35 g/L NaHCO$_3$, the difference $(E_b - E_p)$ has been increased from +370 mV (SCE) to +480 mV (SCE) due to current reversals of 100 μA/cm^2 and 3500 μA/cm^2. To this end, temperature differentials reduce the propensity for spontaneous repassivation of growing pits on stainless steels implanted in the human body. Although the ranges of temperature anticipated from (for example) infections or fever will be much smaller than those studied in this work, temperature differentials from whatever sources in the area of a metallic implant will be detrimental to the corrosion resistance of stainless steels in biofluids and may be a contributing factor to early failures by corrosion or corrosion-assisted fracture [37].

Summary and Conclusions

Normally, Type 316L stainless steels exhibit reasonable corrosion resistance in biological fluids as was demonstrated by the results of Fig. 3. Active corrosion is not apparent at low electrode potentials, and critical current densities for spontaneous passivation are low. Passivation breakdown potentials were shown to be reasonably high in Hanks' physiological solution. All of these factors are beneficial from a corrosion-resistance perspective.

However, the results presented in the foregoing text have shown that local environmental variables (temperature, Cl$^-$, HCO$_3^-$ ions, and pH) can significantly alter electrochemical behavior. It has been shown also that the microstructural variations (grain size, dislocation density) may help to explain differences in behavior between alloys of nominally similar chemical compositions. Variations in stress along the surfaces of metallic implants due to stress concentrations at holes, notches, etc.; variations in microstructure; temperature variations; among others can be shown to promote local redistribution of ions (Cl$^-$, HCO$_3^-$, H$^+$) and the results from the research presented as part of this work have demonstrated that local variations in the concentrations of these ions may adversely influence corrosion resistances. Generally, the following conclusions are drawn from this work:

1. Increasing concentrations of chloride ions adversely affected corrosion resistance by lowering passivation breakdown potentials, marginally lowering corrosion potentials, lowering repassivation potentials, and generally increasing the propensity to pitting attack as evidenced by the increase in size of hysteresis loops from potentiodynamic cycling test.

2. Increasing concentrations of bicarbonate tended to raise the breakdown potentials but also increase the corrosion potential. Thus, in combination with a high chloride concentration, high bicarbonate concentrations may raise the corrosion potentials such that they border on passivation breakdown. The increase in hysteresis loop size on potentiodynamic cycles with increasing bicarbonate concentration shows a lowered resistance to pitting attack and crevice corrosion.

3. Variations in hydrogen ion concentrations (pH) promoted increases in cor-

rosion potentials from acid pH levels to neutral pH; thereafter, however, corrosion potentials were lowered in alkaline solutions to more active values. Decreasing pH caused a lowering of breakdown potentials in the presence of Cl^- and an increase in critical densities for passivation.

4. Increasing temperature tended to raise corrosion potentials, lower breakdown potentials, thereby generally lowering corrosion resistance.

In summary, it is re-emphasized that Type 316L stainless steels generally are corrosion resistant in biological fluids and are suited to application in orthopaedic implants. This is supported by the low incidence of implant failures (<5% from corrosion or corrosion-related fracture mechanisms or both) [37]. However, it is suggested that the origins of the failures that do occur are a consequence of a number of factors acting singly or conjointly to precipitate the failures. It is further suggested that the results presented in the foregoing text demonstrated that localized variations in ionic concentrations and temperature along the surfaces of the implants along with microstructural variations may contribute, in part, to early failure by corrosion of surgically implanted devices.

References

[1] Wickstrom, J. K., Journal of Materials, Vol. 1, No. 2, 1966, pp. 366–372.
[2] Sawyer, P. N. and Srinivasan, S., Medical Engineering, C. D. Ray, Editor in Chief, Yearbook Medical Publications Inc., 1974, pp. 1099–1101.
[3] Cohen, J., Journal of Biomedical Materials Research, Vol. 4, 1970, pp. 223–244.
[4] Mears, D. C., International Metals Reviews, Vol. 22, No. 6, 1977, pp. 122–255.
[5] Weistein, A., Amstutz, H., Pavon, G., and Franceschini, V., Journal Biomedical Materials Research Symposium, No. 4, 1973, pp. 297–325.
[6] Down, G. M. in The Fixation of Fractures Using Plates, Journal of Mechanical Engineering, London, 1974, pp. 24–29.
[7] Sury, P., Corrosion Science, Vol. 17, 1977, pp. 155–169.
[8] Cahoon, J. R., "The Failure of Orthopaedic Implants," presented at Canadian Institute of Mining and Metallurgy Conference on Metallurgists, Quebec City, 26–29 Aug. 1973.
[9] Sonstegard, D. A., Mathews, L. S., and Kaufer, H., Scientific American, Vol. 238, No. 1, 1978, pp. 44–51.
[10] Lisagor, W. B., ASTM Standardization News—Stress Corrosion Cracking, Vol. 3, No. 5, 1975, pp. 20–24.
[11] Ferguson, A. B., Laing, P. G., and Hodge, E. S., Journal of Bone and Joint Surgery, Vol. 42–A, No. 1, 1960, pp. 77–90.
[12] Scales, J. T. in The Fixation of Fractures Using Plates, Journal of Mechanical Engineering, London, 1974, pp. 1–8.
[13] Walker, G. D., Journal of Biomedical Materials Research Symposium, No. 5 (Part 1), 1974, pp. 11–26.
[14] Jacobson, B. and Webster, J. B., Medicine and Clinical Engineering, Prentice Hall, Inc., NJ, 1977, pp. 525–532.
[15] Guyton, A. C., Textbook of Medical Physiology, 5th ed., W. B. Saunders, Philadelphia, 1976, pp. 424–437.
[16] Revie, R. W. and Greene, N. D., Corrosion Science, Vol. 9, 1969, pp. 755–761.
[17] Revie, R. W. and Greene, N. D., Corrosion Science, Vol. 9, 1969, pp. 763–770.
[18] Mueller, H. J. and Greener, E. H., Journal of Biomedical Materials Research, Vol. 4, 1970, pp. 29–41.
[19] Cahoon, J. R. and Paxton, H. W., Journal of Biomedical Materials Research, Vol. 4, 1970, pp. 223–244.

[20] Sandrik, L. and Wragg, L. E., *Journal of Biomedical Materials Research*, Vol. 4, 1970, pp. 275–277.

[21] Bandy, R. and Cahoon, J. R., *Corrosion*, Vol. 33, No. 6, 1977, pp. 204–208.

[22] Jones, R. L., Wing, S. S., and Syrett, B. C., *Corrosion*, Vol. 34, No. 4, 1978, pp. 226–236.

[23] *Annual Book of ASTM Standards*, Part 46, 1979.

[24] *Metals Handbook*, American Society for Metals, 8th ed., Vol. 8, 1973, p. 425.

[25] Kehl, G. L., *The Principles of Metallographic Laboratory Practice*, McGraw Hill Book Company, Inc., 1949.

[26] Foroulis, Z. A., *Journal*, Electrochemical Society, Vol. 113, No. 6, 1966, pp. 532–536.

[27] Greene, N. D. and Saltzman, G. A., *Corrosion*, Vol. 20, 1962, pp. 293t–298t.

[28] Foroulis, Z. A. (Ref. *36*), citing Darken, L., *The Physical Chemistry of Metallic Solutions and Intermetallic Compounds*, Vol. 2, 4G, National Physics Laboratories Symposium, No. 9, London, 1959.

[29] Aragon, P. J. and Hulbert, S. F., *Journal of Biomedical Materials Research*, Vol. 6, 1972, pp. 155–164.

[30] *Annual Book of ASTM Standards*, Part 10, 1979, p. 757.

[31] Ogundele, G. I., Master of Science thesis, The University of Calgary, Sept. 1980.

[32] Uhlig, H. H., *Corrosion Science*, Vol. 7, 1967, pp. 325–339.

[33] Uhlig, H. H., *Corrosion and Corrosion Control*, Wiley, New York, 2nd ed., 1971.

[34] Brown, J. H. U., Jacobs, J. E., and Stark, L., *Biomedical Engineering*, F. A. Davis Company, Philadelphia, 1971, pp. 308–321.

[35] Okamoto, G., *Corrosion Science*, Vol. 13, 1973, pp. 471–489.

[36] Hoar, T. P., *Corrosion Science*, Vol. 7, 1967, pp. 341–355.

[37] White, W. E., Postlethwaithe, J., and Le May, I., *Microstructural Science*, Vol. 4, American Elsevier, New York, 1976, pp. 145–157.

[38] Mattson, E., *British Corrosion Journal*, Vol. 13, No. 1, 1978, pp. 5–12.

Charles C. Irving, Jr.[1]

Electropolishing Stainless Steel Implants

REFERENCE: Irving, C. C., Jr., **"Electropolishing Stainless Steel Implants,"** *Corrosion and Degradation of Implant Materials: Second Symposium, ASTM STP 859,* A. C. Fraker and C. D. Griffin, Eds., American Society for Testing and Materials, Philadelphia, 1985, pp. 136–143.

ABSTRACT: Virtually every manufacturer of AISI Type 316L stainless steel surgical implants uses electropolishing as a surface finishing procedure, but little research has been devoted to the basic chemistry, techniques, and possible benefits of the process. The electrochemical reactions that take place in the electropolishing cell are discussed, as well as the physical techniques used to electropolish stainless steel. Data on the corrosion resistance of electropolished stainless steel is compared to conventional passivation treatments.

KEY WORDS: electropolishing, stainless steels, corrosion resistance, oxide layer, implant materials, degradation, fatigue (materials)

The electropolishing process was first described by Jacquet in 1936 [1].[2] He observed polishing of metallographic copper specimens when an anodic current was applied in orthophosphoric acid. In 1936 to 1937, Faust discovered mixtures of orthophosphoric and sulfuric acids produced a superior polishing effect on stainless steel [2]. He described the metal surface to be highly lusterous and free from scratches and the "piled" layers characteristic of mechanically polished surfaces [3]. These solutions patented in 1943 form the basis for contemporary electrolytes used to electropolish AISI Type 316L stainless steel implants.

The amount of basic research aimed at understanding electropolishing, the techniques involved, and the possible benefits of the process seem less than warranted, considering the wide-spread use in research and industry. Surveying the literature available on electropolishing, one finds only brief references to the increase in corrosion resistance of electropolished surfaces. Only recently has the influence of electropolishing on the corrosion resistance of AISI Type 316L stainless steel been presented [4]. The object of this discussion is to present a brief electrochemical description of the process, industrial techniques used, and data on the corrosion resistance of electropolished stainless steel.

---

[1]President, Digintel, Inc., New York, NY 10006.
[2]The italic numbers in brackets refer to the list of references appended to this paper.

Electrochemistry

A typical polarization curve for stainless steel in a sulfuric-orthophosphoric acid electropolishing solution is shown in Fig. 1 [5]. This curve is characteristic of a metal surface that can be passivated [6]. In the active region (A-B), the surface is aggressively attacked and etching occurs. At the critical current density (B) a thin passive film of slowly dissolving oxides begins to form. The curve enters the passive potential range at the Flade Potential (C) [7]. The dissolution current in this range is approximately 100 times less than critical current density.

As the potential is increased further, the current density begins to rise (D). Surface atoms are oxidized and diffuse into the solution. At high current densities, the dissolution rate is very high and the metal ions are unable to diffuse into the solution rapidly enough to replenish sulfate-phosphate acceptor ions. This is called concentration polarization.

The limiting current density (E) is reached when acceptors into the diffusion layer and metal-acceptors out of the diffusion layer are at a maximum. No more rapid flow of ions can occur in this case [8]. Near the limiting current density, ideal electropolishing conditions are established, that is, the metal surface becomes bright and smooth.

Two discrete reactions at the anode surface are responsible for the mechanism of electropolishing. They are

1. The formation of a thick, viscous diffusion layer that controls smoothing of the surface by dissolution of peaks over 1 μm (macrosmoothing).

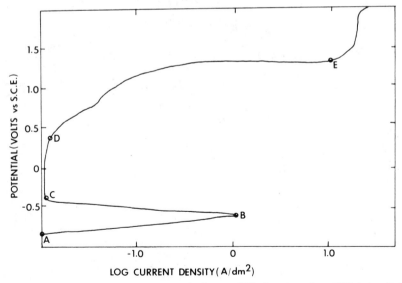

FIG. 1—*Typical anodic polarization curve for iron + 18% chromium alloy, 40°C, in a sulfuric-orthophosphoric acid electrolyte.*

2. The formation of a thin solid film on the surface that controls brightening or dissolution of peaks down to 0.01 μm (microsmoothing) [9].

The viscous film surrounding the anode has a high ohmic resistance that is proportional to the film thickness. Macroscopic peaks are less shielded by the film, therefore receive a higher microcurrent density than the valleys. They dissolve at a faster rate and macrosmoothing occurs, Fig. 2 [2]. The thin solid film is only a few atomic layers thick and is theorized to protect the surface from preferential attack of high energy sites [9]. The random removal of metal ions due to cation vacancies in the film causes brightening of the surface.

At the high potentials involved in electropolishing AISI Type 316L stainless steel, metal ions diffuse into the solution in the highest valent state Eqs 1 to 3 [10].

$$\text{Surface Anode Reaction} \longrightarrow \text{Solution} \longrightarrow \text{Cathode Reaction}$$

$$\text{Fe} \longrightarrow \text{Fe(III)} + 3e \longrightarrow \text{Fe(II)} \tag{1}$$

$$\text{Ni} \longrightarrow \text{Ni(II)} + 2e \longrightarrow \text{Ni(II)} \tag{2}$$

$$\text{Cr} \longrightarrow \text{Cr(VI)} + 6e \longrightarrow \text{Cr(III)} \tag{3}$$

These cations are soluble and migrate in the electrolyte to the cathode, where reduction and precipitation occurs. The dissolved metal content of an electro-

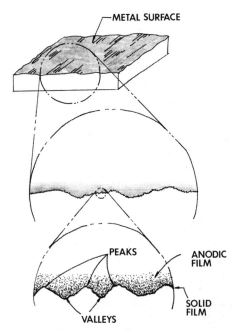

FIG. 2—*Pictorial representation of the anodic film formed during electropolishing.*

polishing solution usually increases with time since the efficiency of the reduction process is less than 100%.

Industrial Physical Techniques

Most electrolytes used in industry to electropolish stainless steel are based upon the sulfuric-phosphoric acid formulas developed by Dr. Faust [2]. Some proprietary solutions include organic compounds such as hydroxy-acetic acid, lactic acid, and diethylene glycol monobuytl ether. The composition of electrolyte usually determines the operating parameters such as temperature, current density, specific gravity, and time, but the techniques illustrated in this discussion are useful for any electropolishing system.

The industrial electropolishing cell is normally a chemical pure lead-lined steel tank, ranging in size from one gallon to several thousand gallons. The tank is electrically insulated from the floor and surrounding environment. Heat is provided by a thermo-regulated lead steam coil or an electric quartz immersion heater. A direct current power supply is used to provide sufficient current and potential for the electropolishing process. The power supply may be as simple as a manual powerstat type or versatile as a solid-state type with automatic constant current, voltage stabilization, ampere-minute timing, etc. Choice of a power supply is usually determined by the degree of process control required. Agitation of the electrolyte is necessary to prevent streaks from the gas evolved on the workpiece.

The agitation should be enough to disburse the gas bubbles, but should not disturb the thick diffusion layer surrounding the anode. Experience has shown that an oscillating anode rod will provide sufficient agitation without disturbing the anodic film.

Copper rods, insulated from the cell, are used to provide power to the electrodes. Cathode materials are normally chemical pure lead or stainless steel. A bare copper fixture can be used to hold the stainless steel implant in most cases, but more often it is coated with an acid-resistant plastisol. Actual contact to the implant is made with a titanium spring clip, using special care not to scratch the implant when attaching it to the fixture.

The implants are cleaned in accordance to ASTM Recommended Practice for Surface Preparation and Marking of Metallic Implants (F 86–76) after they are placed on the fixture. Before entering the electropolish cell, the parts should be relatively dry to prevent water contamination of the electropolishing solution. The fixture is connected to the anode rod, the motor-drive to oscillate the rod is turned on, and the power supply is switched on to the proper operating output. Time for electropolishing is usually determined by the dimensional tolerances and surface finish of the implant prior to electropolishing.

When electropolishing is complete, the power supply is switched off and agitation ceased. The fixture is removed from the solution and a yellow film covering the implants is observed. This is part of the thick, viscous film mentioned earlier. The best method found to remove this high viscosity film is by placing the fixture in a fogging spray of dilute electropolish solution (dragout) [11]. The implants are rinsed in clean water and dried. Once again care must be taken not to scratch the implant when removing them from the fixture.

There is very little documentation concerning quality assurance and process control for electropolishing. In some cases, the recommended practice is to discard the solution when the metal content exceeds the level specified for the electrolyte. The chemistry involved in the process causes an increase in the metal content, a slight decrease in water content from hydrolysis, and increases in water content from water adhering to the work-piece (drag-in) to the tank. A better approach to process control may be to monitor the specific gravity and the metal content. Using this method, decisions to decant part of the solution or add water can be made. Quality assurance for electropolished parts has yet to be determined. Suggested methods include nondestructive visual tests, destructive electrochemical testing, cross-sectioning for microscopic examination, or copper sulfate passivation tests. Clearly, better quality assurance procedures will be determined from additional research in electropolishing.

Corrosion Resistance

The increase in corrosion resistance of electropolished material is of great interest. Aluminum, zinc, stainless steel, carbon steel, and bronze are shown to have a higher resistance to aqueous corrosion than mechanically polished specimens [12]. In 1962, General Dynamics/Astronautics Division in San Diego recommended the use of electropolishing to increase the corrosion resistance of stainless steel in propellant system material to the National Aeronautics and Space Administration. Their observations were based on accelerated and actual seacoast exposure tests [11]. Revie and Greene showed electropolished AISI Type 316 stainless steel to be the most corrosion resistant when compared to a sand-blasted, 1/0 emery polished, and rouge-buffed surfaces [13]. More recently, Sutow has demonstrated the increased corrosion resistance of electropolished AISI Type 316L stainless steel specimens by the more anodic corrosion and breakdown potentials [4].

Several factors contribute to the increase in corrosion resistance of electropolished stainless steel. One of these is undoubtedly the removal of an amorphous, deformed surface called the Beilby Layer. This layer, produced by mechanical finishing, is characterized by crystal fractures and other structural changes. Oxides, polishing compounds, and other materials become embedded in the distorted crystal structure Fig. 3 [12,14]. The conditions established are ideal for the formation of surface corrosion cells. A related factor discussed by Sutow [4] that would affect the corrosion resistance is the surface roughness factor of 1.1 for electropolished and 3.1 for an austenitic stainless steel surface finished with 2/0 emery paper. Faust attributes an increased corrosion resistance to a mildly "anodized" stainless steel surface [14]. A study by Ducrocq et al [15] used secondary ion mass spectroscopy (SIMS) and X-ray photoelectron spectroscopy (XPS) techniques to characterize oxide layers approximately 60 Å deep generated

METAL SURFACE CROSS SECTION

FIG. 3—*Pictorial representation of the deformed surface of a metal subjected to mechanical finishing.*

by different surface preparations of AISI Type 316L stainless steel. Superficial contamination from the polishing abrasion products was found in the first few angstroms near the surface. The thickness of the oxide layer formed by different polishing procedures (600 grit paper, diamond polishing, and electropolishing) are comparable (50 to 70 Å) and correspond with the thickness after passivation. The composition of the films formed by 600 grit paper and diamond polished are comparable. The films formed by dissolution or passivation are enriched with chromium, but only the film formed by electropolishing was near the composition Cr_2O_3. All others were mixtures of iron and chromium compounds. (The electrolyte used in this investigation was a non-aqueous, perchloric acid-ethylene glycol monobutyl ether mixture, not commonly found in industry as a result of the hazardous nature of this solution.)

Seo [16] reports the auger electron spectroscopy (AES) analysis of the composition of AISI Type 316 stainless steel surface films prepared by:

1. mechanical polishing with No. 500 emery paper;
2. chemical etching in a 10% nitric acid, 1% hydrofluoric acid solution, 15 min, 323 K;
3. electropolishing for 50 s in a perchloric-acetic acid solution (1:20), 50 V, 6 × 10^{-2} A/cm^2;
4. electropolishing for 50 s in a sulfuric-orthophosphoric acid solution (2:3), 5 to 8 V, 0.5 to 1.0 A/cm^2;
5. treatment in 10% nitric acid for 30 min at 333 K; and
6. treatment in 30% nitric acid for 1 h at 333 K.

Unfortunately, the depth profiles for the electropolished surfaces were not presented in this discussion. Surface atomic fraction of chromium and ratio of oxygen to total alloying components in the film were given. This analysis showed the electropolished surface had less chromium on the surface than the samples treated with nitric acid. The conditions specified for electropolishing in a sulfuric-orthophosphoric electrolyte are not what one would expect, even though the temperature is not specified. This composition should be operated at approximately 75 to 90°C with a current density of 0.2 to 0.5 A/cm^2. An interesting observation of this study was the oxygen ratio of the surface electropolished in the sulfuric-orthophosphoric acid electrolyte was much higher than the nitric acid treated surfaces.

This study and another by Asami [17] demonstrated that the pitting potential versus saturated calomel reference electrode (SCE) in 3.5% saline solution was in direct proportion to the chromium content in the surface layer. Asami also showed that the corrosion potential increases with increasing chromium content.

Summary

Electropolishing is a valuable finishing process for AISI Type 316L stainless steel implants, but there are many questions pertaining to the basic chemistry and result of the process.

1. What is the actual composition of the oxide film produced by electrolytes of varying concentrations?

2. Is there an electrolyte or operating parameters or both that produce oxide films with maximum corrosion resistance?

3. What are the effects of organic addition agents, commonly found in commercial electropolishing solutions, on the film composition and corrosion resistance?

4. Does passivation in nitric acid solutions after electropolishing change the film characteristics?

Conclusions

Electropolishing is finding wide acceptance as a surface finishing technique, but several basic questions remain open. Basic research is needed to determine these answers. References to electropolishing should include the type of electrolyte and the operating parameters, since there may be a correlation between the operating conditions and the oxide film formed. Anodic polarization curves of both the electropolishing process and the corrosion testing may be helpful. More useful surface analytical techniques such as electron spectroscopy for chemical analysis (ESCA) could provide detailed information about the surface after different electropolishing procedures.

Acknowledgments

I wish to thank Ricsan Industries, Inc., for the generous use of time for research and K. B. Hensel of Hydrite Chemical Company for his technical guidance. A deep appreciation is also due L. G. Lester, Dow Corning Wright, and the manufacturing engineering staff at Richards Medical Company.

References

[1] Jacquet, P. A., *Transactions*, Electrochemical Society, Vol. 69, 1936, pp. 629–655.

[2] Faust, C. L., *Metal Finishing*, Vol. 80, No. 7, 1982, pp. 21–25.

[3] Faust, C. L., U. S. Patent No. 2,334,699, 23 Nov. 1943.

[4] Sutow, E. J., *Journal of Biomaterials Research*, Vol. 14, 1980, pp. 587–595.

[5] Stefanskii, I. S. and Bogoyavenskay, N. V., *Zashchita Metallov*, Vol. 6, 6 Nov., 1970, pp. 707–710.

[6] Kruger, J. in *Corrosion and Degradation of Implant Materials, ASTM STP 684*, B. C. Syrett and A. Acharya, Eds., American Society for Testing and Materials, 1979, pp. 107–127.

[7] Bockris, J. O'M. and Reddy, A. K. N., *Modern Electrochemistry*, Vol. 2, Plenum Press, New York, 1970, p. 1317.

[8] Atkins, P. W., *Physical Chemistry*, Freeman, San Francisco, 1978, p. 971.

[9] Tegart, W. J., McG., *The Electrolytic and Chemical Polishing of Metals in Research and Industry*, Pergamon Press, New York, 1959, p. 2.

[10] Faust, C. L., *Metal Finishing*, Vol. 80, No. 9, 1982, pp. 89–93.

[11] Raney, C., Fin-Tech Electropolishing Service, personal communication.

[12] Parlapanski, M., *Zashchita Metallov*, Vol. 6, No. 2, 1970, pp. 162–165.

[13] Revie, R. W. and Greene, N. D., *Corrosion Science*, Vol. 9, 1969, pp. 763–770.

[14] Faust, C. L. in *Electroplating Engineering Handbook*, A. K. Graham, Ed., Van Nostrand Reinhold Company, New York, 1971, p. 108.

[15] Ducrocq, C., Pivin, J. C., Roque-Carmes, C., and Mathia, T. G., *Journal of Microscopic and Spectroscopic Electronics*, Vol. 6, 1981, pp. 157–167.

[16] Seo, M. and Sato, N. in *Transactions*, Japan Institute of Metals, Vol. 21, No. 12, 1980, pp. 805–810.

[17] Asami, K. and Hashimoto, K., *Corrosion Science*, Vol. 19, 1979, pp. 1007–1017.

Metallic Materials: Pyrolytic Carbon

Ram Kossowsky,[1] *Nir Kossovsky,*[2] *and Manuel Dujovny*[3]

In Vitro Studies of Aneurysm Clip Materials

REFERENCE: Kossowsky, R., Kossovsky, N., and Dujovny, M. *"In Vitro* **Studies of Aneurysm Clip Materials,"** *Corrosion and Degradation of Implant Materials: Second Symposium, ASTM STP 859,* A. C. Fraker and C. D. Griffin, Eds., American Society for Testing and Materials, Philadelphia, 1985, pp. 147–159.

ABSTRACT: Seven surgical aneurysm clips, representing six stainless steels and one cobalt-based alloy, were subjected to a complete metallurgical characterization. The studies involved microstructural analysis, hardness measurements, X-ray diffraction analysis, and measurements of the galvanic passivation and resistance to pitting corrosion. Furthermore, one type of clip was subjected to a rigorous failure analysis. The results will be presented in terms of a comparative evaluation. The clinical significance of *in vitro* galvanic corrosion measurements will be supported by the discussion of the reasons for several failures of one type of aneurysm clip.

KEY WORDS: aneurysm, corrosion, galvanic activity, implant materials, stress corrosion cracking, metallurgy, vascular surgery, X-ray diffraction, biological degradation, fatigue (materials)

The brain is unique as an organ in its tendency to spontaneously bleed within itself. Of all stroke deaths in the United States, one-half are caused by intracranial bleeding. Half of these are secondary to the rupture of intracranial vascular aneurysms.

An aneurysm is characterized by the ballooning of the blood vessel wall (Fig. 1). On microscopic examination, the dome of the aneurysm shows fragmentation of the internal elastic lamina, and degeneration or absence of its smooth muscle wall. Frequently, only a thin fibrous wall may remain and is the most common site of rupture [*1*].[3]

Some estimates are that an intracranial aneurysm lurks in one of every 50 North Americans [*2*]. A first-time bleed may occur in some 28 000 persons in the United States and Canada during the next year, with often devastating results.

---

[1]Manager, Center for Materials Characterization, Westinghouse Research and Development Center, Pittsburgh, PA 15235.
[2]Intern, Department of Pathology, The New York Hospital-Cornell University Medical Center, New York, NY 10021.
[3]Staff member, Department of Neurosurgery, Henry Ford Hospital, Detroit, MI 48202.
[4]The italic numbers in brackets refer to the list of references appended to this paper.

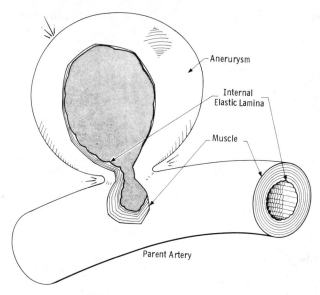

FIG. 1—*Schematic of an aneurysm.*

One form of treatment for this condition is intracranial aneurysm surgery, first reported by Dandy [*3*] in 1938. With the development of microneurosurgery in the early 1970's with concomitant use of the operating microscope, the amount of surgical retraction and exposure of the brain was significantly reduced. Ideally, a metal clip is placed on the neck of the aneurysm, isolating it from the parent artery (see Fig. 2).

Aneurysm clips play a critical role in the control of bleeding; commercially, they are available in a large number of sizes, shapes, and material compositions (Fig. 3). Recently, several papers have appeared in the medical literature that have documented the intracranial failure of one specific type of aneurysm clip, the Heifetz clip [*4,5*]. In this study, we have tested seven commonly used, surgical aneurysm clips that were subjected to complete metallurgical characterization.

Method of Analysis

Seven randomly selected aneurysm clips (types Vari-Angle, Mayfield, Yasargil, Scoville, Vari-Angle McFadden, Pivot®, and Heifetz) were subjected to 10 min of ultrasonic cleaning in an acetone solution followed by an alcohol rinse and dust-free air drying. The clips were then mounted on an aluminum plate and were studied with a Cambridge S-150 scanning electron microscope (SEM) equipped with a computer-controlled Kevex energy dispersive X-ray analysis system (EDS).

The composition of each clip was analyzed in 12 to 18 locations using energy

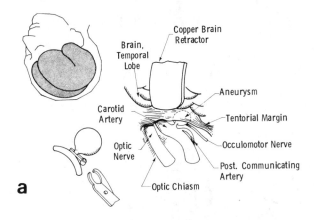

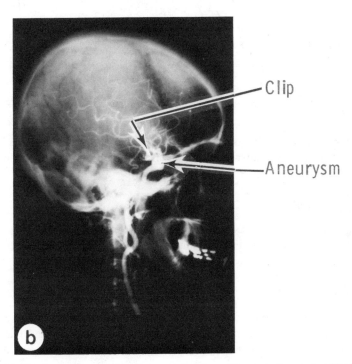

FIG. 2—(a) *Clipping or ligation of the aneurysmal neck provides protection against bleeding, and* (b) *X-ray photograph of clipped aneurysm.*

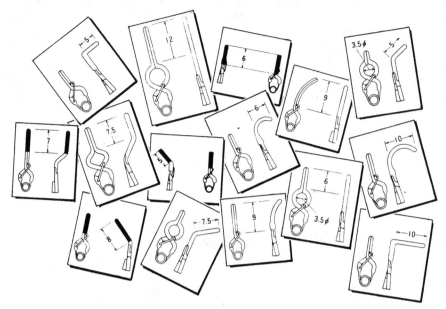

FIG. 3—*Surgical aneurysm clips—a few of the variety of shapes, sizes, and materials.*

dispersive X-ray spectroscopy. The results were obtained using the "standard-less Magic V" computer program, where the results are reproducible within an error margin or ±5% of the analyzed level.

Microstructure

The same clip surfaces that were scanned were then polished and tested on a Philips X-ray Diffractometer to study their metal's crystalline structure and determine the relative proportions of austenitic and martensitic phases.

Hardness

All tested clips were subjected to a hardness evaluation on a Knoops hardness machine. Hardness is an indirect measurement of strength and elasticity that provides a well-documented and reproducible method for assessing the relative strength of materials.

Corrosion Resistance

To evaluate the susceptibility of the various alloys from which clips are constructed, two well-established tests were used. Both these tests rate the alloys on a *relative* basis against a universally accepted standard material. The first, the passivation-reactivation test, is an electrochemical procedure in which each test clip was immersed in an electrolytic solution of 0.5 M H_2SO_4 and 0.05 M

TABLE 1—*Passivation-reactivation test parameters.*

Electrolyte	0.5 M H_2SO_4 + 0.05 M KSCN
Auxiliary electrode	platinum
Reference electrode	standard colomel
Temperature	30°C
Deaeration	continuous N_2 purge
Passivation potential	+350 mV for 4 min
Reactivation scan rate	125 mV/min
Surface finish	1 μm diamond polish

potassium cyanide (KSCN) (Table 1). With the clip functioning in an open circuit as the cathode against a standard calomel anode, a positive potential of 350 mV was established and a protective chromium-rich oxide layer was formed on the clip's surface. This left the clip in the passivated step (Fig. 4). The voltage was then reversed and the current generated by the reactivated test clip was recorded. The peak current (i_{pk}), which is a measure of intergranular corrosion occurring, and the voltage of the peak current (E_{pk}) were noted and compared with accepted standards.

The second, the pitting potential test, is an electrochemical procedure in which each test clip was immersed in lactated Ringer's solution. With the clip again functioning as an anode (Fig. 5), and the circuit open, the free oxidation potential (FOP) was established. The circuit was then closed and the voltage was continuously increased until a sharp current rise was noted. The corresponding pitting

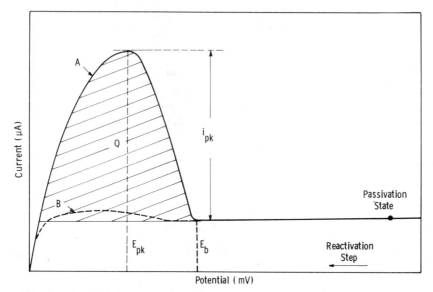

FIG. 4—*Schematic of passivation-reactivation test for material with substantial chromium depletion* (a) *or negligible depletion* (b) *showing breakaway potential* (E_b), *peak potential* (E_pk), *peak current* (i_pk) *and activation change* (Q).

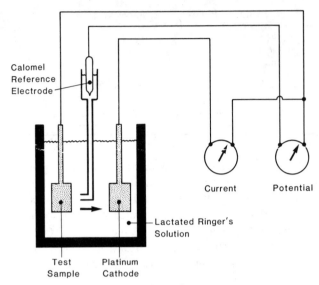

FIG. 5—*Electrochemical test apparatus.*

voltage (E_{pt}) is an indicator of the resistance of the clip to diffuse galvanic attack as well as an indicator of the degree of homogeneity of the material on a microscopic scale. The value was recorded and compared with accepted metallurgical standards. Each clip was then studied by SEM and the extent of the surface pitting was recorded photographically.

Failure Analysis

An eighth clip was acquired postmortem from the base of the skull of a patient who had died following a rupture of an intracranial aneurysm. This same aneurysm was surgically clipped 22 months previously. When recovered, the clip was noted to be in two separate pieces; a portion of one blade was fractured. Failure analysis of this clip included bend-stress assessment, surface examination, and composition determination in the manner just described.

Results

Energy dispersive analysis confirmed each manufacturer's description of the metal type used for each clip. The Vari-Angle clip was made from AISI 17-7PH stainless steel, Mayfield clip from AISI Type 304 stainless steel, Yasargil clip from AISI Type 316 stainless steel, Vari-Angle McFadden clip from MP-35N cobalt alloy, Scoville from EN 58-J (British equivalent of AISI Type 316 stainless steel), and both the Heifetz and Pivot clips from AISI 17-7PH stainless steel. The specific elemental compositions for each alloy are shown in Table 2.

The X-ray diffraction studies revealed varying residual amounts of less than

TABLE 2—*Elemental energy dispersive X-ray analysis of clip materials.*[a]

Clip	Alloy	Al	Ti	Si	S	Cr	Mn	Fe	Ni	Mo	Co
Vari-Angle McFadden	MP35N	1.6	0.8	0.4	0.1	22.3	1.5	0.1	31.4	9.9	31.8
Yasargil	316	1.6	0.8	0.4	0.1	20.1	2.9	65.9	9.1	2.6	0.0
Scoville	316[b]	0.2	0.0	0.4	0.0	16.8	0.0	67.3	9.10	3.2	0.0
Mayfield	304	0.0	0.0	1.1	0.0	19.5	2.0	67.1	10.3	0.0	0.0
Heifetz	17-7PH	1.3	0.0	0.6	0.8	16.8	1.7	72.2	6.7	0.0	0.0
Pivot	17-7PH	1.3	0.0	0.6	0.8	16.8	1.7	72.2	6.7	0.0	0.0
Vari-Angle	17-7PH	1.3	0.0	0.6	0.8	16.8	1.7	72.2	6.7	0.0	0.0

[a]Error range is 10%, not including element with atomic number below sodium.
[b]Designated as EN 58-J stainless steel, a British label that is similar to the AISI Type 316 stainless steel.

TABLE 3—*Hardness and microstructure.*

	Hardness[a]	Martensite, %
Vari-Angle	568	90
Mayfield	614	20
Yasargil	538	0
McFadden	738	0
Scoville	620	0
Heifetz	723	90
Pivot	578	90

[a]Knoops Hardness Number.

desirable martensitic phase in the various clips (Table 3). The Vari-Angle McFadden, Scoville, and Yasargil clips were all 100% austenite. The Mayfield clip was 80% austenite and 20% martensite. The Heifetz, Vari-Angle, and Pivot clips were 10% austenite and 90% martensite.

The microhardness study also showed marked differences. The Vari-Angle McFadden clip was the hardest with a Knoops Hardness Number (KHN) of 738. Heifetz was close with a KHN of 723, then Scoville at 620 and Mayfield at 614. On the softer end were Vari-Angle, Pivot, and Yasargil with 568, 568, and 538 KHN, respectively. The AISI Types 316, 304, and 304 sensitized stainless steels standard samples had hardnesses of 538, 165, and 170 KHN, respectively (Table 3).

The passivation-reactivation test revealed marked differences in the susceptibility of various clips to grain boundary attack. The Yasargil, Vari-Angle McFadden, and Scoville clips were completely resistant with an undetectable critical current density. The Mayfield clip was more sensitive with a critical current density (i_{pk}) of 14 $\mu A/m^2$. The Heifetz, Vari-Angle, and Pivot clips showed the greater sensitivity with a critical current density of 8000 $\mu A/m^2$. The Types 316, 304 and 304 sensitized test standard had critical current densities of 0, 250, and 2600 $\mu A/m^2$, respectively (Table 4). The pitting potential tests revealed that the cobalt-base alloy, MP-35N, had higher immunity to corrosive attack than standard AISI Type 316 stainless steel. The pitting voltage (E_{pt}) for

TABLE 4—*Electrochemical data.*

	Reactivation Voltage	Peak Current, A/m²	Grain Boundary Attack	Pitting Potential, mV
Heifetz 17-7PH	− 50	~8000	severe	+ 70
McFadden MP-35N	not defined	0	none	+ 840
Yasargil 316	not defined	0	none	+ 655
Mayfield 304	+ 250	14	moderate	+ 475
Scoville EN 58-J	not defined	0	none	+ 640
Pivot 17-7PH	− 45	~8000	severe	+ 65
Vari-Angle 17-7PH	− 60	~8000	severe	+ 80

the Vari-Angle McFadden clip was 840 mV, compared to 655 mV for the Yasargil clip, and 640 mV for the Scoville clip. The Mayfield clip, which is constructed from Type 304 stainless steel, had a pitting potential of 475 mV. In marked contrast, the Heifetz, Vari-Angle, and Pivot clips had pitting potentials of less than 100 mV.

The SEM examination of the surfaces correlated well with the results of both the passivation-reactivation test and the pitting potential test. On a relative scale, the Vari-Angle McFadden, Scoville, and Yasargil clips showed no grain boundary attack or crack formation. The Mayfield clip had an occasional evidence of attack, and the Heifetz, Vari-Angle, and Pivot clips had very strong evidence of grain boundary dissolution and incipient crack formation (Fig. 6).

The failure analysis of the recovered Heifetz clip revealed the following. The clip was made from the same alloy as the one we serially examined, that is, 17-7PH stainless steel. The bending stress at the fracture site was high. Assuming that the 2.5-mm-wide and 0.24-mm-thick blades remain 0.25 mm apart [6], the measured load acting on the blades midpoint is 1.4 N. The bend stress at the inner surface of the blade at the location of the observed fracture 10^{-3} m from the apex is

$$\sigma = \frac{6Pl}{bh^2} \quad or \quad \sigma = 568 \text{ MN}/\text{m}^2$$

The yield point of heat-treated 17-7PH steel is approximately 1336 MN/m². In a neutral environment, the bend stress of the clip is relatively insignificant. However, when chemical and electrochemical corrosion occur in concert, the bend stress may approach the critical level.

It is known that 17-7PH steel is a very capricious metal in that the heat treatment needed to give the steel its high yield strength also causes carbides to precipitate at the metal grain boundaries. This, in effect, produces a chemical potential within the steel due to differences in chromium concentration, and it sensitizes the chromium-depleted areas to diffuse corrosion. Because this material has an inherently low fracture toughness and because the stress corrosion produces surface defects and cracks, this material is highly susceptible to stress corrosion catastrophic failure. This impression was confirmed by the pitting potential and reactivation potential data. Finally, SEM micrographs (Fig. 7) showed clear evidence of intergranular, brittle fracture. This is consistent with the balance of the data, and allows us to conclude, with a high degree of confidence, that this particular clip failed as a result of stress corrosion.

Discussion

When intracranial bleeding is a result of a ruptured intracranial vascular aneurysm, there is a 50% chance of mortality expected over the next six to seven years [7]. According to Kassell [8], of the 28 000 North Americans who will

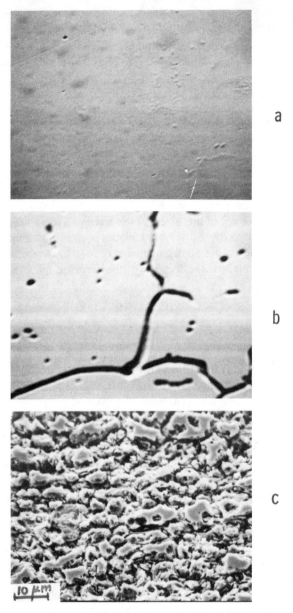

FIG. 6—*SEM micrographs of metal surfaces of Type 304 and 17-7PH steels. Following passi-vation-reactivation test (a) polished surface, Type 304 and 17-7PH, (b) Type 304 after test, (c) 17-7PH after test. (Note: Severe metal disintegration in* (c).)

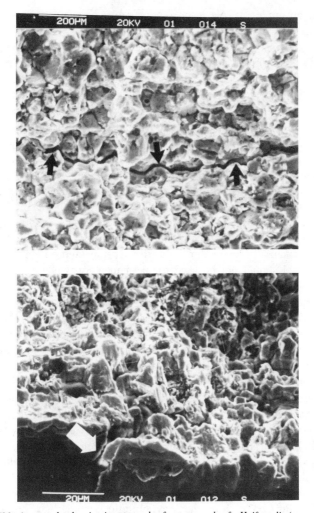

FIG. 7—*SEM micrographs showing intergranular fracture mode of a Heifetz clip (arrows indicate deep secondary cracks).*

experience a hemorrhage for the first time this year, 3000 will die almost immediately, 7000 will die or become disabled due to delayed treatment, 3000 will die or become disabled because of secondary bleeds, 3000 will die or become disabled secondary to vasospasm, 1000 will die or become disabled as a direct result of complications associated with "conservative" nonsurgical treatment, 1000 will die or become disabled as a direct result of complications associated with surgical intervention, and 10 000 or 36% will be "functional survivors." Surgical aneurysm clips play a particularly critical role in the prevention of bleeds in patients who are symptomatic of aneurysms but have not bled. They also play an important role in preventing rebleeding once the initial insult has occurred.

Previous investigators such as Venable [9], Dymond et al [10], Laing [11], and McFadden [12] have studied the nature of host-implant interaction and have provided data on the varying degrees of biocompatibility of the available clips. Based on their work and others, the American Society for Testing and Materials has established guidelines for selecting clips based on their biocompatibility properties (ASTM Practice for Assessment of Compatibility of Metallic Materials for Surgical Implants with Respect to Effect of Materials on Tissue (F 361–80)). They recommend only AISI Type 316 stainless steel, its equivalent. EN-58J, and MP35N cobalt alloy for permanent implantation. Among the clips that we have tested in this study, only the Yasargil, Scoville, and Vari-Angle McFadden clips are made from these metals. We have also actively pursued the question of resistance to stress corrosion failure and our data provides some rational guidelines for the selection of surgical vascular clips. We believe that clips that show a substantially lower threshold to corrosion failure than AISI Type 316 stainless steel should not be used for permanent implantation, such as in the ablation of aneurysms. We believe that the performance of the Heifetz, Vari-Angle, and Pivot clips on the passivation-reactivation test, the pitting potential test, and the post-test surface studies serve as strong indications that these clips are prone to stress corrosion failure. The passivation-reactivation peak current (i_{pk}) of 8000 μA and the pitting potential voltage of 100 mV show that these clips will readily undergo active metal dissolution *in vivo* and thus by metallurgical criteria are at high risk for catastrophic stress corrosion failure. Their use, providing the force provided by the blades is appropriate, should be restricted to temporary applications necessitated by cerebral revascularization or by embolectomy procedures [13]. In marked contrast, the Yasargil, Scoville, and Vari-Angle McFadden are, for practical purposes, electrolytically inert and do not show signs of surface corrosion or crack formation. Thus, these clips appear to be ideally suited for intracranial implantation as far as stress corrosion failure is concerned.

The metallurgical data on the Mayfield clip is borderline and an evaluation from this experimental perspective is difficult. To date, there have been no reports of a metallurgical failure of a Mayfield clip, although several clips have either lacerated the necks of the aneurysm or slipped off [14].

The consequence of an intracranial stress corrosion catastrophic failure of an aneurysm clip can be most profound. The fact that certain clips such as the Heifetz, Vari-Angle, and Pivot show a predisposition to intracranial failure should encourage surgeons to avoid these clips in cases that require long-term implantation. We also feel that the strong correlation between appearance of surface defects as seen by SEM, the results obtained from the passivation-reactivation test, and the pitting potential test strongly suggest that these tests would be ideal preliminary screening devices for manufacturers who are considering a new metal for permanent implantation. Test samples that showed no activity could then be tested more thoroughly with long-term animal implantation followed by meticulous SEM analysis.

Conclusion

In this study, we have described several critical factors that determine a clip's propensity to failure. We have also described the passivation-reactivation test and the pitting potential test that provide an excellent *in vitro* method for testing clips. We hope that this study, in highlighting some metallurgical concerns and applying standard metallurgical techniques, will help in the overall effort to develop better and safer implants.

References

[1] Robbins, S. L. and Cotran, R. S., *Pathologic Basis of Disease,* 2nd Ed., W. B. Saunders, Philadelphia, 1979, p. 1558.
[2] Gunby, P., *Journal,* American Medical Association, Vol. 248, June 1982, pp. 1158–1159.
[3] Dandy, W. E., *Annals of Surgery,* Vol. 107, March 1938, pp. 654–657.
[4] Dujovny, M., Kossovsky, N., and Kossovsky, R., *Journal of Neurosurgery,* Vol. 50, March 1979, pp. 368–373.
[5] Kossowsky, R., Dujovny, M., Kossovsky, N., and Keravel, Y., *Journal of Neurosurgery,* Vol. 57, Aug. 1982, pp. 233–239.
[6] Sugita, K., Hirota, I., Igushi, I., and Mizatani, T., *Journal of Neurosurgery,* Vol. 44, April 1976, pp. 723–727.
[7] Delong, M. and Dujovny, M., *Journal of Neurosurgery,* Vol. 50, Aug. 1979, pp. 662–668.
[8] Kassel, N. F., as quoted in Ref 2.
[9] Venable, C. S., Stuck, W. G., and Beach, A., *Annals of Surgery,* Vol. 105, May 1937, pp. 917–918.
[10] Dymond, A. M., Kaechele, L. E., Jurist, J. M., and Candall, P. M., *Journal of Neurosurgery,* Vol. 33, May 1970, pp. 534–550.
[11] Laing, P. G., *Orthopedic Clinics of North America,* Vol. 4, Feb. 1973, pp. 249–273.
[12] McFadden, J. T., *Journal of Neurosurgery,* Vol. 31, Feb. 1969, pp. 373–385.
[13] Dujovny, M., Kossovsky, N., Kossowsky, R., et al, *Microsurgery,* Vol. 4, Dec. 1983, pp. 124–133.
[14] Carter, C. S., Farwick, D. B., Ross, A. K., et al, *Corrosion,* Vol. 27, 1971, pp. 190–197.

Gordon D. Donald,[1] *David Seligson,*[2]
and Christopher A. Brown[3]

Fatigue of Ball-Joint Rods for External Fixation

REFERENCE: Donald, G. D., Seligson, D., and Brown, C. A., "**Fatigue of Ball-Joint Rods for External Fixation,**" *Corrosion and Degradation of Implant Materials: Second Symposium, ASTM STP 859,* A. C. Fraker and C. D. Griffin, Eds., American Society for Testing and Materials, Philadelphia, 1985, pp. 160–167.

ABSTRACT: Twenty-three Hoffmann ball-joint rods were subjected to sinusoidal loads, each at approximately constant amplitude, between ±400 and ±975 N at a frequency of 5 Hz. Ball-joint rod compliance was plotted against number of cycles for each specimen. The point of component failure was defined as an increase of compliance of 5% from its initial value. Fractographic methods were employed to substantiate a similarity of failure mode in specimens tested and a clear difference from non-tested specimens taken to failure. Experimental loads and number of cycles to failure were well within the clinical range. The fatigue limit of the ball-joint rods is less than 400 N (550 N/mm²). The likelihood of a given component failing is not predictable by inspection or manual manipulation.

KEY WORDS: fatigue (materials), ball-joint rods, external fixation, implant materials

The best prophylaxis against failure is proper clinical judgement in the selection of the device and meticulous technique in its application.[4]

Despite an abundance of literature discussing fatigue failure of metal implants [1–3][5] little or none exists concerning fatigue of external fixation components. Even though corrosive fatigue or stress corrosion cracking [4] plays a significant role in implant failure, stress fatigue is thought to account for the major role [5,6]. Although external fixateurs are not exposed to a corrosive internal environment, loads placed on these devices are as high as metal implants, and one might anticipate stress fatigue. Recent trends in external fixation show increasing

---

[1]Medical student, Burlington, VT 05401.

[2]Kosair professor and chief, Division of Orthopedics, Department of Surgery, University of Louisville School of Medicine, Louisville, KY 40202.

[3]Research assistant, Department of Orthopedics and Rehabilitation, University of Vermont, Hinesburg, VT 05461.

[4]Cohen, J. and Brown, T. in *Trauma Management,* E. F. Cave, J. F. Burke, and R. J. Boyd, Eds., Yearbook Medical Publishers, Inc., Chicago, 1974, p. 927.

[5]The italic numbers in brackets refer to the list of references appended to this paper.

use of half-frame fixateurs thus imposing higher loads on fewer frame components [7–9]. Aggressive physical therapy on the fractured limb contributes further to the loads that fixateurs must endure.

External fixateurs consist of metal frames that connect groups of bone fixation pins. One clinically common device, the Hoffmann fixateur, utilizes ball-joint rods to connect groups of bone fixation pins to the metal frames. These ball-joint rods consist of a vise with a grooved plastic surface for holding the bone fixation pins, and a swivel joint and brazed-on connecting rod for connecting the pin group to the external fixation frame.

Established practice dictates that plates, screws, and other orthopedic implants are not re-used in developed countries [10]. No standards have been set for external fixation components. External fixation pins are no longer re-used in clinical application [8,11] but the high cost of other components often dictate their re-use among patients.

We recently have had a clinical failure of the ball-joint component both of the plastic pin-holding surface and of the brazed vise-rod connection. In this study, the fatigue behavior of the brazed vise-rod junction was examined and correlated with the clinical use of this component.

Materials and Methods

Ball-joint rods (Hoffmann Model B-11[6]) are manufactured from Type 303 stainless steel. The 8.0-mm diameter rod has a 6.0-mm diameter neck region with a flat ground at a slight angle to the longitudinal axis (Fig. 1). The neck region of the rod is force-fit into a 6.0-mm diameter hole in the center of the 53-mm by 17.5-mm by 5-mm vise. The 9.5-mm collar of the vise (Fig. 1) is then brazed to the rod with either a copper or silver-based alloy.

The vise-rod system was tested in an apparatus consisting of a rigid support base to which the ball-joint rods were securely fixed by two screws at either end of the vise. The vise was oriented vertically and the rod horizontally. The support base was in turn fixed to the MTS 810 material testing system table so that the longitudinal axis of the rod was oriented perpendicular to the load cell.

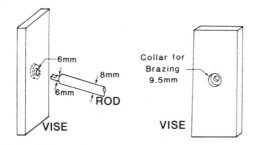

FIG. 1—*Diagramatic representation of Hoffmann ball-joint rod.*

[6]Jacquet-Frerès, Geneva, Switzerland.

TABLE 1—*Test plan: fatigue of ball-joint rods.*

Load, N	920–975	770–815	690–700	585–590	495–500	394–396
Nominal stress (N/mm²)	1300–1380	1090–1150	976–990	828–835	700–707	557–560
Sample size	3	5	3	4	5	3

Twenty-three ball-joint rods were subjected to sinusoidal loads each at approximately constant amplitude, between ±400 and ±975 N at a frequency of 5 Hz (Table 1). Loads were applied at a distance of 30 mm from the clamp rod interface. This distance corresponds to the clinical positioning of the connecting bar bridging a pair of rods in a half-frame fixateur.

Vertical displacements of the rods were measured at the point of loading. Ball-joint rod compliance (displacement/load) was plotted against number of cycles for each specimen. Failure was defined as the point at which rod compliance increased 5% from its value during the first loading cycle. Fifteen specimens were cycled until they failed catastrophically due to either rod fracture or separation at a brazing interface. A standard *S-N* diagram was constructed from a sample of 23 specimens and a fatigue limit was estimated.

Joint properties of all specimens were recorded including type and amount of brazing material, and rotational orientation of the ground flat on the neck in relation to the loading direction.

For those specimens that were cycled until rod fracture occurred, light fractographs were made on which the direction of crack propagation and the crack origins are discernable. Those specimens that were only tested until a 5% increase in compliance was exceeded and did not undergo catastrophic rod failure during cyclic loading were fractured by a single impact. Several uncycled specimens were also fractured by a single impact. Fracture surfaces of these specimens were compared with previous specimens for similarities in evidence of crack origins and propagation.

Results

Initial values of compliance at the onset of testing showed wide variability among specimens although curves of compliance versus number of cycles were remarkably similar. Initially, all specimens demonstrated elastic behavior with compliance remaining constant. As cyclic loading continued, compliance began to slowly increase. Generally, after an increase of 5 to 10% from the initial value, compliance started to increase at a greater rate until the specimen broke or the test was stopped.

A plot of cycles to failure versus load demonstrates an exponential relationship between these variables (Fig. 2). If log of cycles to failure is plotted against load, a linear relationship is evident. The coefficient of linear correlation (r)

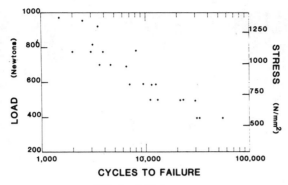

FIG. 2—*S-N diagram.*

calculated for all samples equals -0.92. From Fig. 2, it is apparent that the fatigue limit of the ball-joint rods is less than 400 N (550 N/mm²).

Several sorts of failure occurred. Four of eight specimens tested at loads greater than 770 N failed by separation of the rod from the vise at the brazed interface. The remaining four specimens of this group failed by fracture of the rod at the level of the vise. For the 15 specimens tested at loads less than 700 N, four failed by rod fracture. Five more remained grossly intact, but increased rod compliance and motion at the rod-vise interface could be detected by manual manipulation. The remaining six specimens also remained grossly intact, but without apparent weakness or visible motion.

Physical characteristics of the brazed joint and rotational orientation of the ground flats relative to the loading direction demonstrated no discernable correlation with the mechanism of failure or with fatigue life of a given specimen.

The specimen shown in Fig. 3 fractured by manual manipulation during inspection after 50 000 cycles at a load of 400 N. The loading directions on the fractograph are up and down. The presence of three regions (a, b, and c) that can be discerned on the fractograph are typical of those specimens broken during cycling or by impact after cycling. The uncycled specimens showed only one region similar to Region b, Fig. 3.

The features on the fractograph (Fig. 3) are typical of those commonly associated with reverse bending fatigue fractures [12–14]. Fatigue cracks originated in the regions indicated by the arrows on the circumference of the specimen. The fatigue crack on the lower portion originated in the ground flat. On the upper portion, it is noticed that several fatigue cracks originated along the sharp junction between the neck and the stem, then joined to form a single crack front. The upper and lower fatigue cracks propagated through Regions a and c before the final fast fracture through Region b.

Fracture origins at the junction of the neck and stem, and on the ground flat, were typical of all the observed fractures. These features would be expected to provide the stress concentrations commonly associated with fatigue crack origins [13].

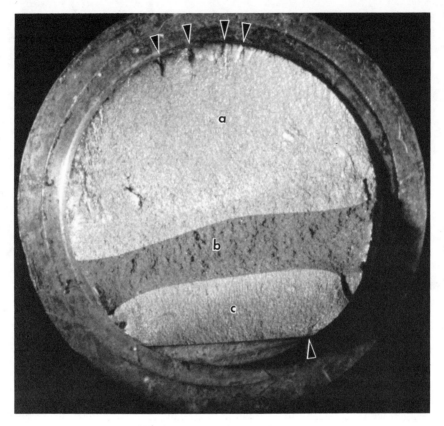

FIG. 3—*Fractograph of failed specimen.*

Discussion

Plates, screws, and other metal implants are not re-used in medical practice [10]. Laurence et al [15] demonstrated that only the strongest of implants tested could withstand the possible forces imposed during weight bearing. With modern surgical metals, corrosion phenomena are too slow *in vivo* to account for device failure before a fracture heals [16]. The excellent corrosion-resistance qualities of Type 316L stainless steel and the titanium alloys are well known [17,18]. Pohler and Straumann showed that clinical breakage of Type 316L stainless steel implants results from fatigue related to fracture instability. Corrosion was not thought to be a major contributing factor to failure of these implants, since fatigue strengths were similar whether conducted in air or Ringer's solution [3].

Clinical failure of a ball-joint rod, which caused painful fracture site motion, led us to examine the mechanical properties of this component. Ball-joint rods were tested to determine if fatigue failure may play a significant role under clinical conditions. Previous material testing literature has amply demonstrated the applicability of *in vitro* testing to clinical *in vivo* effects [5,18]. The specimens examined in this study are constructed of Type 303 stainless steel, a material with relatively less corrosive-resistance than Type 316L stainless steel or titanium alloys. Accordingly, they are utilized in a non-corrosive external environment but undergo similar stresses to implanted plates. Mechanical guidelines currently applied to the use and re-use of plates should then be considered for these components as well.

Sixteen of the ball-joint rods tested in this experiment were previously used clinically. The remaining seven specimens were obtained from components used for demonstration purposes only and had seen no prior significant loading. Although one would expect to see a difference between data from these two groups, scatter was low. This observation may be attributable to several facts. Since the ball-joint rods were used in Vidal-type frames that employ four or more ball-joint rods, loads experienced by individual components for a given degree of limb loading were relatively low. The fractures in which these fixateurs were used were often quite fragmented and unstable so aggressive weight-bearing therapy was not undertaken. Therefore, these samples were seldom subjected to the magnitudes of loading as great as half-frame fixateurs presently used for a variety of fractures. Finally, none of these external fixateur components were re-used. It seems probable that if a group of ball-joint rods re-used several times in clinical situations were tested, a significant difference would be apparent between that group and a group not subjected to previous loading.

Clinical loads imposed on an individual ball-joint rod depend on the number of these components used in fixateur design and the degree of limb loading imposed by therapy and daily activity. The fewer ball-joint rods used and the more aggressive the therapy, the greater the loads on a particular ball-joint rod. If the half-frame fixateur employs only two ball-joint rods with only one connecting rod, the load experienced by each component can approach 2.5 to 3 times body weight [4].

For example, a 70-Kg man with an anterior tibial fixateur who ambulates with the aid of crutches and bears only 25% of body weight on the affected limb may impose a peak force of 430 to 515 N on each ball-joint rod if there is no load sharing by the bone. Clearly, the experimental loads used were well within the realm of clinical loading. Loading frequency for normal ambulation is estimated to be 1 to 3 million cycles per year [4,19]; therefore, it is evident that the number of cycles to failure measured in these tests are obtainable especially if re-use of the components is practiced.

Of the eleven specimens in which rod fracture did not occur, in only five could rod stability be detected by either inspection or manual manipulation. This suggests that the likelihood of fatigue failure occurring in a particular component can not be assessed by inspection or manipulation. Six of the 23 specimens had visible gross defects in the brazing material coupling rod to clamp face. The defects ranged from full thickness voids to shallow surface flaws. Although one might anticipate early failure in these specimens, this proved not to be the case. Rather, the defects appeared to play no detectable role in ultimate fatigue life.

The appearance of the surfaces of the fractures on the specimens broken during cycling and on those impacted after cycling were similar. This indicates that fatigue cracks, undiscernable by an unaided visual examination, had already propagated through more than 50% of the cross section of the part. This contention is supported by the dissimilarity between the previously mentioned fracture surfaces and those created by single impact on uncycled specimens.

Therefore, the surgeon can only estimate the fatigue life of a given component by mechanical testing data and a detailed history of previous loads to which that component had been subjected. Since this history is virtually unknown for a particular ball-joint rod, new components must be employed if fatigue failure is a concern. Device failure in a clinical setting has important legal implications [20].

Perhaps fatigue limits need to be evaluated at loads less than those used in this study, but results indicated that the ball-joint rod fatigue limit is less than 400 N. Since failure would not be anticipated at loads less than the fatigue limit, it must be estimated whether loads imposed on the ball-joint rods will be significantly less than 400 N. If not, it is crucial that new components be used in fixateur construction. Even if loads are estimated to be less than 400 N, the fatigue limit of re-used ball-joint rods may be appreciably lower than predicted if these components were previously exposed to relatively high loads.

Fatigue cracks are known to originate at surface irregularities and geometrical features causing stress concentrations. There are two features on the rod that appear to exacerbate fatigue problems.

The abrupt step from the 8-mm diameter rod to the 6-mm diameter neck coincided with the origin of fatigue cracks in several specimens (Fig. 3). A larger radius of curvature at the junction between the neck and the rod would lessen the stress concentration in this area, thereby promoting greater fatigue life. Such a change in the rod would require a similar change in the plate.

Although there was no correlation between the orientation of the ground flat and fatigue life, cracks were frequently noted to originate on this flat (Fig. 3). Elimination of the ground flat would improve the fatigue fracture life of the rod, however, its elimination may adversely influence the brazing. A smoother surface finish in the ground area would improve the fatigue life, as would increasing the radius of curvature at the flat-neck intersection.

Conclusions

1. Hoffmann ball-joint rods undergo fatigue failure at loads at least as low as 400 N ($550 N/mm^2$).

2. Fatigue life of these components shows an inverse exponential relationship to applied load.

3. Loads at which ball-joint rods were tested are well within the limits of clinical loading conditions.

4. The likelihood of a given component failing is not predictable by inspection of manual manipulation.

5. External fixation appliances are subject to fatigue under load: component elements should be mechanically studied and uniform guidelines evolved to determine re-usability.

References

[1] *Corrosion and Degradation of Implant Materials, ASTM STP 684*, B. C. Syrett and A. Acharyz, Eds., American Society for Testing and Materials, Philadelphia, 1979.

[2] Pohler, O. and Straumann, F., "Characteristics of the Stainless Steel ASIF/AO Implants," *AO Bulletin*, Sept. 1975.

[3] Pohler, O. E. M. and Straumann, F., "Fatigue and Corrosion Fatigue Studies Ion Stainless-Steel Implant Material," *Evaluation of Biomaterials*, Chapter 6, Wiley, New York, 1980.

[4] Fraker, A. C. and Ruff, A. W., *Journal of Metals*, May 1977, pp. 22–28.

[5] Sloter, L. E. and Piehler, H. R. in *Corrosion and Degradation of Implant Materials, ASTM STP 684*, B. C. Syrett and A. Acharyz, Eds., American Society for Testing and Materials, Philadelphia, 1979, pp. 173–192.

[6] Dobbs, H. S. and Scales, J. T. in *Corrosion and Degradation of Implant Materials, ASTM STP 684*, B. C. Syrett and A. Acharyz, Eds., American Society for Testing and Materials, Philadelphia, 1979, pp. 245–258.

[7] Seligson, D. and Pope, M. H., *Concepts in External Fixation*, Grune & Stratton, New York, 1982.

[8] Burny, F. and Bourgois, R., *Acta Orthopedica Belgica*, Vol. 38, 1972, pp. 265–279.

[9] Chao, E. Y., Kasman, R. A., and An, U. N. *Journal of Biomechanics*, Vol. 15, No. 12, 1982, pp. 971–983.

[10] Martz, C., *Journal of Bone and Joint Surgery*, Vol. 38-A, 1956, pp. 827–834.

[11] Green, S. A., *Complications of External Skeletal Fixation*, Charles C. Thomas, Springfield, 1981.

[12] Fellows, J. A., *Metals Handbook: Fractography and Atlas of Fractographs*, Vol. 9, 8th ed., American Society for Metal, Metals Park, OH 1974.

[13] Jacoby, G., *Experimental Mechanics*, March 1965, pp. 65–82.

[14] Lipson, C., *Machine Design*, Sept. 1950, pp. 147–150.

[15] Laurence, M., Freeman, M. A. R., and Swanson, S. A. V., *Journal of Bone and Joint Surgery*, Vol. 51-B, 1967, pp. 754–768.

[16] Cave. E. F., Burke, J. F., and Boyd, R. J., *Trauma Management*, Yearbook Medical Publishers, Inc., Chicago, 1974.

[17] Imam, M. A., Fraker, A. C., and Gilmore, C. M. in *Corrosion and Degradation of Implant Materials, ASTM STP 684*, B. C. Syrett and A. Acharyz, Eds., American Society for Testing and Materials, Philadelphia, 1979, pp. 128–141.

[18] Syrett, B. C. and Davis, E. E. in *Corrosion and Degradation of Implant Materials, ASTM STP 684*, B. C. Syrett and A. Acharyz, Eds., American Society for Testing and Materials, Philadelphia, 1979, pp. 229–242.

[19] Heppenstall, R. B., *Fracture Treatment and Healing*, Saunders, Philadelphia, 1980.

[20] Piehler, H. R., *Corrosion and Degradation of Implant Materials, ASTM STP 684*, B. C. Syrett and A. Acharyz, Eds., American Society for Testing and Materials, Philadelphia, 1979, pp. 328–341.

Miroslav Marek[1]

An Electrochemical Test for the Evaluation of Sulfide Tarnishing of Dental Alloys

REFERENCE: Marek, M., "An Electrochemical Test for the Evaluation of Sulfide Tarnishing of Dental Alloys," *Corrosion and Degradation of Implant Materials: Second Symposium, ASTM STP 859*, A. C. Fraker and C. D. Griffin, Eds., American Society for Testing and Materials, Philadelphia, 1985, pp. 168–178.

ABSTRACT: Tarnishing of gold and silver alloy dental restorations in service is caused mainly by the formation of sulfide films. Since the reactions are electrochemical in nature, electrochemical techniques can be used to examine them. In this study, a practical test of the susceptibility to sulfide tarnishing was designed, based on coulometry at constant potential in an aqueous sodium-sulfide solution. The effects of temperature, potential, and time were examined. The index of susceptibility was defined as the anodic charge per square centimeter of the surface in 0.01 M Na_2S at 37°C and 0.0 V saturated calomel electrode (SCE). The solution must be deaerated to prevent the loss of S^{2-} ions, and the test procedure must ensure reproducible surface conditions and avoidance of spontaneous sulfidation. A series of dental alloys was evaluated using a test period of 1.5 h. The results showed good agreement with reported clinical data.

KEY WORDS: dental materials, sulfur, tarnishing, electrochemical corrosion, gold alloys, silver alloys, corrosion tests, coulometers, implant materials

Dental restorations and appliances made of gold- and silver-based alloys often show discoloration called tarnish, which has been attributed to the formation of sulfide surface films. Although researchers seem to pay more attention to chloride corrosion, dental practitioners, patients, and dental alloy manufacturers are more concerned about the resistance of these materials to sulfide tarnishing.

The high-gold alloys that conform to American Dental Association (ADA) Specification No. 5 and Federation Dentaire International (FDI) Specification No. 7 [1][2] are adequately resistant to tarnishing. As more lower-gold and silver-based alloys are introduced in response to the high price of gold, the need to understand the mechanism and kinetics of tarnishing and to develop techniques for evaluation has become critical.

The best known and very successful testing technique was developed by

---

[1]Associate professor of Metallurgy, Georgia Institute of Technology, Atlanta, GA 30332.
[2]The italic numbers in brackets refer to the list of references appended to this paper.

Tuccillo and Nielsen [2], who designed an apparatus for alternate immersion in an aqueous solution of sodium sulfide. The main disadvantage of the test is the lack of a standard objective technique of quantification of the results.

Many sophisticated techniques have been developed for the study and evaluation of electrochemical corrosion [3]. Most of them take advantage of the fact that the rates of electrochemical processes often can be conveniently and accurately determined by measuring the electrical currents flowing through the electrode. Since the formation of metallic sulfide films is an electrochemical process, there is no reason not to use electrochemical techniques in the investigations. In this study, the attempt was made to develop a practical, electrochemical test procedure for the evaluation of the resistance to sulfide tarnishing. This included the selection of a well-defined environment, the choice of the technique of measurement that would provide a quantitative index of susceptibility, and the examination of the effects of test variables.

Experimental

All electrochemical tests were performed using a temperature-controlled three-electrode glass cell shown in Fig. 1. A standard saturated calomel electrode (SCE) was located outside the cell and connected to it by a salt bridge that ended in a Luggin capillary placed close to the electrode surface. The bridge was filled with distilled water. A 24-carat-gold plate served as the counter-electrode. The specimens were vacuum-cast in an epoxy resin and wet-ground on silicon carbide and Al_2O_3 papers through 0.3 μm grit.

The electrochemical tests consisted of open-circuit potential versus time measurements, and potentiostatic tests. The test solution in the potential-time measurements was a synthetic saliva [4] aerated with a mixture of 10% CO_2 in air. The sulfides were introduced by adding a measured volume of a freshly diluted sodium-sulfide concentrate. All other tests were made using simple aqueous sulfide solutions prepared by diluting either a saturated sodium-sulfide solution or a commercial ammonium-sulfide concentrate (20 to 24% $(NH_4)_2S$) with distilled water. The water was deaerated in the cell by scrubbing with oxygen-free nitrogen for 2 h.

All aqueous sulfide solutions were analyzed for the sulfide concentration by potentiometric titration with lead acetate using an ion selective solid-state electrode[3].

The setup for electrochemical measurements consisted of an electronic potentiostat with a potential meter and a current integrator[4], a logarithmic current converter[5], and X-Y and strip-chart recorders[6].

The samples of dental casting alloys were tested either as-received from the manufacturer, or as-cast using standard dental laboratory procedures. In addition,

[3]Model 94-16, Orion Research, Inc., Cambridge, MA 02139.
[4]Wenking Models 70TS1, PPT69, SSI 70, Sybron/Brinkmann, Westbury, NY 11590.
[5]Model 376, EG&G Princeton Applied Research, Princeton, NJ 08540.
[6]Models SR-207 and SR-206, Heath-Schlumberger, Benton Harbor, MI 49022.

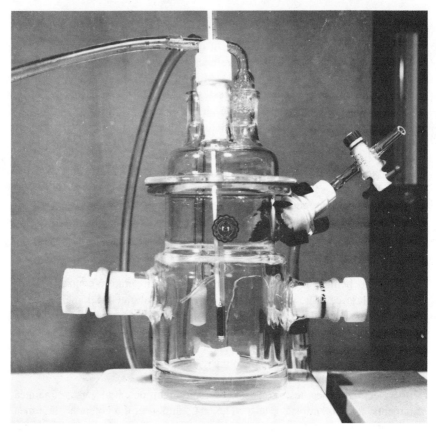

FIG. 1—*Electrochemical cell used in the tests.*

pure silver and gold were tested. The data for the samples are summarized in Table 1.

In the electrode tests in which the anodic current was measured and integrated at an imposed potential, it was important to prevent the sulfidation reactions before the activation of the integrator. This was accomplished by exposing the electrode initially only to distilled water. The electrode was then polarized to the selected potential, all circuits were activated, and sulfide ions were introduced. Since the anodic currents in distilled water were very low, there was no detectable error due to the short (5 to 10 s) polarization period before the introduction of sulfide ions.

Results and Discussion

In any corrosion and tarnishing test, the environment must be well defined. The sulfide compound most commonly used in the preparation of laboratory

TABLE 1—*Metals and alloys used in the test.*

Sample Code	Trade Name	Sample Condition	Major Elements			
			Gold	Silver	Palladium	Copper
MOD	Modulay[a]	cast	77.0	14.0	1.0	7.5
MS	Midas[a]	cast	46.0	39.5	6.0	8.0
F	Forticast[a]	cast	42.0	26.0	9.1	20.9
MX	Maxigold[b]	cast	59.5	26.3	2.7	8.5
MD	Midigold[b]	cast	49.5	35.0	3.5	10.0
MN	Minigold[b]	cast	40.0	47.0	4.0	7.5
MW46	Mowrey #46[c]	cast	42.0	35.0	8.0	15.0
MW120	Mowrey #120[c]	cast	50.0	28.0	4.0	17.0
B	Blaurock[e]	cast	0	89.9	3.4	6.0
PD3	Paladin 3[d]	as-received	0	60.0	25.0	NA[f]
Au	Gold 24-carat[a]	as-received	99.95	0	0	0
Ag	Silver	as-received	0	99.9	0	0

[a]Jelenko/Pennwalt, New Rochelle, NY 10802.
[b]Williams Gold, Buffalo, NY 14214.
[c]W. E. Mowrey Co., St. Paul, MN 55104.
[d]Sterngold, Mount Vernon, NY 10551.
[e]Blaurock Dental Golds, Denver, CO.
[f]NA = not available.

tarnishing test solutions is the sodium-sulfide nonahydrate, $Na_2S \times 9H_2O$, which is highly hygroscopic. The analysis of solutions prepared by dissolving the crystals from five freshly opened bottles showed concentrations of S^{2-} lower than calculated from the mass by 5 to 15%; the error increased to 10 to 25% when the bottles have been in use for two months. This problem was avoided by preparing a saturated solution and diluting it before each test. The analysis showed that the stock solution was 1.8 ± 0.02 M at 25°C; a standard 0.01 M solution selected for the potentiostatic tests was prepared by adding 2 ml of the stock solution to 358 ml of distilled water. Consistent sulfide concentrations also were obtained by diluting the commercial ammonium-sulfide concentrate once the concentration was accurately determined.

The diluted sulfide solutions are not stable. Sodium sulfide, for instance, reacts with aerated water according to the equations

$$2Na_2S + 2O_2 + H_2O = Na_2S_2O_3 + 2NaOH$$

$$Na_2S + H_2O + CO_2 = Na_2CO_3 + H_2S$$

The degradation of solutions of low concentration is very rapid. Figure 2 shows that the S^{2-} concentration of a stirred solution of $(NH_4)_2S$, which was prepared as 0.01 M, dropped to half that value in less than 4 h; after 24 h the S^{2-} concentration was less than 10^{-5} M. When dissolved air was removed by scrubbing with oxygen-free nitrogen, the loss of S^{2-} was only about 4% in 24 h. Therefore, only deaerated solutions were used in the potentiostatic tests.

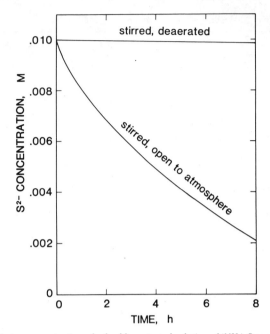

FIG. 2—*Sulfide ion concentration of a freshly prepared solution of $(NH_4)_2S$ as a function of time.*

Figure 3 shows the effect of the addition of sulfide ions to the synthetic saliva on the corrosion potential. The addition of Na_2S to make $0.001\ M$ solution caused a change in the potential in the negative direction by about 0.5 V. The potential then slowly recovered but remained below the original value for at least 24 h.

The change in the corrosion potential is attributed to the anodic reactions that result in the formation of the sulfide surface film. Some of the major reactions and their equilibrium half-cell potentials are shown in Table 2. Because of the highly negative values of the equilibrium potentials, there is a strong tendency for the reactions to proceed spontaneously. Before the addition of the sulfide ions, the alloy exposed to the synthetic saliva is covered with a thin oxide film. When the corrosion potential is stable, the rates of the chemical dissolution of the film, anodic growth of the film, and cathodic reactions (mainly reduction of dissolved oxygen) are all equal. When sulfide ions are added, one or more of the sulfidation reactions proceed in the anodic direction, increasing the total anodic reaction rate. This must be balanced by a similar increase in the cathodic rate, which requires a higher negative overpotential. Therefore, the potential shifts in the negative direction. The change in the electrode conditions is shown schematically in Fig. 4. As the film forms, the reaction rate decreases and the potential changes in the positive direction.

The rate of sulfide film formation depends on the composition of the electrode. There is, however, no single rate of the reaction that can be measured and used to compare the susceptibility to tarnishing. The same tarnishing effect

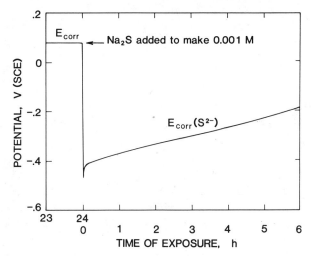

FIG. 3—*Corrosion potential of Alloy PD3 in synthetic saliva [4] showing the effect of sulfides.*

can be achieved by an initially fast reaction that quickly slows down, as well as by a slower reaction that decays more gradually. A multipoint determination of the reaction rate as a function of time at open-circuit potential, although theoretically providing a good measure of the susceptibility, would be very time-consuming, subject to large errors, and generally not suitable as a practical test method.

The reaction rate can be conveniently measured as the anodic current density if the potential is maintained constant using an electronic potentiostat. The current can be integrated over a selected test period so that changes in the reaction kinetics with time are taken into account. The integrated current density (anodic charge) is proportional to the mass of the anodic product that is, in this case, mainly the sulfide film that appears as tarnish. This coulometric technique was used in all further tests.

TABLE 2—*Standard electrode potentials for some electrochemical reactions involving sulfides.*

Reaction	Standard Electrode Potential,[a] V (SCE)
$Ag_2S^b + 2e = 2Ag + S^{2-}$	-0.896
$CuS + 2e = Cu + S^{2-}$	-0.963
$Cu_2S + 2e = 2Cu + S^{2-}$	-1.132
$PtS + 2e = Pt + S^{2-}$	-1.080
$PtS_2 + 4e = Pt + 2S^{2-}$	-0.944
$PdS + 2e = Pd + S^{2-}$	-1.032
$PdS_2 + 4e = Pd + 2S^{2-}$	-0.878

[a]Calculated from thermodynamic data [5]. Potentials on the SCE scale.
[b]Orthorombic α sulfide.

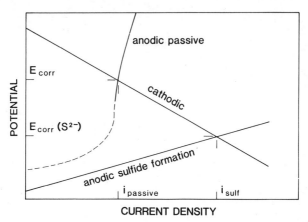

CURRENT DENSITY

FIG. 4—*Schematic illustration of the effect of sulfides on the electrode conditions.*

Figure 5 shows the current density as a function of time for several alloys in 0.01 M Na_2S at 37°C and 0.0 V (SCE). Also included are curves for pure gold and silver. The behavior of all gold- and silver-based dental alloys was basically similar in that the current density reached the peak immediately after the introduction of sulfides and then continuously decayed. It did not increase even in exposures of up to 100 h for any of the alloys, indicating that the sulfide film is at least to some extent protective. The current-time curves for all the dental alloys tested were between the curves for gold and silver, showing that none of the alloys quite equalled the tarnish resistance of pure gold (although some high-gold alloys approached it closely), but all were substantially more tarnish resistant than pure silver.

To standardize the test conditions, the composition and concentration of the solution, the temperature, the potential, and the test period, must be selected. Sodium-sulfide solution was chosen because the concentration could be conveniently controlled by diluting a saturated concentrate. The S^{2-} concentration was arbitrarily selected as 0.01 *M*. Solutions of low concentration are more susceptible to degradation, while high-sulfide concentrations are less relevant to the *in vivo* conditions.

The effect of temperature is shown in Fig. 6 in which the anodic charge is plotted versus the test temperature for one of the alloys. Since the rates of electrochemical reactions usually increase exponentially with increasing temperature, an exponential curve was used to connect the means in Fig. 6. The relationship is a smooth curve that indicates that no sharp changes in the mechanism take place within the range examined. The temperature of 37°C is recommended for future tests, not only because it approximates the oral temperature but also because it can be more conveniently accurately controlled than, for instance, room temperature. Nevertheless, some of the tests in this study were performed at 25°C.

The effect of the potential on the anodic charge is shown in Fig. 7. Again, a

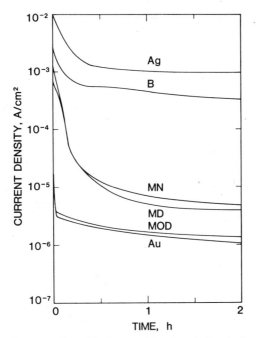

FIG. 5—*Current-time curves for gold, silver, and four dental alloys in deaerated 0.01 M Na_2S at 37°C and 0.0 V (SCE).*

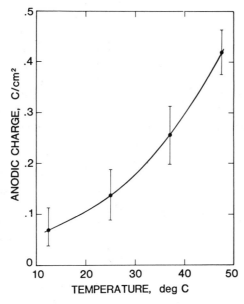

FIG. 6—*Effect of temperature on the 1-h anodic charge in deaerated 0.01 M Na_2S at 0.0 V (SCE); Alloy PD3. Means and ranges of test values are shown.*

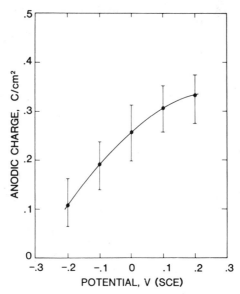

FIG. 7—*Effect of potential on the 1-h anodic charge in deaerated 0.01 M Na_2S at 37°C; Alloy PD3. Means and ranges of test values are shown.*

smooth curve indicates that the potential can be selected within the examined range without a qualitative effect on the reactions. The potential of 0.0 V versus SCE has been selected as a convenient value that, in some cases, allows the use of simpler equipment.

The examination of the current-time curves, such as those in Fig. 5, showed that the initial period of rapidly decaying, high-current density was at most about 1.5-h long. This time was selected as the test period. On the basis of the results of this study, the coulometric index of susceptibility to sulfide tarnishing of noble and semi-noble alloys, T, has been defined as the anodic charge in coulombs per square centimeter of the electrode in a solution of 0.01 M Na_2S at 37°C and 0.0 V (SCE). The subscript indicates the test period in hours, the recommended period being 1.5 h ($T_{1.5}$). Table 3 shows the results of a preliminary series of tests in which the measurements were made at 25°C. It demonstrates that the test is sufficiently sensitive to the differences between the alloys. The ranking of the alloys follows the expected trend, that is, the resistance to tarnishing generally increases with the gold and platinum-group metal content. A study in which the results of this test for alloys in Table 3 and an additional eight experimental alloys were compared with clinical ranking showed very good correlation [6].

The experience with the test to date has shown that the scatter of results is higher than for a similar test used to evaluate chloride corrosion of dental amalgam [7]. The rate of the sulfidation reaction is strongly affected by the presence

TABLE 3—*Results of the coulometric test for a series of dental casting alloys.*

Sample	Mean Anodic Charge,[a] C/cm^2	n^b	σ^c
MOD	0.048	5	0.012
MW120	0.060	5	0.015
F	0.075	5	0.008
MX	0.093	5	0.034
MW46	0.113	5	0.033
MD	0.126	5	0.075
MS	0.175	5	0.021
MN	0.189	5	0.012
B	1.785	5	0.152

[a] Anodic current density in 0.01 M Na_2S at 25°C and 0.0 V (SCE), integrated for 1.5 h.
[b] Number of tests.
[c] Standard deviation.

of the oxide film formed by exposure to air and preexposure in water. For consistent results, the sample preparation and test procedure must be exactly reproduced. Further studies are needed to evaluate the statistical aspects of the test, standardize the procedure, and optimize the test conditions for the best correlation with clinical data.

Conclusions

Electrochemical techniques can be used to study sulfide tarnishing of gold and silver alloys. Aqueous sulfide solutions used in the tests should be analyzed and their preparation standardized to ensure correct concentration. Aerated diluted solutions degrade rapidly; deaerated solutions should be used in controlled potential tests. A coulometric test provides a suitable numerical index of the susceptibility to tarnishing.

Acknowledgments

The cast alloy samples were kindly provided by Drs. K. F. Leinfelder and D. F. Taylor of the Dental Research Center, University of North Carolina.

References

[1] *Guide to Dental Materials and Devices*, 7th ed., American Dental Association, Chicago, 1974.
[2] Tuccillo, J. J. and Nielsen, J. P., *Journal of Prosthetic Dentistry*, Vol. 25, 1971, pp. 629–637.
[3] *1982 Annual Book of ASTM Standards*, Part 10, American Society for Testing and Materials, Philadelphia, 1982.
[4] Wald, F. V. and Cocks, F. H., *Journal of Dental Research*, Vol. 50, 1971, pp. 48–59.
[5] National Bureau of Standards Technical Note 270-3 (1968) and 270-4 (1969), *Selected Values of Chemical Thermodynamic Properties*, Government Printing Office, Washington, DC.
[6] Taylor, D. F., Baldwin, V. H., and Leinfelder, K. F., IADR Program and Abstracts of Papers, Paper No. 1494, *Journal of Dental Research*, Vol. 61, 1982.
[7] Marek, M., *Journal of Dental Research*, Vol. 59, 1980, pp. 63–69.

DISCUSSION

G. I. Ogundele[1] (*written discussion*)—What is the typical pH and sulfide ion concentration at the root of a dental alloy?
Haven't you made the environment a bit too aggressive?

M. Marek (*author's closure*)—Human saliva is nearly neutral. When only sulfides are added to either human or synthetic saliva, the solution becomes alkaline, the pH increase depending on the sulfide ion concentration and the buffering capacity of the matrix. Since the sulfides causing tarnish are mainly introduced in food and drinks, the pH and concentration of sulfides in the periods of sulfidation are highly variable.

The sulfide ion concentration of the test solution was relatively high to accelerate the reactions. There is no reason to believe that the mechanism of tarnishing changes with the concentration of sulfides, and relatively aggressive environments are commonly used in corrosion tests in which only the relative performance is evaluated.

J. Lemons[2] (*written discussion*)—Microstructures, as developed during different casting procedures, can very much influence tarnish due to percolation, etc. Was the casting technique controlled so that microstructural related features did not influence your test results?

M. Marek (*author's closure*)—The microstructure affects the susceptibility to tarnishing, and the described test can be used to evaluate the influence. In this study, however, the emphasis was on the development of the test method and not on the evaluation of the commercial alloys used as samples. Most of the specimens were prepared using standard dental laboratory techniques at the University of North Carolina, while a few were as-received from the manufacturers (see Table 1).

---

[1]University of Calgary, Calgary, Alberta, Canada.
[2]University of Alabama in Birmingham, Birmingham, AL 35294.

Henry Hahn,[1] *Paul J. Lare,*[1] *Russell H. Rowe, Jr.,*[1]
Anna C. Fraker,[2] *and Fred Ordway*[1]

Mechanical Properties and Structure of Ti-6Al-4V with Graded-Porosity Coatings Applied by Plasma Spraying for Use in Orthopedic Implants

REFERENCE: Hahn, H., Lare, P. J., Rowe, R. H., Jr., Fraker, A. C., and Ordway, F., "**Mechanical Properties and Structure of Ti-6Al-4V with Graded-Porosity Coatings Applied by Plasma Spraying for Use in Orthopedic Implants,**" *Corrosion and Degradation of Implant Materials: Second Symposium, ASTM STP 859*, A. C. Fraker and C. D. Griffin, Eds., American Society for Testing and Materials, Philadelphia, 1985, pp. 179–191.

ABSTRACT: The object of this work was to determine mechanical properties of implants with graded porous coatings without reference to the properties of the ingrown bone. Optimum strength of a bone/implant interface consisting of porous metal and ingrown bone requires a gradation from base metal to the original bone. The composite metal-bone interface can be obtained by applying a metal coating of graded porosity, varying from near zero at the substrate surface to more than 50% at the outermost layer, on the original implant. Graded porous coatings of titanium or Ti-6Al-4V were obtained by plasma spraying of selected particle size fractions in three layers of successively decreasing density, the top coat being made with 300 to 850 μm powder. Tensile and shear strengths of the coatings were determined by cementing coated samples face to face with an adhesive resin to simulate ingrown bone. Data from these tests are given. Shear strength values ranged from 5.6 to 9.9 MPa (815 to 1430 psi) and tensile strength values were 5.1 to 25.5 MPa (745 to 3700 psi). Failure occurred within the porous coating and not at the interface between the substrate and the coating. Corrosion fatigue tests in Hanks' solution at 37°C (98.6°F) and a pH of 7.4, with a cyclic, fully reversed, peak torsional shear strain of ±0.01, gave lifetimes comparable to or better than those reported for mill-annealed Ti-6Al-4V, except for the samples that had been sintered.

KEY WORDS: implant materials, orthopedic implants, prostheses, titanium, porous coatings, plasma spray, evaluation, mechanical properties, corrosion fatigue, fatigue (materials), biological degradation

The advantage of a porous metallic coating with open interstices permitting ingrowth of bone for fixation of metal orthopedic implants was demonstrated in

[1]President, vice-president, design engineer, and executive vice-president, respectively, ARTECH Corp., Falls Church, VA 22042.
[2]Metallurgist, National Bureau of Standards, Washington, DC 20234.

1970 [1,2],[3] when bond shear strengths of 12 to 17 MPa (1800 to 2500 psi) were obtained in ovine femora. While the basic concept has shown good promise of obviating the serious shortcomings of fixation by means of acrylic bone cement (and is being used with bone cement as well), the optimization of porous metal coatings continues to offer ample opportunity for progress in prosthesis fixation.

The objectives of surface treatment of an orthopedic implant designed for bone ingrowth are

1. biocompatibility,
2. rapid initial ingrowth of bone for stabilization, and
3. maximum ultimate bond strength within a reasonable time.

Objective 1 is achievable by proper selection of the coating material; the present work deals with titanium alloys, whose biocompatibility is well established [3–5]. Objective 2 is achieved by suitable adjustment of the pore size of the coating, for example, by selecting an appropriate size range of metal particles from which it was made. Objective 3, however, demands consideration not merely of a coating layer with ingrown bone but rather of the entire structural system consisting of the metal implant, the coating layer, and the bone itself. The compressive, tensile, flexural, and torsional forces acting on the implant are transmitted to the bone through compressive, tensile, and shear stresses in the coating layer. Tension and shear, which tend to break the continuity of the ingrown bone, are the more serious problems.

The type of interface structure that maximizes the resistance to tensile or shear failure at the interface between a rigid material and a more readily deformable material is that of the roots of a tree—or of a tooth: a graded interface composed mostly of the rigid phase where it meets that phase and of the more complaint phase at the other boundary, with a continuous variation of composition in between. In the present work, a graded porous surface for an orthopedic implant (patent pending) was achieved by plasma spraying of metal powder with control of process variables such as carrier gas flow, powder feed rate, plasma arc current, spraying distance, and angle of incidence. The object of the work was to evaluate mechanical properties of the titanium alloy coating system itself, without reference to the effects of bone ingrowth, as a point of departure for studies in which bone growth is a variable.

Materials

Coatings were prepared from two types of powder, spherical Ti-6Al-4V and ground titanium sponge.[4] The Ti-6Al-4V powders were of two particle size ranges: (1) Consolidated "prealloyed spherical Ti-6Al-4V" powder, 45 to 600

---

[3]The italic numbers in brackets refer to the list of references appended to this paper.
[4]Mention of commercial products in this paper is for reference only and does not imply endorsement by the authors or their organizations.

μm (-30 $+325$ mesh, that is passing a 30 and retained on a 325 mesh screen)[5] and (2) Nuclear Metals "P.R.E.P. Ti-6Al-4V powder", 425 to 850 μm (-20 $+40$ mesh).[6] The ground titanium sponge, from Reactive Metals,[7] had a particle size range of 300 to 850 μm. The Consolidated powder was screened to obtain a <180 μm (-80 mesh) fraction for the base coat and a 180 to 250 μm (-60 $+80$ mesh) fraction for the intermediate coat. The Nuclear Metals and Reactive Metals powders were used as received for the top coat. Substrates were fabricated from commercial Ti-6Al-4V alloy sheet or bar stock in the annealed condition.

Procedure

Plasma Spraying

Appropriate areas of fabricated test specimens were grit blasted to a uniform texture with coarse aluminum oxide abrasive. The prepared areas were then coated by plasma spraying in three stages. A 130-μm (0.005-in.) base layer of dense metal, sprayed with the <180-μm (-80 mesh) Consolidated powder fraction, was first applied to improve adhesion, followed by a 380-μm (0.015-in.) porous layer of the 180 to 250-μm (-60 $+80$ mesh) Consolidated powder fraction. Finally, a low-density layer 1.1 to 1.4 mm (0.045 to 0.055 in.) thick was sprayed with 300 or 425 to 850 μm (-20 $+40$ or $+50$ mesh) powder. The plasma spray parameters for the porous coatings were adjusted to leave the larger powder particles at most partly melted. The settings found to produce the desired densities and pore sizes, and used for the test specimens, are indicated in Table 1. It was found necessary in spraying the top coat to reduce the heating of the powder either by extending the spray nozzle orifice 9 mm ($11/32$ in.) forward from the standard location or enlarging it by milling two 3-mm ($1/8$-in.) slots 3 mm ($1/8$ in.) long, extending radially from the periphery of the round orifice on opposite sides. The extended nozzle was used for spraying the Reactive Metals titanium sponge powder and the enlarged orifice for the Nuclear Metals Ti-6Al-4V powder.

Test Specimens

Lap Shear—The coating to be tested was applied to a 25-mm (1-in.) length at the end of each of two 25 by 76 by 5.5-mm (1 by 3 by 0.215-in.) plates of titanium alloy having a 12.70 or 14.22-mm (0.500 or 0.560-in.) hole 12.7 mm (0.50 in.) from the uncoated end. Torr Seal® adhesive resin[8] was mixed, applied to both coated areas, and pressed into the porous surface with a spatula. The two pieces were then clamped, with the coated faces together, in a fixture that established their alignment and spacing. The opposed faces of the substrate were

[5]Consolidated Astronautics Division, United International Research, Inc., Hauppauge, NY 11787.
[6]Nuclear Metals Inc., Concord, MA 01742.
[7]Reactive Metals, Inc., Niles, OH 44446.
[8]Varian Associates, Palo Alto, CA 94303.

TABLE 1—*Plasma spray conditions for graded porosity titanium coatings.*

	Base Coat	Intermediate Coat	Top Coat	
			Titanium Sponge	Ti-6A1-4V
Coating thickness,				
mm	0.13	0.38	1.6 to 1.8	1.8 to 1.9
(in.)	(0.005)	(0.015)	(0.065 to 0.070)	(0.070 to 0.075)
Plasma current, A (d-c)	600	520	460	440
Gas flow rates, standard ft³/h[a]				
Total	He 90,[b] H₂ 10	He 90,[b] H₂ 10	Ar 80,[b] H₂ 20 to 25	N₂ 80, H₂ 20 to 25
Carrier gas (feeder)	3	8	8	18
Feed rate, g/s	0.2	0.6	0.3	0.3
Gun distance,				
m	0.10 to 0.15	0.20 to 0.25	0.10	0.10
(in.)	(4 to 6)	(8 to 10)	(4)	(4)
Angle of incidence	0°	45°	75 to 85°	75 to 85°

[a]One cubic foot per hour equals 7.86×10^{-6} m²/s.
[b]Flow rate indicated on gage calibrated for N₂.

4.8 mm (0.19 in.) apart and the highest peaks of the porous coating were nearly in contact. An excess of adhesive squeezed out as the fixture clamp was tightened and was sanded off after curing. The resin was cured by heating in an oven at 60°C (140°F) for at least 1.5 h.

Tension—Cylinders of titanium alloy 28.6 mm (1.125 in.) in diameter and 19 mm (0.75 in.) long, with an axial UNC ¼-20 or UNC ⅜-16 threaded hole 15 mm (0.6 in.) deep on one end, were coated on the other end while rotating at 40 to 50 rpm. The adhesive was applied to the coated ends as for the lap shear specimens and the two cylinders were clamped in a machinist's V-block while the coated faces were held together by finger pressure. Curing of the specimen clamped in the V-block and removal of excess resin were performed as for the lap shear specimens.

Corrosion Fatigue—Cylindrical rods of Ti-6Al-4V, 76 mm (3 in.) long and 6.5 mm (0.25 in.) in diameter, were machined to a diameter of 5.0 mm (0.197 in.) over a 12.7-mm (0.5-in.) length at the center. This reduced center section, which was to be exposed to the solution, was polished with successively finer abrasives down to 0.05-μm aluminum oxide to produce a smooth, scratch-free surface. The center sections of specimens to be plasma coated were not mechanically polished. The graded coating was applied to the portion of reduced diameter while the rod was rotating at 40 to 50 rpm.

Sintering

Coated specimens to be sintered were placed in a Centorr® vacuum furnace with a tantalum resistance element and a single tantalum heat shield. After evacuation to 5×10^{-5} torr, the furnace was heated to 870°C (1600°F), held for 16 h, and allowed to cool to room temperature before air was readmitted. Early in the heating period, the pressure in the vacuum chamber increased to a maximum of 9×10^{-5} torr owing to gradual outgassing of its contents. Thereafter, the pressure decreased to 8×10^{-6} torr. The specimens were supported in the furnace by thin sheets of aluminum oxide ceramic. The lap shear specimens were set on their long edges, the tension specimens were set on end with the coated ends uppermost, and the corrosion fatigue specimens were laid on the ceramic with their axes horizontal. The sintering temperature was chosen to maximize the sintering rate while keeping the specimens safely below the β transus.

Testing Procedure

Lap Shear and Tension Tests—Lap shear and tension specimens were tested on an Instron Model TTC machine of 45-kN (10 000-lb) capacity at a crosshead speed of 0.25 mm (0.010 in.)/min. The lap shear specimens were mounted in clevis grips similar in principle to those described in the ASTM Test for Strength Properties of Adhesives in Shear by Tension Loading in the Temperature Range from -267.8 to $-55°C$ (-450 to $-67°F$) (D 2557-72) except that spacers on the clevis pins were used, rather than doublers attached to the specimens, to ensure correct alignment of the bond plane. The tension specimens were mounted on 250-mm (10-in.) lengths of threaded steel rod, which were attached to standard grips used with threaded-end tension specimens.

(*top*) Titanium sponge top coat.
(*center*) Ti-6Al-4V top coat.
(*bottom*) Ungraded standard coating.

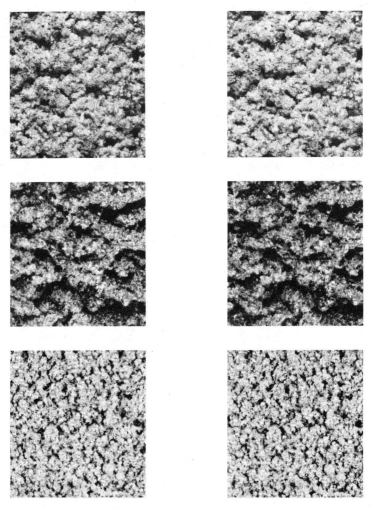

FIG. 1—*Stereo photographs of plasma coated surfaces* ($\times 5$). *For complete stereo viewing instructions see Ref 8. The following process may facilitate viewing without optical aids: (1) Place the eyes close to the photographs and rotate the page slowly until the blurred images merge. (2) Move the page slowly away until the merged images focus.*

Corrosion Fatigue Tests—The tests were performed with the specimen bathed in flowing Hanks' solution [6,7] at a temperature of 37°C (98.6°F) and a pH of 7.4. This environment simulates the corrosive effect of body fluids. The specimens were stressed in an apparatus [7] that applies a fully reversible cyclic twist to one end, while the other end is fixed with respect to rotation but free to move axially. The shear strain imposed by the test apparatus is $\gamma = r\theta/L$ where r is the radius of the specimen prior to coating, L is the length of the test section, and θ is the angle of twist. The radius of the specimen was measured in the uncoated condition. The added thickness of solid metal was less than 40 μm (0.001 in.), although the total thickness of the porous coating was approximately 1 mm (0.04 in.). The peak value of the shear strain was ±0.010 in all the tests. The frequency of the cyclic twist was 1 Hz. Each specimen was run until it had endured more than 10^6 cycles, or it failed, and the number of cycles was recorded.

Metallographic Sections

After testing, shear and tension specimens were sectioned, infiltrated with epoxy casting resin filled with Al_2O_3 powder, and polished through graded boron carbide grits and finally 3 and 0.05-μm Al_2O_3 in vibratory polishers. The polished sections were etched with aqueous 2% HF + 3% HNO_3.

Results

Structure

The external appearance of the coatings with top coats of the titanium sponge powder and of the Ti-6Al-4V powder is shown by the stereoscopic photographs in Fig. 1, together with an ungraded standard coating having only the base coat and intermediate coat. It is clear from the photographs that the top coatings, containing particles with a larger size range, have cavities for bone ingrowth that are considerably larger at the mouth than those in the ungraded coating.

The micrographs shown in Fig. 2 indicate the typical appearance of the metallographic sections. Because the cavities in the coatings are variable in cross section and tortuous in three dimensions, the planar sections do not reveal clearly their effective diameters. The spherical shapes of the larger powder particles are often preserved in the coatings, but many of the particles were at least partly molten on impact, and are greatly deformed. The melting contributes to adhesion of the particles and resultant strength of the coating. The maximum particle size of the Ti-6Al-4V coating is clearly larger than that of the titanium sponge coating.

Strength

All test specimens broke within the porous coating, below the portion reinforced by penetration of the adhesive, rather than in the adhesive layer between coated faces. This result would be expected inasmuch as the base metal and the portion of the coating reinforced by the Al_2O_3-filled epoxy resin are both stronger

(*top*) Titanium sponge top coat.
(*center*) Ti-6Al-4V top coat.
(*bottom*) Ungraded standard coating.

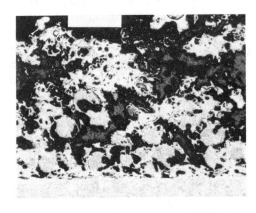

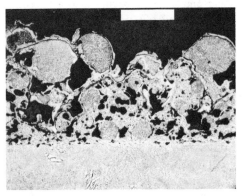

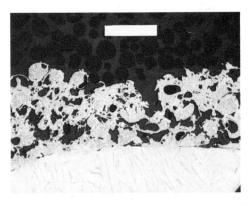

FIG. 2—*Metallographic sections of plasma coated surfaces, infiltrated with Al₂O₃-filled epoxy resin. (Length of white bar indicates 500 μm.)*

than the porous coating alone. The test results are listed in Table 2. The average shear and tensile strengths were generally in the range 7 to 14 MPa (1000 to 2000 psi). The test results represent satisfactory performance in a simulation of ingrowth that is not yet complete, although they do not show conclusively that any one of the three types is superior.

Corrosion Fatigue

The results of corrosion fatigue tests on samples of the two types of graded coatings, sintered and unsintered, are given in Table 3 together with results from testing of specimens representing successive stages in the preparation process: original machined and polished surface; grit blasted; and with base and intermediate coats only (that is, a standard ungraded coating). The results for these samples are plotted in Fig. 3 along with the fatigue endurance curve obtained [9] for another batch of mill-annealed Ti-6Al-4V.

Some unsintered samples had greater fatigue endurance than that which would correspond to mill-annealed material. This may be due to effects of grit blasting and coating. The result for Sample M, which was only machined and polished, is also above the curve. The base metal from which the samples were prepared may have superior properties over the material that was used to establish the

TABLE 2—*Shear and tension test results.*

	Titanium Sponge Top Coat				Ti-6A1-4V Top Coat	
	Unsintered,		Sintered		Sintered,	
	MPa	(psi)	MPa	(psi)	MPa	(psi)
	LAP SHEAR					
	9.86	(1430)	6.65	(965)	9.31	(1350)
	6.90	(1000)	5.62	(815)	8.41	(1220)
	7.45	(1080)	8.00	(1160)	11.27	(1635)
	8.96	(1300)	7.72	(1120)	9.96	(1445)
	8.27	(1200)	9.38	(1360)	13.82	(2005)
	8.76	(1270)	...	...	11.17	(1620)
Average	8.36	(1213)	7.47	(1084)	10.66	(1546)
standard deviation	±1.07	(156)	±1.42	(206)	±1.90	(275)
	TENSION					
	14.02	(2033)	13.22	(1917)	28.45	(4127)
	5.14	(745)	25.51	(3700)	...	...
	6.25	(906)	...	...	...	...
	8.74	(1268)	...	...	...	...
Average	8.54	(1238)	19.36	(2808)	...	...
standard deviation	±3.95	(573)	±8.69	(1260)	...	...

TABLE 3—*Corrosion fatigue test results for 0.01 shear strain.*

Specimen (ELI Ti-6A1-4V, annealed)	Number of Cycles to Failure
M = polished through 0.05-μm A1$_2$0$_3$	>10^6
F = grit blasted	>10^6
G = grit blasted	>10^6
L = two-layer standard coating	>10^6
J = two layer standard coating, sintered	9.3 × 10^5
K = two layer standard coating, sintered	9.4 × 10^5
D = graded coating with titanium sponge top coat	>10^6
C = coated as in D, sintered	2.2 × 10^5
B = graded coating with Ti-6A1-4V top coat	4.8 × 10^5
A = coated as in B, sintered	7.5 × 10^4

curve, although both were samples of the ELI grade Ti-6Al-4V alloy. Grit blasting and coating were not shown to affect the fatigue life adversely. They may even be beneficial. It does appear, however, that the sintering operation reduces the fatigue life. Oxygen contamination or traces of oxygen in the vacuum furnace or in the coating, producing a thin surface layer of higher oxygen content, could lead to this loss of fatigue life.

All of the unsintered specimens except one were "runouts" having fatigue lives exceeding 10^6 cycles. Observations of these specimens, which had endured

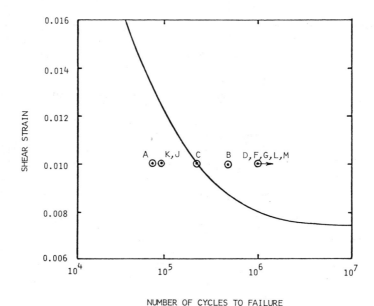

NUMBER OF CYCLES TO FAILURE

FIG. 3—*S-N curve for Ti-6Al-4V [9] with data points from tests of Ti-6Al-4V specimens with porous coatings. Tests were conducted in fully reversible torsion in Hanks' saline solution at 37°C, at a pH of 7.4 and a frequency of 1 Hz. Data points are identified by letters referring to specimens listed in Table 3.*

over one million cycles without failure, indicated that no damage to the coating or the substrate had occurred. Light microscopy examination of the surface showed no evidence of cracks. Spherical particles still were visible on the surface as well as the deformed particles that were partly or completely melted in the plasma spraying process. Viewing the edge of the coated area, where the metal substrate and coating were visible together, revealed that the interface was intact. Additional evidence for the integrity of the coating was the absence of any residue or particles in the bottom of the environmental test cell. Light microscopy observation of failed specimens did not reveal cracks other than at the failure site.

Test results from a larger number of samples will be needed to establish the optimum compromise between ultimate strength and fatigue life, corresponding to the most desirable sintering treatment.

Conclusions

It has been established previously that arc plasma spraying is a useful method for applying metal porous coatings to metal substrates of various types. Porous coated metal surfaces can be advantageous for use in surgical implant fixation, both with and without cement. This paper dealt with arc plasma sprayed porous coatings on Ti-6Al-4V alloy; the following conclusions can be drawn from this work.

1. Arc plasma spraying can be used to apply porous coatings of either titanium sponge or Ti-6Al-4V to Ti-6Al-4V substrates.
2. The strength of the interface bond formed between the coating and the substrate is high. Failure in lap shear tests always occurred within the coating.
3. Parameters of the coating process can be varied to give a coating with either a large or a small pore size or to give a coating having a graduated pore size.
4. Control of plasma spraying techniques and subsequent heat treating are important to prevent oxidation and also to prevent microstructural changes in the substrate metal.
5. Application of the porous coating has not been shown to reduce the corrosion-fatigue life of the substrate if the coated specimen is not sintered. Sintering can lower the fatigue endurance.

References

[1] Hahn, H. and Palich, W., *Journal of Biomedical Materials Research*, Vol. 4, 1970, pp. 571–577.
[2] Hahn, H., "Bone Implant," U.S. Patent 3 605 123, 1971.
[3] Laing, P. G., Ferguson, A. B., Jr., and Hodge, E., *Journal of Biomedical Materials Research*, Vol. 1, 1967, pp. 135–149.

[4] Laing, P. G. in *Corrosion and Degradation of Implant Materials, ASTM STP 684*, B. C. Syrett and A. Acharya, Eds., American Society for Testing and Materials, 1979, pp. 199–211.
[5] Wheeler, K. R., Karagianes, M. T., and Sump, K. R. in *Titanium Alloys in Surgical Implants, ASTM STP 796*, H. A. Luckey and F. Kubli, Jr., Eds., American Society for Testing and Materials, 1983, pp. 241–254.
[6] Hanks, J. H. and Wallace, R. E. in *Proceedings*, Society for Experimental Biology, Vol. 71, 1949, pp. 196–200.
[7] Imam, M. A., Fraker, A. C., and Gilmore, C. M. in *Corrosion and Degradation of Implant Materials, ASTM STP 684*, B. C. Syrett and A. Acharya, Eds., American Society for Testing and Materials, 1979, pp. 128–143.
[8] Fellows, J. A. in *Metals Handbook*, 8th ed., Vol. 9, American Society for Metals, Metals Park, Ohio, 1974, pp. 281–282.
[9] Imam, M. A., Fraker, A. C., Harris, J. S., and Gilmore, C. M. in *Titanium Alloys in Surgical Implants, ASTM STP 796*, H. A. Luckey and F. Kubli, Jr., Eds., American Society for Testing and Materials, 1983, pp. 105–119.

DISCUSSION

L. Gustavson[1] (*written discussion*)—Why did you elect to use a torsional fatigue test mode rather than axial or bending fatigue loading? Do you feel that 10^6 cycles is a sufficient runout point rather than 10^7 cycles?

H. Hahn, P. J. Lare, R. H. Rowe, Jr., A. C. Fraker, and F. Ordway (*authors' closure*)—The torsion fatigue testing mode is more severe than either axial or bending fatigue loading. The stress distribution in a bending mode is more complex than the stress distribution in an axial or torsional mode. The maximum stress in the torsion test is on the surface of the specimen and decreases to zero at the center of the specimen whereas the stress distribution in the axial test is uniform throughout the cross-sectional area. Data from these different testing modes are not comparable directly, but if the tests are conducted under the same type of control such as load or displacement control, the ranking of materials, in terms of fatigue strength, should be the same.

Many of the tests listed as a runout at $>10^6$ actually ran to 1.2×10^6 cycles before the tests were stopped. The decision to stop the tests at this point was made on the basis of time. There may not be any difference in the results if the tests are run to 10^7 cycles, but probably it would be better to do so. One way to accomplish this in a shorter time would be to increase the frequency from 1 Hz to 2 Hz for future tests.

Walter P. Spires, Jr.[2] (*written discussion*)—Do you, in fact, have a three-dimensional interconnected porosity? How can you quantify it and what percentage of porosity is obtained? How do you control it?

H. Hahn, P. J. Lare, R. H. Rowe, Jr., A. C. Fraker, and F. Ordway (*authors' closure*)—Yes, the porosity, at least in the portion of the coating where the

[1] Howmedica, Inc., Rutherford, NJ 07070.
[2] Dow Corning Wright, Arlington, TN 38002.

sprayed particles are largely unmelted, is interconnected as one would expect in a more or less closely packed aggregate of spheres. The appearance of the surface in scanning electron micrographs and the forms visible in metallographic sections support this opinion.

We do not have a mathematical description adequate to characterize such a surface structure; consequently, no suitable experimental technique has been defined for the characterization. The percentage of porosity clearly varies as a function of distance above the original substrate surface, being close to zero near the substrate and going to 100% at the elevation of the highest peak. The pore size within aggregates is approximately proportional to the diameter of the sprayed particles, assuming that the fractional melting of the particles is approximately the same for the different particle sizes used. The size of spaces between adjacent peaks increases with distance from the substrate. Unfortunately, the methods of porosimetry and surface area measurement applicable to homogeneous bodies are not suitable for graded porous coatings.

The reproducibility of the graded porous coating depends on constancy of the spraying conditions, including gas compositions and flow rates, arc current, powder composition, particle size and shape, feed rate, and specimen distance, angle, and traverse rate. Developing the set of optimum conditions is a process of multiparameter refinement that may be performed by systematic analytic methods but is most effectively carried out with the intuitive guidance of a skilled operator.

Metallic Materials: *In Vivo* and *In Vitro*

Katharine Merritt[1] and Stanley A. Brown[1]

Biological Effects of Corrosion Products from Metals

REFERENCE: Merritt, K. and Brown, S. A., **"Biological Effects of Corrosion Products from Metals,"** *Corrosion and Degradation of Implant Materials: Second Symposium, ASTM STP 859*, A. C. Fraker and C. D. Griffin, Eds., American Society for Testing and Materials, Philadelphia, 1985, pp. 195–207.

ABSTRACT: Metallic implants inserted into the body will eventually undergo some degradation by a variety of mechanisms including abrasion and corrosion. The biological effects of the material released into the tissue is a subject of much concern and investigation. We have been investigating the ability of the metallic implants upon degradation to stimulate metal sensitivity reactions, and studies have been undertaken in humans and animals to investigate the tissue response to metallic implants in the presence of metal sensitivity. It is apparent that many patients have a sensitivity to the metals in the implants and that the presence of metals in a sensitive animal or human will elicit inflammatory responses and sometimes the formation of foreign body giant cells. This may have an adverse effect on the performance of the implant with pain, swelling, and tissue necrosis at the site and, in some cases, loosening of the implant. Studies on implant site infections revealed that infection rates are altered by the presence of this tissue response.

Metal sensitivity reactions are not to the small metal ions but to a complex of metal ions and host tissue. The nature of the binding of the metal ion to tissue or cells, the distribution of the metal ion or metal complexes in the body, and the biological responses to these complexes are of concern. We have undertaken *in vitro* and *in vivo* studies on the binding of metal ions as metal salts or as corrosion products, and it is evident that metals bind mostly to albumin. The ability of these different metals to bind to red cells and white cells varies with chromium with a valence of $6+$ binding most strongly to cells. The chrome from corrosion products binds strongly to cells and thus appears to be a chromium $6+$.

Metal salts or corrosion products injected intramuscularly are rapidly distributed in the body with cell binding again being seen mostly with chromium $6+$. Cobalt is not distributed as rapidly in the body and apparently binds to the tissue at the site of the intramuscular injection. The interaction of serum and cells with chromium on the surface of the implant may markedly effect its corrosion and the biological activity of the complexes formed. Some of the tissue and protein binding may be altered by shifts in the pH in the tissue during inflammatory responses or in infection.

KEY WORDS: stainless steels, metal sensitivity, metal-protein complexes, metal-cell complexes, sensitivity, infection, corrosion, implant materials, biological degradation

Implants used for maintenance or restoration of health and body function have been of various designs and materials. This report will focus on the *in vivo*

[1]Associate professors of Biomedical Engineering, Department of Biomedical Engineering, Case Western Reserve University, Cleveland, OH 44106.

biological responses to orthopaedic implants and the *in vitro* and *in vivo* responses to corrosion products of stainless steel or cobalt-chrome alloys. Metallic implants for orthopaedic use may be designed for short-term function or for the life of the patient, as in total joint replacements. With anticipated function of greater than 20 years in a weight-bearing situation, concern has developed for the biological response to the implants and any corrosion or wear products. That metal loss does occur from these implants is well recognized and the subject of various symposia [1–3].[2] Examination of tissue surrounding metallic implants that have failed or that have been removed because of morbidity associated with the implant, has revealed the presence of acute inflammatory responses with the accumulation of polymorphonuclear leukocytes (PMNS); more chronic responses with the accumulation of monocytes, macrophages, and lymphocytes or even plasma cells; and, finally, long-term chronic responses with the accumulation of giant cells and evidence of necrosis [4,5].

Studies on tissue retrieved from the site of implants have revealed that metal is present, but analysis of the amount of corrosion of the implant and amount of metal in the tissue does not reveal a strong correlation; and the composition of the alloy and the composition of the metal in the tissue often does not correlate [6,7]. The observation that metal can be detected in remote tissues including hair, urine, and peripheral blood [8] indicates that much of the metal released from the implant is transported away from the site. In addition, analysis of the histopathology of the tissue at the site of the implant does not always correlate with the degree of corrosion of the implant or amount of metal in the tissue, and indicates differences in host responses [6].

Detailed studies on total hip replacements that were removed because of pain, loosening, and signs of inflammation revealed that there was a correlation between the presence of a cutaneous sensitivity response to one or more of the metal components and the morbidity necessitating the removal [9,10]. In addition, these cases and other cases of sensitivity responses to metallic implants revealed that there may be manifestations of a sensitivity response occurring at some distance from the implant [11–14]. This has led to much controversy and confusion regarding the role of the implant in the development of the symptoms.

We became interested in the problem of sensitivity reactions to orthopaedic implants and undertook studies in order to understand the mechanism of this biological response to orthopaedic implants. This report will discuss (1) the study on the sensitivity response to metals in the human using an *in vitro* test for leukocyte migration inhibition factor (LIF) as the assay, (2) the use of the rabbit to model and study these sensitivity responses, (3) infection studies in mice and hamsters to elucidate the role of sensitivity reactions in variations in infection rates, (4) studies on the cell and protein binding of metal salts and corrosion products to understand the antigen involved in sensitivity reactions, (5) studies on distribution of metal salts and corrosion products in the body in order to

---

[2]The italic numbers in brackets refer to the list of references appended to this paper.

understand the mechanism of systemic responses, and (6) preliminary studies on the effect of pH on protein binding to further elucidate some of the biological responses to implants and corrosion products.

Methods

Metal Salts

Nickel: certified nickel(ous) chloride, contaminating cobalt (Co) 0.04%. Cobalt: certified cobalt chloride, contaminating nickel (Ni) 0.02%. Chromium: certified chromium(ic) chloride, certified ACS potassium dichromate.[3]

Corrosion Products

Corrosion products were generated from the fretting corrosion of Type 316L (ASTM Specification for Stainless Steel Bars and Wire for Surgical Implants (Special Quality) (F 138-76)) plates and screws[4] or MP 35N plates and screws (ASTM Specification for Wrought Cobalt-Nickel-Chromium-Molybdenum Alloy for Surgical Applications (F 562-78))[5] for four weeks in saline or in 10% calf serum using the fretting corrosion simulator previously described [15]. These products were resuspended prior to use and the resulting suspension used in the biological assays.

Leukocyte Migration (LIF) Test

This test was performed using the method in agarose as previously described [16]. Briefly for our procedure, we use Click's medium in agarose with Hepes and NaHCO, and newborn calf serum. For the test, the buffy coat from 10 to 15 ml of blood is collected and centrifuged at 500 g for 20 min. The pellet is washed and then resuspended with 100 μl of buffer. The cell suspension is then pipetted in 20 μl quantities into test tubes containing the test solutions. The first tube contains 10 μl of saline and 10 μl of the patient's serum as a positive control for migration. The second tube contains saline and phytohemagglutinin (PHA) as a control for LIF production. The remaining tubes contain 10 μl of serum and 10 μl 0.05% nickel chloride, cobalt chloride, and potassium dichromate, respectively, to test for metal sensitivity. The tubes are incubated for 30 min at 37°C then 10 μl samples are pipetted in duplicate from each tube into the 3 mm-diameter wells cut into the agarose plates. The plates are incubated in a candle jar or carbon dioxide (CO_2) incubator. The next morning, the plates are removed from the incubator, inverted, and examined with a standard light microscope with a $\times 4$ objective and $\times 20$ eyepieces. One eyepiece contains a micrometer disk with 25 boxes (five down by five across). The number of cells in each box is counted to a maximum of 50. The sum of the cells in the five

---

[3]All required from Fisher Scientific Co., Pittsburgh, PA 15219.
[4]Zimmer, Inc., Warsaw, IN 46580.
[5]Sulzer Bros, Inc., New York, NY 10166.

boxes next to the well is added to 2 × the sum of the second row, 3 × the sum of the third row, and so forth for the five rows. Migration fully filling the boxes is termed "total" and given a score of 3750. The two sides of the well with the most migration are counted and averaged. Migration in the wells with the metal salts must be less than 50% the migration of the control well with buffer to state that the patient has a sensitivity to that metal. For a patient to be termed a nonmigrator, migration in all the wells including the control well with the buffer must be less than 250. In addition, control cells known to migrate must have migrated better than 2000 in that batch of agarose plates.

Animals

Mature New Zealand white rabbits were used in some experiments. Golden hamsters, 120 to 150 g, and Swiss white mice 20 to 25 g, were used in the other animal experiments. All were given food and water *ad lib.* Animals to be made sensitive were injected with metal salts in complete Freund's adjuvant followed by monthly injections of aqueous solutions of the metal salts.

Blood

Human, rabbit, and hamster blood was collected in heparin and separated into serum, red cells, and white cells using 2% dextran 500 T[6] to sediment the cells [16].

Histologic Analysis

Tissue from the implant site of the human or tissue containing the implant from animals was fixed in formalin. It was then embedded in polyethylmethacrylate as previously described [17]. The implant, if present, was electrolytically dissolved away [18], and the tissue thin sectioned on a Jung microtome. The tissue was analyzed for the presence of inflammatory cells, plasma cells, giant cells, necrosis, and blood vessels, connective tissue, and bone [16,17].

Infection Study

Mice or hamsters were made sensitive to the metal salts by injection. Then, small metallic implants were inserted subcutaneously using a tube and trocar technique [19]. Control and sensitized animals were then left for five to ten weeks at which time a known inoculum of bacteria from 10 to 10 was injected into the site of the implant. The animals were sacrificed five days later and the implants and surrounding tissue were removed, ground in glass beads and saline, and the resulting solution plated on Trypticase Soy Agar for bacterial counts to determine the incidence of implant site infections in the presence and absence of metal sensitivity.

[6]Pharmacia Inc., Piscataway, NJ 08854.

Analysis of Metal Content

The amount of nickel, cobalt, and chromium on the cells or in the serum was determined using a Perkin Elmer 403 [20] or a Perkin Elmer 2380 Atomic Absorption Spectrophotometer with a graphite furnace.

Studies of the Effect of pH on Biological Activity

Solutions of Bovine Albumin (Miles Fraction V) and of Bovine Gamma Globulin (Miles Fraction II) were prepared at a concentration of 0.5 g% in physiologic saline. The pH was then adjusted with 1 N NaOH or 1 N HCl. Albumin was used at its isoelectric point (pH 4.5), acid to its isoelectric point (pH 3.0), and basic to its isoelectric point (pH 8.0). Gamma globulin was used at its isoelectric point (pH 8.0), acid to its isoelectric point (pH 5.9), and basic to its isoelectric point (pH 11.0). Fretting corrosion of Type 316L stainless steel was then studied in these different solutions and the amount of corrosion was monitored by weight loss of the plates and screws, nickel in solution, and chromium in solution [15].

Results

Studies on Metal Sensitivity in Patients with Orthopaedic Implants

LIF Results—The data on 629 patients receiving metallic implants revealed that 25% were sensitive with 18% sensitive to nickel, 15% sensitive to cobalt, and 3.5% sensitive to chromium. These numbers add to greater than 25%, because some were sensitive to more than one of the metal salts. Some of these patients had had prior implants, and in others the history of prior implants was unknown. The presence of previous implants as well as the increased sensitivity of the LIF test may account for the increase in the incidence of sensitivity in these patients when compared to the incidence of sensitivity determined by skin testing [21].

The data on 283 patients having metal implants removed revealed that only 43% were nonsensitive with 29% sensitive and 27% were nonmigrators in the LIF test indicating that they were sensitive to one or more of the metal salts and actually reacting to the presence of the implant [6,15,16,22]. Thus of the 283 patients having implants removed, 56% demonstrated sensitivity to one or more of the metal salts. When the data on the retrieved implants was analyzed, it was apparent that 55 were cobalt-chrome appliances and that 74% of these patients were sensitive or sensitive and reacting, and that there were 187 stainless steel (316 or 316L) inplants and 57% of these patients were sensitive or sensitive and reacting. The results are summarized in Table 1.

Histologic samples were available on some of these patients that revealed signs of acute inflammation in some and more chronic inflammation or foreign body giant cells evident in others. These patients were manifesting a tissue response to the presence of the implant (Fig. 1).

TABLE 1—*The incidence of sensitivity in patients receiving implants and having implants removed. The sensitivity associated with the two alloys of which the removed implants were manufactured is also shown.*

	No Sensitivity, %	Sensitive, %	Reacting (nonmigrator), %
629 implantations	75	25	0
283 removals	43	29	27
	TYPES REMOVED		
187 stainless steel	43	37	20
55 cobalt chrome	26	36	38

Animal Studies on Metal Sensitivity

In order to model the problem of sensitivity reactions to orthopaedic implants, rabbits were made sensitive to nickel, cobalt, or chromium. The rabbits were sacrificed at 3, 6, 9, and 12 weeks after implantation of stainless steel (ASTM F 138-76) or cobalt-chromium (ASTM Specification for Wrought Cobalt-Nickel-Chromium-Molybdenum-Tungsten-Iron Alloy for Surgical Implant Applications (F 563-78)) bone screws and the histologic response to the bone screws determined. The LIF tests were done on these rabbits at approximately one month intervals. The LIF testing revealed that all the injected rabbits became sensitive to the metals and that about 50% of these animals extended the sensitivity to a condition of sensitive and reacting. Histologic analysis revealed inflammatory and giant cell reactions around the implant. In addition, areas of necrosis were

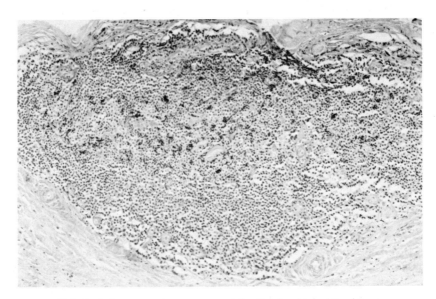

FIG. 1—*Inflammatory tissue response at the site of an implant in a human.*

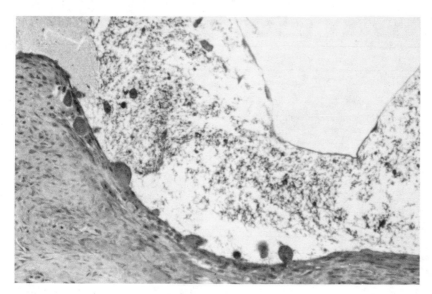

FIG. 2—*Area of necrosis at a bone screw interface in a sensitive and reacting rabbit.*

noted, especially in the animals that were sensitive and reacting [16,17] (Fig. 2).

The observation that some patients with a metal sensitivity reaction extended this reaction to a condition in which there was no migration in any of the wells including the control well has led to some concern about the immunologic responses of these individuals. Studies have shown that the PMNS and monocytes from these patients are incapable of spontaneous migration and do not respond to chemotactic factors. Blastogenesis in lymphocytes appears normal or elevated [22].

Infection Studies

The presence of this migratory defect in patients and in rabbits bearing implants led us to believe that these individuals might have an altered ability to combat infection and that there might be a change in the incidence of implant site infections. The results are shown in Table 2. The studies in mice that had been immunized for ten weeks and then bore the implants for nine weeks before injection with Staphylococcus aureus revealed that immunization with nickel or cobalt chloride did increase the infection rate but there was no significant difference ($P = 0.7$) [23]. The results from the hamster study revealed that this response is time dependent and that early in the response the acute inflammatory response resulted in a significant decrease in infection rate in sensitized animals whereas the late chronic response resulted in an increase in infection rate.

TABLE 2—*Implant site infections in immunized and control animals. Staphylococcus aureus was injected at the implant site five days prior to sacrifice.*

	Number Infected		
	Immunized	Unimmunized	
MICE			

Ten weeks of immunization with metal salts followed by nine weeks of implantation of steel or cobalt-chromium implants.

	Immunized	Unimmunized	
316L steel	26/43 (60%)	22/47 (47%)	$p = 0.75$
Co-Cr	20/43 (46%)	18/46 (39%)	$p = 0.95$
Total	46/86 (53%)	40/91 (44%)	$p = 0.6$
HAMSTERS			

Short term: five weeks of immunization and five weeks of steel implants.

	37/103 (36%)	52/97 (54%)	$p = 0.02$

Long term: eight weeks of immunization and ten weeks of steel implants.

	33/61 (53%)	29/59 (47.6%)	$p = 0.5$

Cell and Protein Binding of Metal Salts and Corrosion Products

Because the *in vitro* testing and the animal modeling used metal salts but the *in vivo* response was to the alloy or the corrosion products from the alloy, studies were initiated using corrosion products in comparison to metal salts. The first series of experiments were undertaken to determine the cell and protein binding of the salts and corrosion products. Analysis of human and rabbit blood after the addition of metal salts or corrosion products revealed that most of the metal was in the serum with some binding to cells. Cobalt bound to cells but the most extensive binding to cells was seen with the chromium with a valence of 6+ [24]. It did not matter whether the salts and the corrosion products were added to whole blood that was then separated into the components or whether they were added to the components of blood already separated. Thus, the presence of serum did not affect the binding to either the red blood cells or the white blood cells. Analysis of the serum by isoelectric focusing and by gel filtration revealed that the bulk of the metal was bound to albumin with some binding to gamma globulin [20]. The results are summarized in Table 3.

Studies on the effect of serum and serum proteins on fretting corrosion had revealed that albumin was the protein that most affected the corrosion [25]. The two most likely reasons why this might be the case were (1) the relative concentration and (2) the pH. Thus, these studies were repeated using 0.5 g % albumin and gamma globulin. This is the concentration that would mimic the concentration of albumin in 10% serum (which is our standard test solution for fretting corrosion) but would be five times the concentration of gamma globulin normally present. The addition of albumin or gamma globulin to USP saline for injection[7] results in a pH of around 6.0 that is basic to the isoelectric point of albumin but acid to the isoelectric point of gamma globulin. The addition of

[7]Abbott Laboratories, North Chicago, IL 60064.

TABLE 3—*Protein and cell binding of metal salts added to blood.*

	Albumin	Globulin	Red Blood Cells	White Blood Cells
Nickel	+	...	+	...
Cobalt	+	...	+	+
Chromium 3+	+	...	+	+
Chromium 6+	+	±	+	+

serum to saline at a concentration of 10% serum results in a pH of about 6.5. Again, this is a pH basic to the isoelectric point of albumin but acid to the isoelectric point of gamma globulin. Thus, we were studying one protein with negative charges and the other protein with positive charges.

The results of our study with the same concentration of both proteins at different pH's revealed that for both of them a pH acid to the isoelectric point gives corrosion closest to that of saline alone and a pH basic to the isoelectric point results in corrosion closer to 10% serum. Neither protein alone completely mimics the effects of serum.

Transport of Metals in the Body

The final series of experiments to be described were designed to answer the question on distribution of metals from the site of an implant. The animals were injected intramuscularly into the thigh area with corrosion products or the individual metal salts. The animals were bled at 0 time, 2, 4, 6, 24, 48, and 96 h [25]. Analysis of the blood components (serum, red blood cells, and white blood cells) revealed that the metal was rapidly distributed into the blood with high levels reached at 2 h. Results with the metal salts revealed that the bulk of nickel, cobalt, and chromium 3+ was in the serum, and there was a little cell binding by nickel. On the other hand, chromium with a valence of 6 was found on the cells. The corrosion products from stainless steel and cobalt-chrome behaved similarly to the metal salts with the bulk of the nickel and cobalt in the serum phase. There was an additional late binding of nickel to red blood cells. The chromium was found largely on cells indicating that it was behaving as a chromium 6+ rather than a chromium 3+.

Discussion

It is evident that there are a wide range of biological responses to corrosion and wear products from implants. This report has focused only on metallic implants of stainless steel and cobalt-chrome and has not addressed the question of other materials. Biological responses to wear, degradation, and corrosion products from other materials are of importance and need to be addressed.

Our studies have indicated that there is a major change in the migratory capacity of the white blood cells in the peripheral blood of patients who are bearing orthopaedic implants. There is a loss of spontaneous migration of these cells and an inability to respond to chemotactic factor. This would mean that these patients would be unable to mount a strong defense mechanism at a distant site, such as a site of infection. However, studies on these patients have not indicated a higher infection rate and the real consequences of this migratory defect remain to be elucidated. The presence of this migratory defect has been associated with increased pain at the implant site and loosening of the implant, or local bone resorption. In addition, the presence of cutaneous symptoms has also been more common in patients with the migratory defect.

Studies in rabbits have indicated that the presence of a metal sensitivity does alter the histologic response to stainless steel and cobalt-chrome screws. The presence of the migratory defect was associated with a more intense inflammatory response with more giant cells and necrosis present. *In vitro* studies with the rabbit cells have indicated that phagocytosis and killing are normal. Thus, cells could not migrate into the site at which they were needed but those that were there would be functional.

Studies on infection rates at implant sites in mice and hamsters have revealed some interesting findings. Animals sensitive to the metal and bearing the implant for a short period of time have a decreased infection rate compared to unimmunized controls. If they were sensitive and the implant was in place for a longer period of time, the infection rate was greater than that in the control. This would indicate that the acute inflammatory response would be protective, whereas the chronic inflammatory response with giant cells and necrotic tissue would provide a histologic response supportive of growth of bacteria and the lack of migration of cells might prevent other cells from entering the site to combat the infection. These studies are done in soft tissue and are really not studies on biological responses to corrosion products but do indicate the host response to the implant. Studies in hamsters immunized with corrosion products from stainless steel and then implanted subcutaneously with stainless steel implants are barely underway and the data would indicate that the short-term implants have a decreased infection rate in the animals immunized with the corrosion products. The long-term animals have not yet been challenged.

The presentation of patients with signs of sensitivity reactions at some distance from the implant and the observation that there is a defect in function of peripheral white blood cells indicated that there are systemic responses to local implants. This would indicate that there is distribution of corrosion products from the site of the implant. Studies *in vitro* were designed to assess the cell and protein binding of these products. All the metals from salts or from corrosion products bound to albumin. This is the protein in the serum in highest concentration and also is the blood protein present in highest concentration in the tissue fluid. Thus, the observation that the metals will bind to this protein provides an explanation for the rapid dissemination of the metal through the body. The host responses of sensitivity could therefore occur wherever the immune system encountered the antigen and this could be in almost any dermal site and in some deep tissues. The role of pH in fretting corrosion has been discussed in a preceding paper in this volume [26]. In terms of the biological response, inflammatory responses and some infections tend to lower the pH at the site. This would decrease the effect of proteins on the fretting corrosion and result in increased corrosion. Some infections raise the pH that might alter the albumin/globulin binding of the metals. It is therefore important to look at the effects of pH.

The observation that some of these metals are highly cell binding explains some of the disparate findings in tissue analysis. Cobalt *in vitro* was highly cell binding but was not found on circulating cells in the hamsters injected intramuscularly and then bled. This would indicate that the cobalt bound to cells at the site and did not enter the peripheral blood. This would account for high levels of cobalt found in the tissue around cobalt-chrome implants [6,7]. Chromium with a valence of 6+ also bound strongly to cells but also entered the circulation indicating some transport away from the site. Nickel seemed to be more readily transported away from the site and into the peripheral blood. These studies were not undertaken with the intent of studying the metabolism of the metals as discussed by others [27,28], but were designed to determine which cells and proteins were involved in metal binding and might serve as the real antigen in stimulation of the immune responses.

Attempts to identify the active antigen are being undertaken in the laboratory. Animals sensitive to metals are being challenged *in vivo* with the different components. *In vitro* LIF tests are underway with the different components. The major limitations to these studies are that when one injects a substance into the body, cell and protein binding can occur and modify the response. Similarly, the LIF test *in vitro*, as it is currently done, involves the use of media containing serum. We are in the process of developing the LIF test in serum-free media and hope to define more accurately the corrosion product and cell-protein interactions that lead to adverse biological responses to implants. Once these processes are understood, we can better understand how to detect those patients who are apt to manifest an adverse response and select products that do not pose a hazard.

References

[1] *Implant Retrieval: Material and Biological Analysis*, A. Weinstein, D. Gibbons, S. A. Brown, and A. W. Ruff, Eds., National Bureau of Standards, Special Publication 601, in press.

[2] *Corrosion and Degradation of Implant Materials*, ASTM STP 684, B. C. Syrett and A. Acharya, Eds., American Society for Testing and Materials, Philadelphia, 1978.

[3] *Mechanical Properties of Biomaterials*, G. W. Hastings and D. F. Williams, Eds., *Advances in Biomaterials*, Vol 2, Wiley, New York.

[4] Williams, D. F. and Meachim, G. A., "Combined Metallurgical and Histological Study of Tissue-Prosthesis Interactions in Orthopaedic Patients," *Journal of Biomedical Materials Research*, Vol. 5, No. 1, 1974.

[5] Willert, H. G., Buchhorn, G. H., and Semlitsch, M., "Recognition and Identification of Wear Products in the Surrounding Tissues of Artificial Joint Prosthesis," *Tribology of Natural and Artificial Joints*, J. H. Dumbleton, Ed., Elsevier, New York.

[6] Simpson, J. P., Geret, V., Brown, S. A., and Merritt, K., "Retrieved Fracture Plates: Implant and Tissue Analysis," *Implant Retrieval: Material and Biological Analysis*, A. Weinstein, D. Gibbons, S. A. Brown, and A. W. Ruff, Eds., National Bureau of Standards, Special Publication 601, in press.

[7] Williams, D. F. and Roaf, R. in *Implants in Surgery*, Saunders, London, 1973, pp. 246–256.

[8] Coleman, R. F., Herrington, J., and Scales, J. T., *British Medical Journal*, Vol. 1, 1973, p. 527.

[9] Evans, E. M., Freeman, M. A. R., Miller, A. J., and Vernon-Roberts, B., *Journal of Bone and Joint Surgery*, Vol. 56B, 1974, pp. 626–642.

[10] Benson, M. K. D., Goodwin, P. G., and Boostoff, J. *British Medical Journal*, Vol. 4, 1975, pp. 374–375.

[11] McKenzie, A. W., Aitken, C. V. E., and Risdill-Smith, R., *British Medical Journal*, Vol. 4, 1967, p. 36.

[12] Barranco, V. P. and Soloman, H., *Journal*, American Medical Association, Vol. 220, 1972, p. 1244.

[13] Kubba, R. and Champion, R. H., *British Journal of Dermatology*, Vol. 93, No. 11, 1975, p. 41.

[14] Halpin, D. S., *Journal of Bone and Joint Surgery*, Vol. 57B, 1975, p. 451.

[15] Brown, S. A. and Merritt, K., *Journal of Biomedical Materials Research*, Vol. 15, 1981, pp. 479–488.

[16] Merritt, K. and Brown, S. A., *Acta Orthopaedica Scandinavica*, Vol. 51, 1980, pp. 403–411.

[17] Merritt, K. and Brown, S. A., *Advances in Biomaterials*, Vol. 4, 1982, pp. 85–93.

[18] Brown, S. A. and Simpson, J., *Journal of Biomedical Materials Research*, Vol. 13, 1979, pp. 337–338.

[19] Merritt, K., Shafer, J. W., and Brown, S. A., *Journal of Biomedical Materials Research*, Vol. 13, 1979, pp. 101–108.

[20] Merritt, K., Sharkey, N. A., and Brown, S. A., "Serum and Cell Binding of Nickel," *Transactions*, 7th Annual Society of Biomaterials, Vol. IV, 1981, p. 16.

[21] Rudner, E. J. et al, *North American Dermatological Survey. Contact Dermatitis*, Vol. 1, 1975, pp. 277–280.

[22] Merritt, K. and Brown, S. A., "Loss of Random and Chemotactic Cell Migration in Humans and Rabbits," *Proceedings*, 80th Meeting of American Society of Microbiology, 1980.

[23] Merritt, K. and Brown, S. A. in *Transactions*, Society of Biomaterials, Vol. 3, 1979, p. 122.

[24] Merritt, K. and Brown, S. A. in *Transactions*, 8th Annual Society of Biomaterials, Vol. V, 1982, p. 73.

[25] Brown, S. A. and Merritt, K., *Advances in Biomaterials*, Vol. 4, 1982, pp. 195–202.

[26] Brown, S. A. and Merritt, K., this publication, pp. 105–116.

[27] Woodman, J. L., Black, J., and Jimenez, S. in *Biomaterials 1980*, G. Winter, D. Gibbons, and H. Plenk, Eds., Wiley, New York, 1980, pp. 245–250.

[28] Smith, G. K. and Black, J., this publication, pp. 223–247.

DISCUSSION

G. I. Ogundele[1] (*written discussion*)—What is the average age of your human patients?

Did you do any statistical analysis on the range of ages of your human patients that were sensitive and non-sensitive to metal implants?

K. Merritt and S. A. Brown (*authors' closure*)—We have not determined the

[1]University of Calgary, Calgary, Alberta, Canada.

average age. The age range is 16 to 99. The bulk of the reactions to internal fixation devices have been in the young adult. The reactions to joint replacements have occurred in the elderly.

The sensitivity status does not seem to be associated with age but we have not analyzed this statistically.

Ralpy Kafesjian[2] (*written discussion*)—Is the amount of iron released from the stainless steel implants sufficient to cause problems (in addition to the nickel and chromium)?

K. Merritt and S. A. Brown (*authors' closure*)—Significant amounts of iron are released from the stainless steel implants. The amount of iron can markedly affect infection rates with some infections enhanced and some inhibited by large amounts of iron. Sensitivity reactions to iron are unknown and iron is a major constituent of normal tissue. Certainly, iron overload can be a problem and this is probably best addressed by Dr. Smith and Dr. Black in their presentation and published work.

[2]Medical Specialties, American Hospital Supply, Irvine, CA 92715.

Linda C. Lucas,[1] Larry J. Bearden,[2] and Jack E. Lemons[3]

Ultrastructural Examination of *In Vitro* and *In Vivo* Cells Exposed to Elements from Type 316L Stainless Steel

REFERENCE: Lucas, L. C., Bearden, L. J., and Lemons, J. E., "Ultrastructural Examination of *In Vitro* and *In Vivo* Cells Exposed to Elements from Type 316L Stainless Steel," *Corrosion and Degradation of Implant Materials: Second Symposium, ASTM STP 859*, A. C. Fraker and C. D. Griffin, Eds., American Society for Testing and Materials, Philadelphia, 1985, pp. 208–222.

ABSTRACT: The ultrastructural features of *in vivo* and *in vitro* cells exposed to Type 316L stainless steel corrosion product solutions were examined. The *in vivo* cells were from fibrous tissue capsules formed around stainless steel rods implanted in the back muscles of New Zealand white rabbits. Aqueous solutions of the stainless steel alloy were injected at the implant-to-capsule interface to increase the concentration of alloy constituents bathing the tissues. The *in vitro* cells were cultured human gingival fibroblasts also exposed to stainless steel solutions of similar concentration. The ultrastructural features of both *in vitro* and *in vivo* cells were examined by transmission electron microscopy.

The following observations were made for both the *in vitro* and *in vivo* cells:

1. The severity of the cellular response increased when the concentration of the solutions of stainless steel exposed to the cells increased.

2. As the severity of the cellular response increased, the cells showed a decrease in rough endoplasmic reticulum, an increase in cytoplasmic vacuolation with altered organelles, a more diffuse plasma membrane, and a decrease in cellular organization.

KEY WORDS: transmission electron microscopy, stainless steels, implant materials, biocompatibility, ultrastructures of cells, tissue response, biomaterials

The host response to surgical implant alloys is often investigated utilizing two approaches: (1) the examination of tissues from humans and laboratory animals exposed to implant materials (*in vivo* analyses) [1–7][4] and, (2) studies of *in vitro* cellular responses (*in vitro* analyses) [8–17]. Both of these approaches provide information on the biological responses to surgical materials. The histological examination of tissue adjacent to surgical implants can provide information on

---

[1]Assistant professor, Department of Biomedical Engineering, The University of Alabama in Birmingham, AL 35294.

[2]Consultant, Birmingham, AL 35294.

[3]Professor and chairman, Department of Biomaterials, School of Dentistry, The University of Alabama in Birmingham, Birmingham, AL 35294.

[4]The italic numbers in brackets refer to the list of references appended to this paper.

the thickness of the fibrous capsule, the type of cells present, and the tissue response to the degradation products released from the implant. This type of information is normally on a macroscopic level and indicates the overall tissue response to the material. Numerous variables in addition to implant composition contribute to the observed tissue responses. This often makes it difficult to interpret the results.

The effect of released metallic constituents at the cellular level has been investigated by exposing cultured cells to alloys and/or components of these alloys. Pappas and Cohen [13] exposed varying concentrations of Type 316L stainless steel and cobalt-chromium-molybdenum (Co–Cr–Mo) (Vitallium)[5] powders to cultured cell lines. While both of the alloy powders elicited toxic responses, the responses to the Co–Cr–Mo powders showed more toxicity. Campbell, Meirowsky, and Hyde [10] exposed metal disks of stainless steel and Co–Cr–Mo alloys to cultured fibroblasts. They reported that the growth rates of the cells were not altered in response to these alloys. Exposing the alloys in the form of powders to cultured cells, as was done by Pappas and Cohen, increased the surface area of material exposed to the cells and, therefore, increased the quantity of ionic constituents released to the cells. Consequently, they observed more toxic responses in their investigations. Elemental salt solutions of the elements comprising the alloys have also been added to cultured cells. Daniel et al [11] exposed cultured rat fibroblasts to solutions of cobalt chloride and found that concentrations as low as 5 μg/mL Co^{+2} killed all the cells within 48 h. Bearden and Cooke [18] exposed fibroblasts to cobalt and nickel chloride solutions within a concentration range of 7.5 to 30 μg/mL. At all cobalt concentrations, they observed depressed growth rates for the cultures. Nickel solutions enhanced the cellular growth rate at the lower concentrations of 7.5 μg/mL, but depressed the growth rate at 30 μg/mL.

While the ability of cultured cells to grow and multiply is one factor that has been monitored with cell cultures, another factor is the ability of the cells to continue functioning normally in the presence of degradation products. Rae [14] recorded levels of lactic dehydrogenase (LDH) and glucose–6–phophate dehydrogenase (G6PD) in cell cultures of marine macrophages after exposure to metallic particulates. When cell cultures were exposed to particulates of cobalt, nickel, and the cobalt-chromium alloy, a release of LDH was observed, indicating cellular damage. A lower level of G6PD was also observed that indicated a low phagocytic capacity of the cells. Chromium, molybdenum, and titanium had no effect on the activity of either enzyme. These types of *in vitro* experiments can provide insight into the effect of degradation products at the cellular level.

In vitro cell cultures have also been utilized to investigate the biological response at the ultrastructural level. Bearden [9] exposed fibroblast cell cultures to cobalt chloride solutions and examined the cellular ultrastructures with transmission electron microscopy. Many of these cells had lost their ultrastructural

[5]Registered trademark of Howmedica Inc., Rutherford, NJ 07070.

organization; some contained filamentous material and had breached plasma membranes. These were compared to ultrastructural events described by Trump et al [19] for *in vivo* cells exposed to trauma. The events outlined by Trump represented irreversible events beginning with a normal resting cell (Stage 1) and ending in cellular death (Stage IV). The subsequent *in vitro* study showed that cellular ultrastructural damage induced *in vitro* by cobalt or nickel was similar to the events Trump described for cells under *in vivo* trauma conditions. This comparison suggested further studies of the ultrastructural alterations resulting from exposure to prosthetic materials.

The objective of the present investigation was to examine the ultrastructural features of both *in vitro* and *in vivo* cells exposed to solutions of Type 316L stainless steel alloy. The *in vitro* cells were cultured human gingival fibroblasts; the *in vivo* cells were from the fibrous capsules formed around implanted rods of stainless steel. The cellular responses to the stainless steel solutions were observed and correlations were made between the *in vitro* and *in vivo* conditions.

In Vivo and *In Vitro* Methods

In Vivo *Methods*

In vivo cells were obtained from fibrous capsules formed around cylindrically shaped (0.5 cm in diameter and 1 cm in length) specimens of Type 316L stainless steel (ASTM Specification for Stainless Steel Bars and Wire for Surgical Implants (Special Quality) (F 138–76)) implanted in the back muscles of New Zealand white rabbits. Six samples were implanted for each rabbit. Prior to implantation, the alloys were mechanically and chemically cleaned, and passivated according to ASTM Recommended Practice for Surface Preparation and Marking of Me-

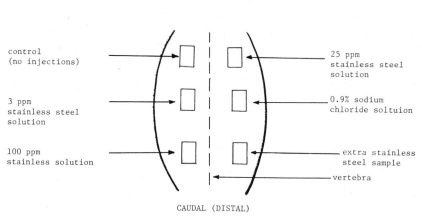

FIG. 1—*Injection sites for the Type 316L stainless steel solutions.*

tallic Surgical Implants (F 86–76). The specimens remained *in situ* for a period of six weeks during which time an intact fibrous tissue capsule formed around each of the implants.

To increase the quantity of corrosion products at the implant-capsule interface, solutions of stainless steel alloy were injected with a 22-gage needle at the implant-fibrous capsule interface after the sixth week. Figure 1 shows the sites for all injections. The solutions of stainless steel represented the chemical composition of the alloy as specified in ASTM F 138–76. To make these solutions, the following iron, nickel, chromium, molybdenum, and manganese atomic absorption (AA) standards were used:

Iron (Fisher AA standard); ferric chloride in dilute HCI.

Chromium (Banco AA standard); K_2CrO_4 in distilled water.

Nickel (Fisher AA standard); nickel metal in dilute HNO_3.

Molybdenum (Fisher AA standard); molybdic anhydride in dilute aqua regia.

Manganese (Fisher AA standard); manganese nitrate in distilled water.

The chemical composition of Type 316L stainless steel (based only on the five elements iron, chromium, nickel, molybdenum, and manganese) is 64% iron, 18.5% chromium, 13.5% nickel, 2.2% molybdenum, and 1.8% manganese. Based on this chemical composition, a 200-ppm solution of the alloy containing iron, nickel, chromium, molybdenum, and manganese was made from the preceding AA standards. Serial dilutions of the alloy solution were made with a 0.9% sodium chloride solution to make stainless steel solutions of 100, 25, and 3 ppm. The pH of all solutions was adjusted with sodium bicarbonate to a neutral pH. These solutions were injected at the implant-capsule interface as shown in Fig. 1. Each interface received two injections, one on week six and the second on week seven. Thus, the local quantity provided to the tissue was 200, 50, and 6 mg of stainless steel. Since the alloy solutions were prepared with a 0.9% sodium chloride solution, and the atomic absorption standards were also chloride based solutions, one capsule interface for each rabbit was injected with saline. The control for these experiments was a fibrous tissue capsule that was not injected with either the alloy solution or the saline solution.

On week eight, the implants were retrieved and the tissue adjacent to the implants immediately fixed with a 2.5% glutaraldehyde solution. The tissue sections were cut into very small pieces due to the low diffusion rate of the glutaraldehyde. These tissue samples remained in glutaraldehyde for a 2½-h period followed by emersion in a 0.1 M cacodylate buffer at pH 7.35 for 5 h to remove residual glutaraldehyde. After tissue fixation, the specimens were prepared for transmission electron microscopy (TEM) (Phillips, 300 EM) analyses [17]. All specimens were embedded in a low viscosity Spurr[6] medium. Thin sections were made with an ultratome and subsequently stained with uranyl acetate and lead citrate for examination.

---

[6]Ladd Research Labs, Burlington, VT 05402.

In Vitro *Methods*

The *in vitro* cells were human gingival fibroblast (HGF) primary cell cultures. The gingival tissue for the cell cultures was extracted from patients receiving periodontal treatment. Due to human use protocols, the medical and dental histories of the patients cannot be obtained. *In vitro* cells for the tissue culture were grown in a Waymouth's medium with 10% calf serum and maintained in a 5% CO_2 continuous flow incubator at 37°C. The cultured cells were exposed to solutions of stainless steel also made from atomic absorption standards and based on the chemical composition of the alloy. Concentrations of 3, 25, and 100 ppm were exposed to the cells for a 24-h period. Cells exposed only to Waymouth's growth medium with 10% calf serum served as a control. All solutions exposed to the cultured fibroblasts were adjusted to a neutral pH. After exposure to the stainless steel solutions, the cells were trypsinized, fixed with 2.5% glutaraldehyde for 1 to 2 h, and stored in a 0.1-*M* cacodylate buffer. The samples were prepared for TEM analyses and embedded in low viscosity Spurr medium in the same manner as had been done for the *in vivo* cells. Thin sections were prepared and stained with uranyl acetate and lead citrate. Ultrastructural analyses were made with the TEM at an accelerating voltage of 60 kV. All cells within the thin sections were observed. Representative photomicrographs at × 10 000 magnification were made for comparisons.

Ultrastructural alterations were studied as a function of the alloy solution concentration for the *in vitro* cultured cells and *in vivo* cells from the tissue capsules. Cellular alterations were recorded photographically at magnifications of × 10 000 for each alloy solution. Cellular organelles such as the mitochondria, plasma and nuclear membranes, and rough endoplasmic reticulum (RER) were

TABLE 1—*Phases of cellular necrosis for* in vivo *cells.*

Phase (observed events)	Concentration, ppm	Frequency, %
Normal-to-Early Phase		
Normal quantity and morphology of	control	80
the RER, some dilated RER, some	3	50
detached ribosomes, continuous	25	7
membrane, normal cytoplasmic	100	0
vacuolation, normal mitochondria.		
Middle Phase		
Decreased quantity of RER, increased	control	20
cytoplasmic vacuolation, dense	3	33
staining around nuclear membrane,	25	36
swollen mitochondria, diffuse plasma	100	28
membrane, membrane blebs.		
Late Phase		
No RER, extensive cytoplasmic	control	0
vacuolation, extensive clumping around	3	17
nuclear membrane, swollen mito-	25	57
chondria, loss of cellular organization.	100	72

examined. Changes in these ultrastructures were compared to a sequence of ultrastructural alterations recognized by Trump et al [*18*] as irreversible events eventually leading to cellular necrosis. The ultrastructural alterations were observed, and the frequency with which these events occurred was recorded for each concentration. This information was collected for both the *in vitro* and *in vivo* cells.

Results and Discussion

In Vivo *Results*

The ultrastructural features observed in the *in vivo* cells are shown in Table 1. The events recorded range from normal or only minor deviations of normal cellular ultrastructure to extensive alterations resulting in cellular death. The cellular events were divided into three phases: normal-to-early, middle, and late.

Cells in the normal-to-early phase displayed either normal ultrastructures or ultrastructures with only slight deviations from normal morphology. The cells contained a normal quantity of RER. While most of the RER had a normal morphology, cells in a very-early stage of necrosis sometimes exhibited evidence of endoplasmic dilation and detached ribosomes. The plasma membrane was either intact (normal phase) or had only a few diffuse areas (early phase). The remaining cellular organelles listed were typical of normal and healthy fibroblasts. An example of a normal fibroblast is shown in Fig. 2. The overall spindle shape of the cell and its nucleus are typical of normal fibroblasts. The area around

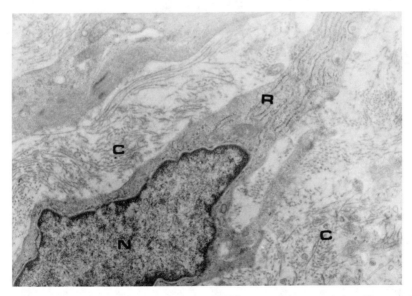

FIG. 2—In vivo *cells (original magnification* × *10 000)—control site: normal phase* (R = *rough endoplasmic reticulum,* N = *nucleus, and* C = *collagen fibers*).

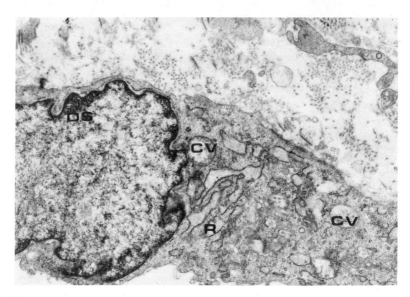

FIG. 3—In vivo *cells (original magnification × 10 000)—100-ppm site: middle phase (R = rough endoplasmic reticulum, CV = cytoplasmic vacuolation, and DS = dense staining around nuclear membrane).*

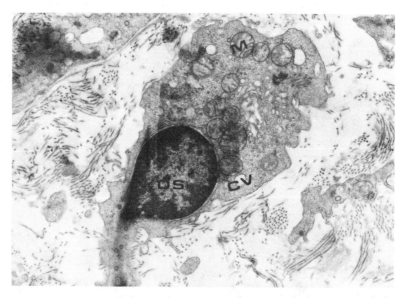

FIG. 4—In vivo *cells (original magnification × 10 000)—25-ppm site: late phase (CV = cytoplasmic vacuolation, M = mitochondria, and DS = dense staining around nuclear membrane).*

the cell contained a significant amount of collagen fibers. The events representative of this phase were seen in 80% of the control cells, 50% of the cells exposed to 3-ppm alloy solutions, 7% of the cells exposed to 25-ppm alloy solutions, and 0% of the cells exposed to 100-ppm alloy solutions.

Cells in the middle phase exhibited more cellular alterations than those in the normal-to-early phase. An example of these cellular events is shown in Fig. 3. These alterations included decreased amounts of RER with the RER having a dilated appearance, cytoplasmic vacuolation, dense staining around the nuclear membrane, and a diffuse plasma membrane often containing membrane blebs (small extrusions from the membrane). These features were seen in 20% of the control cells, 33% of the cells exposed to 3-ppm alloy solutions, 36% of cells exposed to 25-ppm solution, and 28% of cells exposed to the 100-ppm solution.

Cells in the late phase exhibited extensive cellular alterations. An example of these cellular events is shown in Fig. 4. These included a total loss of RER, cytoplasmic vacuolation, extensive clumping around the nuclear membrane, swollen and rounded mitochondria, and a loss of cellular organization. These characteristics were observed in 0% of the control cells, 16% of the cells exposed to 3-ppm alloy solutions, 57% of the cells exposed to 25-ppm alloy solutions, and 73% of the cells exposed to 100-ppm alloy solutions.

Examination of the cells from the fibrous capsules revealed several trends. When a higher concentration of corrosion products was injected into the capsule area, a severe tissue response was observed. However, some overlapping of the cellular responses and solution concentrations did occur. As shown in Table 1, the normal-to-early phase was observed, primarily in cells receiving a 0- to 3-ppm stainless steel corrosion product solution. The late phase was seen most often in cells receiving 25- to 100-ppm injections. The increased severity of the tissue response was indicated by a decrease in the amount of RER, an increase in cytoplasmic vacuolation, an increased number of swollen mitochondria, and a general loss of cellular organization. Clearly, a cell experiencing structural changes of this magnitude will have extreme difficulty in functioning normally.

Several factors contribute to the response previously described. The concentration of the stainless steel solutions injected at the implant-capsule interface is one of these factors. Another factor would be the valence states of the ionic constituents in the stainless steel solutions. Of significant interest would be the valence state of the chromium ion since several studies have demonstrated that Cr^{+6} can be particularly toxic [17,20]. The chromium atomic absorption standard utilized for this study was made of K_2CrO_4 dissolved in distilled water. Due to the chemistry of the atomic absorption standard, the valence state in the solution was probably +6. In studying the tissue responses to ionic constituents released from implant alloys, the ionic constituents used in the experiment should be similar to the actual products released. Current studies have reported that the chromium ion released as a result of the corrosion phenomena has a valence state of +6 [21]. Since chromium can exist in a variety of states, the possibility

TABLE 2—*Phases of cellular necrosis for* in vitro *cells.*

Phase (observed events)	Concentration, ppm	Frequency, %
Normal-to-Early Phase		
Normal RER, normal mitochondria,	control	100
continuous plasma membrane, some	3	50
cytoplasmic vacuolation.	25	0
	100	0
Middle Phase		
Decreased quantity of RER, increased	control	0
cytoplasmic vacuolation, dense	3	50
staining around the nuclear membrane,	25	50
swollen mitochondria.	100	0
Late Phase		
No RER, extensive cytoplasmic	control	0
vacuolation, total loss of cellular	3	0
organization.	25	50
	100	100

of several chromium valence states also existed; however, these other valence states will not be near as toxic as the Cr^{+6}.

The results from the *in vivo* study also depend on the concentrations of stainless steel solutions that remain at the implant-capsule interface. There exists the possibility that some of the solutions could have migrated from the interface at the time of injection or that the ionic constituents in the solution were not evenly distributed around the implant. This may be one reason for some of the variable responses listed in Table 1. Overall, the responses listed in Table 1 demonstrated a correlation between the cellular responses and the concentration of stainless steel solutions injected at the implant site. The same correlation was observed for the *in vitro* experiments and will be discussed in the next section.

In Vitro *Results*

The ultrastructural features observed for the *in vitro* cultured cells are summarized in Table 2. Similar trends were observed in the *in vitro* cells as were seen with the *in vivo* cells. However, with the *in vitro* cells the correlation between the extent of the cellular response and the concentration of the stainless steel solution added to the cell cultures was more direct than for the *in vivo* cells. Low concentrations of alloy solutions elicited milder responses, whereas the higher concentrations generally elicited more severe responses. These trends can be seen in Figs. 5 through 8.

Figure 5 shows a cultured HGF exposed only to Waymouth's complete medium and represented a normal *in vitro* cell. The cytoplasm contained a significant amount of RER with only a small amount of vacuolation. In general, the amount of RER in the normal *in vitro* cells was generally less than the amount of RER in the normal *in vivo* cells. Also, the cultured cells had a more rounded appearance

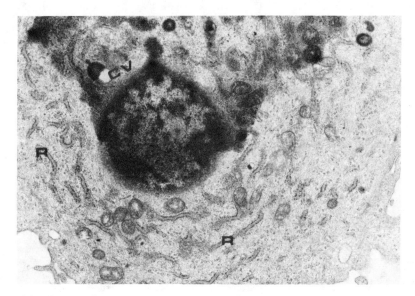

FIG. 5—In vitro *cells (original magnification × 10 000)—control cell: normal phase (R = rough endoplasmic reticulum and CV = cytoplasmic vacuolation).*

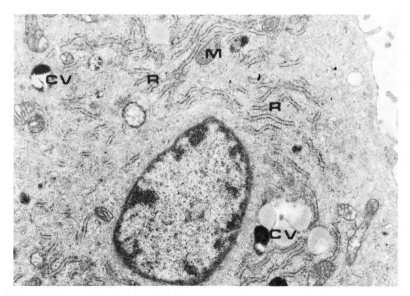

FIG. 6—In vitro *cells (original magnification × 10 000)—3-ppm exposure: early phase (R = rough endoplasmic reticulum, CV = cytoplasmic vacuolation, and M = mitochondria).*

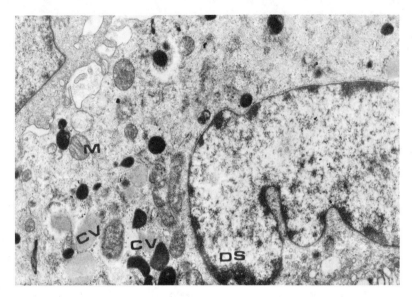

FIG. 7—In vitro *cells (original magnification* ×10 000)—25-ppm *exposure: middle-to-late phase* (*CV* = *cytoplasmic vacuolation, M* = *mitochondria, and DS* = *dense staining around nuclear membrane*).

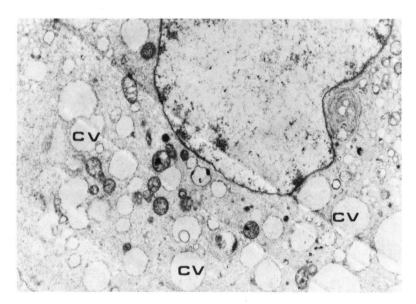

FIG. 8—In vitro *cells (original magnification* ×10 000)—100-ppm *exposure: late phase* (*CV* = *cytoplasmic vacuolation*).

than the *in vivo* cells. This may have been due to trypsinizing the cells prior to fixation. This will be evident in all the *in vitro* photomicrographs.

The cell shown in Fig. 6 was exposed to a 3-ppm alloy solution. The amount of RER was somewhat less than in the control cell. Also, the RER had a discontinuous appearance. Some cytoplasmic vacuolation was seen and some mitochondria had a slightly swollen appearance. These features represented an early phase of cellular necrosis.

Figure 7 shows a cell exposed to a 25-ppm alloy solution. In this cell, there was a significant decrease in RER and a definite increase in cytoplasmic vacuolation with dark staining pigment. The nucleus contained some dense staining around the nuclear membrane and the mitochondria had a swollen and spherical appearance. This represented a middle-to-late phase of cellular necrosis. The cell shown in Fig. 8 was exposed to a 100-ppm alloy solution and was representative of the late phase. In this cell, there was extensive vacuolation throughout the entire cytoplasm. Such cells exhibited very little ultrastructural detail.

The features exhibited by the *in vitro* cells were similar to those seen in the *in vivo* cells from the fibrous capsule around the implants. For both types of cells, increasing the concentration of alloy solutions resulted in reduced quantities of RER, swollen mitochondria, increased vacuolation, diffuse plasma membranes, and general loss of cellular organization. This sequence of ultrastructural changes is very similar to that observed in traumatized cells that are undergoing necrosis.

As with the *in vivo* portion of this study, several factors including the concentration of the stainless steel solutions contributed to the *in vitro* cellular responses. Again, the valence state of the ionic constituents in the stainless steel solutions was one of these factors. Since the same atomic absorption standards were used, the valence of chromium was again mostly $+6$, the most toxic chromium state. Chromium was a significant contributor to the observed toxic responses.

Since the *in vitro* experiments were conducted in a closed system, the concentration of stainless steel exposed to the fibroblasts remained at the site. A different situation existed for the *in vivo* fibroblasts of the fibrous capsules. For the *in vivo* experiments, the ionic constituents in the solutions could migrate from the sites. For the *in vitro* experiments (Table 2), a more direct correlation between cellular response and solution concentrations existed than did for the *in vivo* experiments (Table 1).

Summary and Conclusions

Ultrastructural features were examined for *in vivo* and *in vitro* cells exposed to solutions of Type 316L stainless steel. The *in vivo* cells were from fibrous tissue capsules formed around implanted stainless steel rods in the back muscles of New Zealand rabbits. Injections of stainless steel alloy solutions were made at the implant-to-capsule interface to increase the quantity of alloy constituents exposed to the tissues. The *in vitro* cells were cultured human gingival fibroblasts exposed to similar concentrations of solutions of stainless steel. The ultrastructural features of both the *in vitro* and *in vivo* cells were examined by transmission electron microscopy.

The following observations were made for both the *in vitro* and *in vivo* cells:

1. The severity of the cellular response increased when the concentration of the solutions of stainless steel exposed to the cells increased. However, this dose-response correlation was stronger for the *in vitro* cells.

2. As the severity of cellular response increased, the cells showed a decrease in rough endoplasmic reticulum, an increase in cytoplasmic vacuolation with altered organelles, a more diffuse plasma membrane, and a decrease in cellular organization.

As shown in this study, the transmission electron microscope (TEM) can be used for both *in vitro* and *in vivo* studies. Through these studies, trends in cellular responses were observed for both the *in vitro* and *in vivo* cells. Ultrastructural analyses could provide a means for enhancing other such studies and hopefully increase our understanding of the effect of degradation products on biological tissues.

References

[1] Beder, O. E. and Gilbert, E., *Surgery,* Vol. 39, 1955, p. 420.

[2] Beder, O. E., Stevenson, J. K., and Jones, T. W., *Surgery,* Vol. 41, 1956, p. 1012.

[3] Emneus, H., Stenram, U., and Baecklund, J., *Acta Orthopaedica Scandinavica,* Vol. 30, 1960, p. 226.

[4] Ferguson, A. B., Akahoshi, Y., Laing, P., and Hodge, E. S., *Journal of Biomedical Materials Research,* Vol. 44-A, 1962, p. 323.

[5] Ferguson, A. B., Laing, P. G., and Hodge, E. S., *The Journal of Bone and Joint Surgery,* Vol. 42-A, 1960, p. 77.

[6] Laing, P. G., Ferguson, A. B., and Hodge, E. S., *Journal of Biomedical Materials Research,* Vol. 1, 1967, p. 135.

[7] Mears, D. C., *The Journal of Bone and Joint Surgery,* Vol. 48-B, 1966, p. 567.

[8] Baumhammers, A., Langkamp, H. H., Matta, R. K., and Kilbury, K., *Journal of Periodontology,* Vol. 49, 1978, p. 592.

[9] Bearden, L. J., "The Toxicity of Two Prosthetic Metals (Cobalt and Nickel) to Cultured Fibroblasts," Ph.D. thesis, Clemson University, May 1976.

[10] Campbell, E., Meirowsky, A., and Hyde, G., *Annals of Surgery,* Vol. 114, 1941, p. 472.

[11] Daniel, M., Dingle, J. T., Webb, M., and Heath, S. C., *British Journal of Experimental Pathology,* Vol. 44, 1963, p. 163.

[12] Mital, M. and Cohen, J., *The Journal of Bone and Joint Surgery,* Vol. 50-A, 1968, p. 547.

[13] Pappas, A. M. and Cohen, J., *The Journal of Bone and Joint Surgery,* Vol. 50-A, 1968, p. 535.

[14] Rae, T., *The Journal of Bone and Joint Surgery,* Vol. 57–B, 1975, p. 444.

[15] Sisca, R. F., Thonard, J. C., Lower, D. A., and George, W. A., *Journal of Dental Research,* Vol. 46, 1967, p. 24B.

[16] Lucas, L. C., Lemons, J. E., and Buchanan, R. A. in *Transactions,* Eighth Annual Meeting of the Society for Biomaterials, Orlando, FL, 1982, p. 72.

[17] Lucas, L. C., "Biocompatibility Investigations of Surgical Implant Alloys," Ph.D. thesis, University of Alabama in Birmingham, Birmingham, AL, 1982.

[18] Bearden, L. J. and Cooke, F. W., *Journal of Biomedical Materials Research,* Vol. 14, 1980, p. 289.

[19] Trump, B. F., Valigorsky, J. M., Dees, J. H., Mergner, W. J., Kim, K. M., Jones, R. T., Pendergrass, R. E., Garbos, J., and Cowley, R. A., *Human Pathology,* Vol. 4, 1973, p. 89.

[20] Black, J., *Biological Performance of Materials—Fundamentals of Biocompatibility,* Marcel Dekker, Inc., New York, 1981.

[21] Wortman, R. S., Merritt, K., Brown, S. A., and Millard, M. M. in *Transactions,* Ninth Annual Meeting of The Society for Biomaterials, Vol. 6, 1983, p. 105.

DISCUSSION

James M. Anderson[1] *(written discussion)*—Did you carry out analyses to locate and determine the concentration of metal ions within the cells?

Did you see a zonal effect (distance from the implant) in the fibrous capsule regarding cellular necrosis?

L. C. Lucas, L. J. Bearden, and J. E. Lemons (authors' closure)—To answer your first question, no, this was not done. However, in future projects of this type, this should be done.

For the transmission studies, we examined the fibroblasts immediately adjacent to the specimen and did not determine if there was a zonal effect in the capsule. We did, however, see a zonal effect with some of the histological sections. Muscle tissues adjacent to the fibrous capsules were affected by the metallic constituents in varying degrees. These histological responses did display a zonal or gradient effect.

Eugene P. Lautenschluger[2] *(written discussion)*—We have heard over and over again that the corrosion of solid metal stock, even though of identical chemical composition, can be significantly altered by such phenomena as degree of cold work, history of heat treatments, phase morphology and distribution, and grain size. While your metal salt test solutions have the same ratios of major metal components as found in Type 316L stainless steel, they do not possess any of the aforementioned solid phenomena. Would you therefore care to comment on the relevancy of your test solutions as a model of corrosion products from a solid?

L. C. Lucas, L. J. Bearden, and J. E. Lemons (authors' closure)—When a surgical alloy is implanted, a corrosion process occurs whereby metallic ions are released. It is true that the factors mentioned by the reviewer can definitely affect the rates of corrosion release and may also affect the types of specific ionic constituents that are released. However, the fact remains that these metallic

[1]Department of Pathology, Case-Western Reserve University, Cleveland, OH 44106.
[2]Department of Biological Materials, Northwestern University, Chicago, IL 60611.

ions are released. Once these constituents are released, it is also important to evaluate the effect of the corrosion products on the biological tissues. Numerous investigators have found corrosion products intracellularly. The objective of this study was to evaluate the effect of the released metallic ions at the cellular ultrastructure using transmission electron microscopy. By studying these phenomena, we believe that a better understanding of cellular responses might be obtained.

Gail K. Smith[1] and Jonathan Black[1]

Estimation of *In Vivo* Type 316L Stainless Steel Corrosion Rate from Blood Transport and Organ Accumulation Data

REFERENCE: Smith, G. K. and Black, J., **"Estimation of *In Vivo* Type 316L Stainless Steel Corrosion Rate from Blood Transport and Organ Accumulation Data,"** *Corrosion and Degradation of Implant Materials: Second Symposium, ASTM STP 859,* A. C. Fraker and C. D. Griffin, Eds., American Society for Testing and Materials, Philadelphia, 1985, pp. 223–247.

ABSTRACT: A model utilizing implanted Type 316L stainless steel in laboratory rabbits was developed to investigate the systemic transport and distribution of constituent iron and chromium corrosion products. Results utilizing flameless atomic absorption spectrophotometry indicate that levels of circulating iron and chromium in the blood-transport compartment reflect the implanted surface area of Type 316L stainless steel. A similar positive correlation was found between implanted Type 316L stainless steel surface area and liver (reticuloendothelial) accumulations of iron and chromium. The kidney, having no reticuloendothelial capacity, demonstrated no constituent element accumulations.

Approximate corrosion rates of Type 316L stainless steel were calculated by: (1) superimposing the observed plasma iron elevations on transport compartment size and plasma iron kinetics, and by (2) evaluating storage site accumulations of iron and chromium over time. All calculations are consistent with Type 316L stainless steel's reported corrosion rate as measured by more precise *in vitro* and *in vivo* analytical techniques (20 to 300 ng/ cm^2/day).

Although the clinical significance of these findings is presently indeterminable, it is clear that the biological system as a whole must be considered when assessing the long-term biocompatibility of implanted Type 316L stainless steel.

KEY WORDS: corrosion, biocompatibility, animal model, systemic effects, iron metabolism, chromium, trace elements, organ accumulations, atomic absorption spectrophotometry, implant materials, stainless steels

Type 316L stainless steel (316L) is presently the most widely used surgical implant alloy. Although corrosion resistant by conventional engineering standards, 316L has a finite corrosion rate *in vivo* [1–6][2] resulting in the release,

---

[1]Assistant professor and associate professor, respectively, Department of Surgery, School of Veterinary Medicine and Department of Orthopaedic Surgery, School of Medicine, University of Pennsylvania, Philadelphia, PA 19104.

[2]The italic numbers in brackets refer to the list of references appended to this paper.

locally and systemically [7], of constituent ions and corrosion complexes after implantation. This phenomenon, as predicted from thermodynamic theory [8], is supported by both experimental [7,9–17] and clinical experience [18–24].

The primary constituents of 316L are iron, chromium, and nickel, all of which are biologically essential trace elements under precise physiological regulation and function. Accordingly, implantation of a device containing 316L affords a long-term *in vivo* source of reactive trace elements, any of which may produce profound systemic sequelae when introduced at altered levels or by nonphysiological means. Systemic effects, anticipated and observed, include metabolic, immunologic, or carcinogenic phenomena [25].

More than 20 years have passed since the classic work of Ferguson and co-workers [7] that clearly demonstrated the potential for corrosion products from 316L implants to be distributed and systemically deposited at remote sites. Though their findings raised the question of the potential adverse interaction of corrosion products with the biochemistry of the implant recipient, surprisingly little attention has been directed toward this concern to date. To our knowledge, only three other investigations on the systemic distribution of 316L constituents appear in the literature. One is by Owen et al [26] reporting no increase in the level of chromium in the hair following stainless steel metal-on-plastic total hip replacement. The second is a study by Michel et al [27] on the influence of metal implants on trace element contents in human and mammalian tissue and organs. Analysis of lymphatic tissue, liver, and kidneys of rabbits with implants demonstrated a tendency toward enrichment of chromium and nickel. The third is a very recent report by Merritt and Brown investigating the distribution of 316L corrosion products manufactured *in vitro* [28].

Investigating the real or potential systemic sequelae of corrosion products from implanted 316L necessitates an awareness of normal trace element metabolism and the induced alterations in this metabolism with excess exogenous administration. Unfortunately, for most trace elements (including essential ones) such information is scant and incomplete. The absence of specific studies in this regard requires surveying literature from relatively distant disciplines (for example, data from toxicological, nutritional, biochemical, or industrial exposure studies). Unfortunately, experimental toxicological studies are typically of short-term duration, employing rapid, high-dose metal-salt administration either *per os*, intravenous, or subcutaneous, and thus of limited value [29–31]. Such studies fail to reproduce the unique situation created by implanted metals that allow a slow, continuous release of metal ions into the system. Similarly, results of industrial exposure surveys [32,33], though often of a more long-term nature, typically relate to changes in the exposed skin or respiratory epithelium subsequent to chronic contact or inhalation of noxious gases or particulate material. Here again, the discrepancies in routes of administration and distribution are obvious.

In general, the important factors in determining host response to implants are: (1) the *rate* of corrosion product release into the system (surface-type, surface-area dependent); (2) the nature of the released corrosion products; (3) the mode and efficiency by which the corrosion products are transported, metabolized, or excreted; (4) the sites of distribution or accumulation; and (5) the specific sequelae, local or systemic, of corrosion product release.

With the intent of clarifying some of these points, an investigation was conducted to study the systemic transport and distribution of constituent iron and chromium from 316L implants in laboratory rabbits and to estimate implant corrosion rate from organ accumulation data and alterations in transport-compartment kinetics. Hopefully, the many questions resulting from this study will stimulate continued interest in this area of research.

Methods

An investigation into the systemic distribution of 316L corrosion products necessarily begins with a suitable *in vivo* source of corrosion products. Ideally, the experimentally implanted metal should be in bulk form, that is, not particulate, and have surface characteristics indistinguishable from commercially available 316L surgical devices. Accordingly, two forms of implantable 316L were utilized. Steinmann pins manufactured of surgical 316L were selected as control (standard) material to simulate the clinical implantation of any device of 316L specifications. Steinemann pins were fashioned such that experimental animals would receive a standard surface area relative to body weight (SA/BW) that was derived from the human equivalent of a 70-kg man receiving a standard total hip prosthesis of 200 cm², that is, 2.90 cm²/kg. Henceforth, this SA/BW ratio will be disignated "1×" or "standard man" equivalent.

The second experimental corrosion product source consisted of 55 μm diameter microspheres[3] manufactured from a bulk electrode satisfying 316L specifications. The high surface area per unit volume of this material ($\sim$144 cm²/g) provided a means to implant in experimental animals, multiples of the standard man SA/BW ratio specifically, 1×, 10×, and 100× (2.9 cm²/kg, 29 cm²/kg, and 290 cm²/kg, respectively). The objective was to amplify or exaggerate within the constraints of the investigation, the system's exposure to corrosion products originating from implanted 316L.

A nominal particle size of 55 μm was small enough to provide a large implant surface-area to volume ratio, yet of sufficient size to preclude transport from the site of implantation either by cellular phagocytosis or bulk lymph flow. The excellent sphericity of the powders enabled a simple optically-based calculation of surface areas as a function of weight (144 cm²/g). Moreover, the surface of a spherical powder, 55 μm in diameter, far better approximates the surface characteristics of a bulk 316L device than does a powder of irregular configuration, for example, splat-cooled or mechanically fractured powders. Indeed, with respect to surface area, the microsphere surface can be approximated simply

[3]Nuclear Metals, Inc., Concord, MA.

as an equivalent *planar* area, since the calculated increase in surface energy as a function of radius of curvature is negligible for microspheres of this size.

Pilot investigations to characterize the Steinmann pin and spherical powders included metalography, X-ray fluorescence, and X-ray diffraction and revealed both forms to have comparable compositions, but to have microstructures reflecting the method of processing; Steinmann pins had a highly cold-worked microstructure while the microspheres were largely as-quenched. Owing to technical and investigational constraints, no attempts were made to assess discrepancies in surface characteristics or relative corrosion rates, although the utility of such information is clear.

The New Zealand white rabbit was selected as the laboratory animal model to standardize the new methodology with respect to the ASTM Recommended Practice for Experimental Testing for Biological Compatibility of Metals for Surgical Implants (F 361-72) [34] soft tissue protocol and other similar rabbit implant models. Rabbits are of a size such that surgery can be accomplished on a macroscopic scale and their prominent ear vasculature affords easy access for repeated blood collection. Moreover, rabbits are not meal eaters but rather consume their food *ad libitum* over the day or night. This behavior eliminates, or at least diminishes, the often wide variation in blood trace element levels commonly observed in meal eaters relative to their intake [35]. Rabbits, upon delivery, were examined and allowed two weeks to equilibrate relative to their new diet and environment before the experiment was begun. During the experiment, rabbits were maintained in galvanized cages with stainless steel feeding dishes and watering spouts. Diet consisted of commercial rabbit ration (Wayne Medicated) and water. Supplements of carrots and kale, which are normally fed once weekly, were eliminated from the diet to avoid a corresponding weekly fluctuation in trace element levels. Also, variations in individual rabbit preference for these supplements were thought to be a potential source for non-uniform dietary intake in the colony.

Seventy male New Zealand white rabbits received implants of 316L in either rod or particulate form in two anatomic sites, the paraspinal musculature or the femoral medullary canal. Table 1 lists the six experimental groups and their respective implant type, site, and SA/BW ratio. Group 1 consisted of rabbits receiving intramuscular Steinmann pins pursuant to the ASTM F 361-72 soft tissue protocol. Implanted SA/BW ratio was 2.90 cm^2/kg (1 ×, standard man equivalent). Group 2 rabbits had an equivalent 1 × surface area of Steinmann pins implanted bilaterally in the femoral medullary canal. Rabbits in Groups 3, 4, and 5 received 1 ×, 10 ×, and 100 ×, respectively, of spherical powder unilaterally in the femoral medullary canal. Group 6 rabbits served as control; seven rabbits underwent sham surgery with no metallic implantation and seven rabbits were left totally unoperated. All surgeries were performed with strict adherence to sterile technique. Anesthesia was conducted similarly for all animals undergoing surgery. A neuroleptanalgesic agent (Innovar-Vet) in combination with atropine was administered intramuscularly as pre-anesthetic drug. Subse-

TABLE 1—*Listing of experimental groups with corresponding implant site and type.*

Group	n =	Implant	Site	SA/BW (1 × = 2.9 cm²/kg)
1	10	pin	muscle	1 ×
2	10	pin	bone	1 ×
3	15	powder	bone	1 ×
4	10	powder	bone	10 ×
5	10	powder	bone	100 ×
6 (control)	15	...	...	0

quent epidural anesthesia of 2% lidocane in the lumbosacral space provided complete flaccidity and analgesia in the caudal portion of the animal and lasted between 1½ to 2 h following administration. Anesthesia by this means was safe (mortality < 2%) and resulted in a smooth, uneventful recovery.

Blood samples were drawn from each rabbit preoperatively and at 4, 8, 20, 24, and 28 weeks postoperatively. Specimens were heparinized, centrifuged twice, and the clear plasma supernatant stored in clean polystyrene tubes at −72°C to await eventual iron and chromium trace element analyses. Blood sample volume was minimized (less than 2 mL) to reduce the drain upon the rabbits erythropoietic capacity and thereby, avoid the potential antagonism of the blood-letting process on the levels of circulating implant constituents, particularly iron.

Radiographs were taken immediately postoperatively and at the end of the seven-month implantation period to determine any gross changes in implant position or untoward tissue reaction.

At the end of the seven-month implantation period, rabbits were euthanized by means of a large intravenous bolus of barbiturate and autopsy was performed. Copper-bronze instruments were used to avoid contamination of tissue with stainless steel constituents. Sections of the kidney and left lateral liver lobe were resected and stored at −72°C in hinged-lid polyethylene containers. The left portion of the liver was sampled because less splanchnic circulation is supplied to this side of the liver.

Iron and chromium concentrations of liver, kidney, and plasma samples were measured by means of flameless atomic absorption spectrophotometry (Perkin-Elmer Model 360 double-beam spectrophotometer with HGA-2100 graphite furnace accessory). Plasma iron concentration was determined by the ultra-micro-method of Olsen et al [36]. Plasma chromium concentrations and tissue and iron and chromium concentrations were determined by modifying the wet digestion techniques of Davidson and Secrest [37].

The analytical techniques provided ample sensitivity (44 × 10⁻¹² g for 1% absorption) with a practical detection limit of approximately 2 pg (with appropriate signal amplification). The wet digestion methods permitted measurement of ultralow concentrations (ng/mL) of iron and chromium in a variety of organic matrices of small sample size. Pilot research to check the precision and recovery of iron and chromium analytical techniques revealed a maximum coefficient of variation of 8% with a recovery yield of >96% of known additions.

All experimental parameters, that is, plasma and tissue iron and chromium concentrations, were expressed as means with corresponding 95% confidence intervals. Where significance between control or experimental groups were suspected, the method of Student was employed to determine the level of confidence or p value. The null hypothesis was rejected if $p \leqq 0.05$.

To determine the significance of regression coefficients between experimental and control groups or between any group and zero, the method of least squares was invoked [38].

Results

Plasma Iron Analysis

Mean plasma iron concentrations relative to implant surface area and duration appear in Table 2. Rabbit plasma iron concentration, like that in humans [39], decreased with increasing rabbit age over the course of the experimental period. Owing to this continuous base-line diminution, fluctuations in mean plasma iron concentrations (attributable to implant corrosion) were not amenable to serial comparisons. Accordingly, data from the experimental surgical groups were evaluated relative to the control group within the same blood collection period only. Throughout the experiment, no significant differences in either plasma iron or chromium concentrations were detected between sham-operated controls and non-operated controls and, therefore, both subpopulations were combined as one control group.

Pre-operative mean plasma iron concentrations relative to the prospective surgical groups were uniform and within the 95% confidence interval of the Group 6 control rabbits (1.72 ± 0.32 μg/mL iron). An apparent exception is Group 2, which demonstrated a reduced (though not significant) mean plasma iron concentration. This group was, by necessity, added to the rabbit population two weeks after the other experimental groups, that is, just prior to surgery, and did not have time to equilibrate relative to the new diet and environment. Data, therefore, may reflect a residuum of the previous environment at the breeding colony.

No significant plasma iron elevations were noted until 20 weeks postoperatively when Group 5 (100×) demonstrated a 14% increase over mean control plasma iron concentration. Although all of the other experimental groups exhibited mean plasma iron concentrations above control levels, none of the differences were statistically significant.

Blood samples at 24 weeks postoperatively were collected in the midst of an on-going laboratory-animal-handler's strike, during which period numerous breaks in control were encountered. This collection could legitimately be excluded from experimental findings. However, the results serve to emphasize the need for tight experimental control when conducting studies of this type. Mean control

TABLE 2 — *Mean plasma iron concentrations (μg iron/mL plasma).*

| | | | Time | | |
Rabbit Group	Pre-operative Plasma	8 Weeks Postoperative Plasma	20 Weeks Postoperative Plasma	24 Weeks Postoperative Plasma[a]	28 Weeks Postoperative Plasma
Group 1 (1×, intramedullary pins)	1.88 ± 0.30	1.89 ± 0.47	1.52 ± 0.35	2.03 ± 0.27	1.34 ± 0.33
Group 2 (1×, intramedullary pins)	1.45 ± 0.26[b]	1.65 ± 0.27	1.50 ± 0.21	1.71 ± 0.24[c] (↓ 17%)	1.25 ± 0.17
Group 3 (1×, powder)	1.84 ± 0.19	1.52 ± 0.18	1.57 ± 0.18	1.94 ± 0.24	1.26 ± 0.14
Group 4 (10×, powder)	1.57 ± 0.25	1.63 ± 0.27	1.51 ± 0.28	1.66 ± 0.16[c] (↓ 19%)	1.44 ± 0.22
Group 5 (100×, powder)	1.87 ± 0.46	1.54 ± 0.16	1.67 ± 0.11[c] (↑ 14.4%)	2.01 ± 0.19	1.53 ± 0.17[c] (↑ 12.5%)
Group 6 (control + sham)	1.72 ± 0.32	1.53 ± 0.14	1.46 ± 0.16	2.06 ± 0.16	1.36 ± 0.09

[a]Laboratory-animal-handlers strike on.
[b]Group added late, not steady state.
[c]Significantly different from control; ($p < 0.05$) same time period.

plasma iron was 2.06 ± 0.16 μg/mL, breaking the heretofore observed trend of diminishing plasma iron levels with age. Subsequent total protein measurements of the same blood samples were found to be elevated, suggesting that the rabbits during this period were suffering from mild dehydration. This impression is supported by the fact that Group 5 plasma (collected on the same day) demonstrated a similar propensity toward hemoconcentration. Moreover, groups 2 and 4, collected together on a different day, have comparable plasma iron values, both of which are significantly less than control. In general, no consistent patterns in plasma iron concentration, attributable to implant corrosion, were recognized. If such patterns indeed existed, they were obscured by the numerous breaks in experimental protocol during this period.

Control mean plasma iron concentration of blood drawn 28 weeks postoperatively was 1.36 ± 0.9 μg/mL, resuming the recognized trend toward diminishing plasma iron levels with age. Experimental group means were distributed about the control mean and, therefore, did not reveal the clear-cut pattern in plasma iron elevations as demonstrated by the 20-week blood samples. Group 5 ($100\times$), however, again exhibited a significant 12.5% elevation in plasma iron concentration over the control mean.

Regarding plasma iron hemostasis throughout the experiment, Table 2 reveals several interesting patterns. Excluding the poorly controlled 24-week blood collection, control mean plasma iron concentrations decreased gradually from a pre-operative value of 1.72 ± 0.32 μg/mL to a pre-sacrifice value of 1.36 ± 0.09 μg/mL. However, a regression of concentration on time using the method of least squares revealed this gradual diminution *not* be significant (slope not different from zero $p < 0.05$). In general, as the experiment progressed, analytical results, irrespective of experimental group, clustered more closely about their respective group means. The largest spread in data was observed in the pre-operative blood collection, apparently because animals had not as yet achieved steady state relative to their new diet and environment. For all rabbit groups individually, the least square obtained-slopes of plasma iron concentration versus time were not significantly different from zero, nor significantly different one from another.

Plasma Chromium Analysis

Results of plasma chromium analysis appear in Table 3, including means and 95% confidence intervals. Fluctuations in base-line or control mean chromium concentration from one blood collection to another precluded a meaningful running comparison of results over the experimental duration. Within-run data analysis, therefore, was conducted, comparing plasma chromium values with corresponding control values from the same blood collection period.

Mean plasma chromium drawn pre-operatively from the control group of rabbits measured 13.1 ± 0.93 ng/mL. Experimental groups prior to surgery demonstrated mean plasma chromium values within the 95% confidence interval

SMITH AND BLACK ON STAINLESS STEEL CORROSION RATE 231

TABLE 3—Mean plasma chromium concentrations (ng chromium/mL plasma).

| | | | Time | | | |
Rabbit Group	Pre-operative Plasma	4 Weeks Postoperative Plasma	8 Weeks Postoperative Plasma	20 Weeks Postoperative Plasma	24 Weeks Postoperative Plasma	28 Weeks Postoperative Plasma
Group 1 (1×, intramedullary pins)	12.5 ± 0.94	12.5 ± 1.45	lost group	10.9 ± 0.57	11.9 ± 0.53[a] (↑ 15.5%)	19.1 ± 2.30[a] (↑ 17%)
Group 2 (1×, intramedullary pins)	11.6 ± 0.95[a,b] (↓ 11.4%)	11.5 ± 0.52	12.0 ± 0.61	11.3 ± 0.41	10.7 ± 0.30	16.1 ± 1.04
Group 3 (1×, powder)	12.8 ± 0.88	11.6 ± 0.49	12.2 ± 0.77	11.9 ± 0.58[a] (↑ 6.2%)	12.1 ± 0.58[a] (↑ 17.5%)	16.3 ± 0.49
Group 4 (10×, powder)	13.9 ± 1.31	11.1 ± 0.36	11.5 ± 0.79	11.5 ± 0.90	12.8 ± 1.25[a] (↑ 24.3%)	16.7 ± 1.64
Group 5 (100×, powder)	12.8 ± 1.35	13.3 ± 1.15[a] (↑ 16.7%)	14.1 ± 0.90[a] (↑ 12.8%)	13.6 ± 1.66[a] (↑ 21.4%)	12.7 ± 1.81[a] (↑ 23.3%)	18.0 ± 1.55[a] (↑ 10.4%)
Group 6 (control + sham)	13.1 ± 0.93	11.4 ± 0.43	12.5 ± 0.86	11.2 ± 0.40	10.3 ± 0.70	16.3 ± 0.72

[a]Significantly different from control; ($p < 0.05$) same time period.
[b] = Group added late, not steady state.

of control values. An exception was Group 2 (1×) that, as discussed in the plasma iron analysis section, was added to the rabbitry later than the other experimental groups and was not given sufficient time to equilibrate prior to blood collection. Group 2 rabbits displayed a significantly lower (11.4%) plasma chromium concentration versus control; and again tighter experimental control was clearly indicated.

In the four-week blood collection, Group 5 plasma demonstrated a significant 16.7% elevation ($p < 0.01$) in chromium concentration. Accompanying the elevation was an increased spread of data about the mean value, possibly indicating either preferential corrosion in some rabbits in the group or a relative individual inability to regulate excess exogenous chromium.

The pattern at four weeks was duplicated in the eight-week postoperative blood collection that exhibited a 12.8% elevation of Group 5 plasma chromium concentration over control. The similarity between the four-week and eight-week results suggested the attainment of plasma steady-state relative to implant chromium release.

At 20 weeks postoperatively, mean control plasma chromium concentration measured 11.2 ± 0.40 ng/mL and Group 5 plasma chromium again was slightly elevated (21.4%) over control. Group 3, as well, displayed a significant though marginal (6.2%) increase in plasma chromium concentrations. No trends were noted in the other experimental groups.

As discussed in reviewing plasma iron results, the 24-week postoperative blood collection was plagued by numerous breaks in control stemming from circumstances surrounding an on-going laboratory-animal-handler's strike. Despite the situation, however, plasma chromium patterns seem not so obviously affected as corresponding iron levels. The underlying reasons for this phenomenon are open to speculation. At 24 weeks postoperatively, mean plasma chromium concentration registered 10.3 ± 0.7 ng/mL, while four of the five experimental groups demonstrated ($p < 0.01$) significant elevations over this value. The degree of plasma chromium elevation appeared related to implanted stainless steel surface areas though not strongly.

The final blood collection at 28 weeks yielded significant elevations in Group 1 rabbit plasma (17%) and again in Group 5 plasma (10.4%). Other experimental groups exhibited no trend toward elevation.

Patterns in plasma chromium levels over the course of the experiment were not so well defined as in the corresponding iron data. Control mean plasma chromium concentration with time demonstrated a trend toward gradually diminishing values that reversed abruptly at 28 weeks postoperatively. Regression analysis, however, of control chromium concentration on time revealed no significant difference between this slope and zero.

Regarding experimental groups specifically, Group 5 (100×) mean plasma chromium concentration was significantly elevated over controls for all postoperative blood sample collections. Only Group 2 (1×, intramedullary pins) showed no plasma chromium increase during the course of the seven-month study.

All remaining experimental groups exhibited sporadic plasma chromium elevations particularly for the last two months of the experimental period. Least-square slopes of concentration over time for the experimental groups, individually, were not significantly different one from another or from zero.

It should be noted that blood samples from Groups 5 ($100 \times$) and 6 (control) were collected, prepared, and analyzed in parallel. Respective data, then, from these two groups reflect optimum experimental control and should receive commensurate credibility. Data from the other experimental groups, it can be argued, suffer uncertainties stemming from blood collection on days other than Groups 5 and 6. On the other hand, if day-to-day fluctuations are indeed a random occurrence, one would expect as many negative deviations from control plasma values as positive. The fact that this is not the case suggests that the plasma chromium elevations as presented in Table 3 are real.

Tissue Trace Element Analysis:
Kidney and Liver, Iron and Chromium Accumulations

Results of kidney iron and chromium analyses were normalized with respect to dry weight and are summarized in Table 4. No significant differences were observed between any of the experimental groups and controls. In general, control kidney iron concentrations were two to three orders of magnitude larger on a weight basis than kidney chromium concentrations.

Results of liver analyses from experimental and control groups appear in Table 5. Group 5 ($100 \times$) rabbits demonstrated a 30% elevation ($p < 0.02$) in liver iron concentration over control values. None of the other experimental groups displayed any tendency toward elevation. Liver chromium concentrations reflected an ability on the part of the liver to accumulate excess chromium. Group 4 ($10 \times$) liver samples displayed mean chromium concentration 65% ($p < 0.02$) over control concentrations. Group 5 liver chromium concentrations were 85% ($p < 0.01$) elevated. The other experimental groups demonstrated no tendency toward increased liver chromium accumulations.

Discussion

Today, implants manufactured from 316L are surgically applied with increasing frequency and with expanding application. As emphasized, all 316L implants in a biological environment are subject to corrosion mechanisms of various sorts. Corrosion products are released into the biological environment at a rate dependent on the surface area of the bulk implant, as well as the size, surface area, and surface energy of wear products emanating from multi-component metallic devices. For example, the fretting products produced at the interface between the screw and plate of a fracture fixation device may exhibit accelerated corrosion rates as a function of their small size, as well as their highly cold-worked nature. Moreover, many of these products are small enough to be transported either via bulk lymphflow or by phagocytic mechanisms to remote sites where corrosion may continue. This systemic distribution of corrosion products and corrosion

TABLE 4—Kidney iron and chromium concentrations (% change).

316L Constituent	Group 6, Control Kidney Concentration	Group 1, Intramuscular Pin, 1×	Group 2, Intramedullary Pin, 1×	Group 3, Powder, 1×	Group 4, Powder, 10×	Group 5, Powder, 100×
Iron	0.36 ± 0.04 μg iron/mg dry weight	NS[a]	NS	NS	NS	NS
Chromium	1.97 ± 0.58 ng chromium/mg dry weight	NS	NS	NS	NS	NS

[a]Not significantly different from control.

TABLE 5—Liver iron and chromium concentrations (% change).

316L Constituent	Group 6, Control Liver Concentration	Group 1, Intramuscular Pin, 1×	Group 2, Intramedullary Pin, 1×	Group 3, Powder, 1×	Group 4, Powder, 10×	Group 5, Powder, 100×
Iron	0.65 ± 0.07 μg iron/mg dry weight	NS[a]	NS	NS	NS	↑30.8%
Chromium	1.87 ± 0.65 ng chromium/mg dry weight	NS	NS	NS	↑64.2%	↑85%

[a]Not significantly different from control.
[b]p < 0.05.

particles constitutes a source of concern, particularly in light of the advent of porous surface implants. Because of its relatively inferior corrosion resistance, stainless steel is not presently used for such purposes; however, a porous form of the F-75 cobalt-chromium alloy has been approved for surgical implantation. The development of this type of coating has been prompted by the capacity of the bone to grow into interstices approximately 100 μm in size, thereby affording rigid bone to implant fixation without the need for an interposed layer of bone cement. Such a coating increases the implanted surface area by one to three orders of magnitude.

It was the purpose of the present study to examine the effect of implanted 316L surface area on the distribution of corrosion products at various sites throughout the body. It should be noted that the 55 μm microspheres used in this investigation are too large to be transported systemically as particles, and do not have the corrosion potential of highly-worked wear particles. Therefore, the results of this investigation normalized to surface area, likely represent a conservative estimate of the corrosion potential of an actual multicomponent stainless steel device used in clinical application.

Judging from the apparently satisfactory performance of 316L in clinical service, one might readily conclude that systemic complications stemming from stainless steel implantation are unimportant, if not nonexistent. This impression is misleading and at least partially responsible for the absence of controlled prospective investigations in this regard. In actual fact, no controlled retrospective or prospective studies correlating systemic disease in the presence of an implant has yet been performed. Considering the apparent lack of concern regarding this question, perhaps it is true that 316L implants occasion no adverse systemic sequelae. More likely, however, systemic effects are real, but are unrecognized against the general background of increasing disease in the aging patient. Conceivably, the presence of a metallic implant could accelerate or intensify a preexisting problem. For example, it is generally recognized that the incidence of neoplasia, arthritic disorders, and heart disease, all increase with age. The important question then, is whether populations of patients with implanted metallic devices have a statistically higher incidence, earlier age of onset, or increased severity of these maladies than control-matched populations.

A logical approach to evaluating potential systemic effects begins with an understanding of the corrosion process and the dissemination of corrosion products throughout the body. Having this information, one can then pinpoint various organs or organ systems and focus on pathological changes, large or small, occasioned by 316L implantation. Studies of this sort necessitate a strict adherence to experimental control. Human populations with attendant wide variations in age, diet, life-style, ethnic background, and geographical location typically exhibit correspondingly wide variations in organ or plasma trace element levels. Consequently, a tightly controlled animal model, incorporating age, diet, environment, and sex, is essential. Information derived from such a model though perhaps not directly applicable to human beings is, nevertheless, helpful to initiate a basic understanding of the problem and its magnitude. As base line date, the results of plasma, liver and kidney iron, and chromium analyses in the control animal population of this study were in acceptable agreement with data in the literature [40–42].

Corrosion and Iron Metabolism

The simplest model to describe the iron metabolic system is shown in Fig. 1, with modification to account for the presence of an exogenous iron source, for example, a corroding stainless steel implant [39]. In the erythroid circuit, transferrin-bound iron leaves the plasma to bind with red blood cell precursors in the bone marrow whereupon hemoglobin synthesis is accomplished. (Transferrin is a specific iron-binding protein responsible for the intravascular transport of iron.) Mature red cells leave the marrow and circulate in the vascular pool until senescence and death. Iron derived from hemoglobin catabolism in the reticuloendothelial system refluxes rapidly to the plasma upon demand.

The extravascular iron compartment represents transferrin-bound iron in interstitial fluid and lymphatics. The pool exchanges freely with tissue compartment T_1 to T_Z containing iron enzymes, myoglobin, and storage iron.

Regarding iron metabolism, the foregoing research addressed two fundamental questions: (1) is the plasma iron compartment, as determined through measurements of plasma iron, altered by the presence of implanted stainless steel?; and (2) does remote tissue storage of iron reflect implant corrosion?

Results of plasma iron determinations on rabbits having high surface areas of implanted stainless steel indicate that indeed, the plasma iron compartment is moderately altered; that is, plasma iron concentration is increased relative to implantation of high surface areas of stainless steel. In answer to the second question, elevated iron concentrations were found in the livers of rabbits having high surface areas of implanted 316L. It appears then, that corrosion at the implant site is modifying steady-state iron equilibrium, such that a net transport of iron from the implant site to the plasma and eventually to storage sites is occurring. This of course assumes that the basic model itself has not been

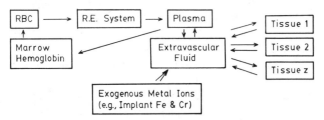

FIG. 1—*Modification of the basic iron-metabolism model to account for the introduction of corrosion products from 316L stainless steel implants. Results indicate that iron and chromium elevations occur in both the plasma-transport compartment and the reticuloendothelial compartment with sparing of the nonerythroid and nonreticuloendothelial tissue compartment* (T_1, T_z).

inherently altered by the mere presence of a chronically corroding metallic implant.

An experimental elevation of mean plasma iron concentrations can be interpreted to mean that more iron is entering the plasma-iron transport pool than leaving it (assuming constant corrosion rate) until a new steady state is achieved. More specifically, when plasma iron introduction (rate) exceeds the plasma iron "turnover rate" (PIT = total amount of iron leaving the plasma per day), plasma iron concentration rises. This relationship, however, is not as clear-cut as it first sounds. Finch et al have shown that the PIT in human subjects remains relatively constant (approximately 0.75 mg/100 mL of blood per day) despite wide variations in serum iron levels [43]. On the other hand, plasma iron clearance, the measurement of diminishing tracer ^{59}Fe activity following injection, shows wide variation relative to the body's demand for iron. For example, patients with iron deficiency anemia clear their plasma of ^{59}Fe much more rapidly than those with iron overload or hemochromatosis. Therefore, a finding of clinically elevated plasma iron, as in this investigation (assuming, of course, rabbit iron metabolism is similar to man), may suggest not only that plasma iron turnover is exceeded by the corrosion process, but that plasma iron clearance as well, in response to chronic overload, is depressed.

Results of this experiment have shown that plasma iron was elevated 12 to 15% in the $100 \times$ group of rabbits at 20 and 28 weeks postoperatively. Coupling this with available information from man on plasma iron turnover, an indirect calculation of stainless steel corrosion rate at the implant site can be made. Assuming a PIT in human subjects of 0.75 mg/100 mL of blood per day, a 4-kg rabbit would scale down to having a PIT of 3 mg of iron daily. The finding of a 12 to 15% elevation in plasma iron (in the $100 \times$ group) over control, equates to an additional 0.4 mg of iron being turned over each day. Again, this assumes constant corrosion rate, constant PIT, and no major suppression of plasma iron clearance or absorption. Dividing the additional iron turnover per day by the total surface area of implanted microspheres (1160 cm^2) yields a corrosion rate of 345 ng of iron/cm^2/day. Considering that iron comprises only 65% of 316L and assuming uniform corrosion of all stainless steel constituents, this figure would calculate to a 316L corrosion rate of 531 ng/cm^2/day. It is assumed in making this calculation that iron that corrodes from an implant passes through the plasma enroute to its eventual destination.

It is noteworthy to compare this corrosion rate with others in the literature (Table 6) that were determined both *in vitro* and *in vivo* by sensitive electrochemical techniques. All reported rates as shown for 316L fall approximately within an order of magnitude above or below the calculated corrosion rate. It must be emphasized that whereas the cited corrosion rates are direct electrochemical measurements, the calculation performed here constitutes, at best, a rough indirect estimate of corrosion rate. The potential errors in this treatment are large. For example, there is no evidence to suggest that in rabbits, PIT is either constant or equivalent to the value used in this calculation. Additionally,

TABLE 6—*316L corrosion rates in ng/cm²/day measured* in vivo *and* in vitro.

Material	in vitro	in vivo	Technique
316L	...	20 to 70[47]	polarization resistance
CrNiMo	...	30 to 120[47]	polarization resistance
CrNi	...	100 to 300[47]	polarization resistance
316L	...	1000[3]	charge curve analysis
316L	9500[48]	...	linear polarization
SAE1018	...	30 000 to 100 000[49]	linear polarization
CrNiMo	90 to 7000[50]	...	linear polarization
316L	...	1.5 to 500 (present study)	plasma dynamics and tissue accumulations

in vivo corrosion rate or the release of corrosion products from the extravascular pool to the plasma may be sporadic rather than constant. Further, should some of the iron corrosion complex stay within the implantation site or be transported to remote sites by means other than the plasma, one would expect the figure of 500 ng/cm²/day to be a low estimate. If, on the other hand, the effects of chronic iron overload, as studied by Finch [44] are manifesting, one would anticipate a depressed plasma iron clearance with consequent back-up in the plasma. In this case, the calculated corrosion rate estimate would be in excess of the real value.

A similar calculation to the foregoing one can be performed using data from deterministic models of iron metabolism. Sharney et al [45] and Najean et al [46] have modeled the intercompartmental transfer of iron and report that in a 70-kg man, 4.7 to 5.1 mg of iron enters the plasma daily. This figure scaled down to a rabbit on a weight basis is approximately 0.28 mg of iron per day. A similar calculation to the previous one yields an estimated microsphere corrosion rate of approximately 40 to 60 ng of iron/cm²/day. This figure is within the range of Steinemann's [47] determinations for 316L corrosion rate using techniques of polarization resistance.

Considering these low estimated corrosion rates, it is not surprising that the 1 × and 10 × rabbit groups did not demonstrate significant iron elevations relative to background. That is, even assuming that the higher corrosion rate of 500 ng/cm²/day is accurate, the additional iron added to the system in a 10 × rabbit is 40 μg/day or a 1.3% increase over the plasma iron turnover (for 1 × only 0.1%). This value is not sufficient to stand out from biological and experimental variability. Again, it should be emphasized that the foregoing treatment is an exercise based on several assumptions, any or all of which may be in error.

Tissue Iron Concentrations

Results from analyses of kidney and liver iron concentrations for the various experimental groups are consistent with a model of chronic low-level iron overload. This is in contrast to high-level iron overload such as hemochromatosis. In either hemochromatosis or excessive iatrogenic iron administration, iron enters the system so rapidly that it exceeds the iron-binding capacity of the blood and forms an unbound and readily diffusible fraction. Excessive iron or ferric salts at neutral pH are extensively hydrolyzed and tend to form polynuclear aggregates in plasma that bind nonspecifically to several plasma proteins. This nontransferrin-bound iron is not under physiological control but rather diffuses nonspecifically into the parenchymatous tissues of the body (T_1 to T_Z, Fig. 1) causing its known toxic effects [51].

In contrast, evidence from this study suggests that corrosion of even large surface areas of 316L does not exceed the plasma transferrin's capacity to transport iron to remote sites. The absence of a diffusible fraction is supported by the result that in the $100 \times$ rabbit group, iron was found to accumulate preferentially in the liver rather than the kidney. If nonspecific parenchymatous distribution of iron-containing corrosion products were occurring, one would expect both organs to show accumulations. Selective liver accumulations, therefore, reflect the liver's reticuloendothelial capacity to bind and store excess iron.

Minimum corrosion rates of the implanted 316L can be calculated from the known organ accumulation of implant constituents over the seven-month implantation period. The $100 \times$ rabbit group demonstrated a mean iron concentration of 0.85 μg iron/mg of liver dry-weight, while the mean control liver iron concentration was 0.65 μg iron/mg dry weight. Considering that the average 4-kg rabbit liver weighs approximately 150 g, and that dry weight is roughly 33% of wet weight, the absolute increase in iron content of liver over the seven-month period was roughly 1 mg or an accumulation rate of 4.8 μg/day. Dividing this figure by the total surface area of implanted 316L (1160 cm^2), the corrosion rate calculates to 4.1 ng iron/cm^2/day, or 6.3 ng 316L/cm^2/day. Understandably, this figure is much lower than those calculated previously, since the liver contains roughly only one third of the total body iron stored. It does, however, demonstrate that a 30% increase in liver iron storage is a realistic figure in terms of known implant corrosion rates and plasma iron kinetics.

Chromium Corrosion and Metabolism

Chromium's distribution and metabolism in the biological system is not nearly so quantitatively investigated as that of iron. Indeed, no model has yet been proposed to fully explain chromium's physiological role. Further, unlike plasma iron, plasma chromium levels are not reliable indicators of body chromium status. Results from this study, therefore, are not as yet amenable to evaluation relative to known systemic chromium distribution or kinetics, but must be interpreted on face value to reflect happenings at the implant site, specifically the corrosion process. Current knowledge suggests a model, whereby chromium leaves the implant site, is complexed specifically to transferrin or nonspecifically to albumin and then is transported by the plasma to sites of storage as yet undetermined (though probably the reticuloendothelial system) or is excreted by the kidney or bile, particularly in response to high intravenous injection of chromium salts [29,31].

Distribution and storage models for chromium depend exquisitely upon the assumed valence of the released metal ion. The general assumption [52] has been that chromium (III) predominates under balanced conditions, but more recent studies by Merritt and Brown [28] and by Wapner et al [53] suggest that a significant portion of chromium from in vivo corrosion is the more biologically active form, chromium (VI).

For the 100× groups of rabbits, the fact that chromium was consistently elevated postoperatively would suggest that chromium's release into the intravascular pool is exceeding compensatory excretion and storage. The relatively constant magnitude of chromium elevations from 4 to 28 weeks postoperatively implies a constant corrosion rate and achievement of steady state relative to the biological system. On the other hand, sporadic elevations in plasma chromium concentrations in the other experimental population would suggest that this is not the case (that corrosion rate is not constant), since biological control of chromium has long rate constants.

Alternatively, these sporadic deviations can be explained perhaps by deviations in experimental protocol from one experimental group to another. To optimize experimental control, blood samples from the 100× rabbit group were always collected at the same time (within 2 h) on the same day. Unfortunately, constraints of time prevented all other experimental groups from being handled in a similar fashion, and therefore, samples from other groups were collected on days other than the control and 100× groups. Perhaps, then, what we are observing, in fact, is a situation of implant corrosion superimposed on the normal day-to-day variations in plasma chromium levels or metabolism. Interestingly, these normal variations never resulted in an experimental group having a plasma chromium concentration significantly less than the control value, despite the difference in collection day. This would support the view that implant corrosion alone is responsible for the observed sporadic plasma chromium elevations.

Examination of the plasma chromium levels reveals that no elevation above 25% was observed irrespective of implanted 316L surface area. From this, one may postulate the existence of a "threshold" phenomenon at chromium levels above 25%. It is conceivable that the compensatory physiological mechanisms to regulate excess chromium do not operate until levels are 25% above normal, or that chromium above this level constitutes a readily diffusible fraction that is excreted in the kidney or bile. Such a phenomenon is not without biological precedent. Glucose, for example, above its physiological threshold in the blood is "spilled" through the renal glomerulus into the urine, causing glycosuria as in diabetes mellitus. In the case of chromium, exceeding the postulated threshold would result in chromium's rapid elimination from the intravascular fluid space to sites as yet undetermined.

In summary then, the findings of sporadic chromium elevations in the 1× and 10× groups and consistent elevations in the 100× group over the seven-month implantation period suggest that plasma chromium levels reflect implanted 316L surface area and corrosion. The threshold phenomenon as postulated here is in marked contrast to the estimated plasma steady-state chromium (III) concentrations as reported by Onkelinx [31] using results from the radioactive ^{51}CrCl wash-out studies in rats. Onkelinx has divided chromium metabolism into three theoretical compartments, the kinetics of which is best approximated by the mathematical expression:

$$S_1 = A_1 e^{-a_1 t} + A_2 e^{-a_2 t} + A_3 e^{-a_3 t}$$

where S_1 = plasma ^{51}Cr concentration; t = time post-injection; and A_1, A_2, A_3, a_1, a_2, a_3 = constants evaluated from data. This relationship, however, does not take into account a threshold phenomenon as postulated in the present study, predicted that at steady state, plasma chromium levels would be several times the base-line physiological level. It should be mentioned in addition, however, that the rate of experimental chromium infusion was approximately 1 μg/kg/h, or a figure at least ten times the expected corrosion rate in the 100× rabbit group employed in this investigation. Therefore, whether Onkelinx's results are applicable to the implant corrosion process is open to question. The pitfalls of infusion experiments to simulate implant corrosion are many. Obviously, in this experiment, the rate of administration, though slow by design, still exceeds physiological levels and the actual *in vivo* corrosion process (by two to three orders of magnitude).

Tissue Chromium Concentrations

Results of tissue analyses reveal several interesting points when comparing trace element data from control liver and kidney. First, the kidney having negligible reticuloendothelial capacity, demonstrated steady-state iron concentrations one-half those of the liver. This is not surprising, considering the liver's prominent role in normal iron homeostasis and the kidney's lack of reticuloendothelial components. Secondly, the finding that control liver and kidney chromium concentrations in non-loaded rabbits are essentially identical, would suggest that chromium homeostatic mechanisms are common to both types. Alternatively, it may imply that chromium is distributed equally throughout all cell types of the kidney and liver and that, ordinarily, liver is not a major storage site of chromium. Nevertheless, the fact that experimental liver chromium concentrations selectively increased in the 10× and 100× groups over kidney levels, suggests the specificity of the storage capacity of liver's reticuloendothelial system. Furthermore, the magnitudes of liver's chromium elevations, that is, 64% for the 10× group and 85% for the 100× group, demonstrates the storage capacity to be a function of implanted 316L surface area. The obvious lack of proportionality of implanted 316L to liver chromium accumulations could again reflect the existence of a threshold phenomenon where high plasma chromium levels result in excessive loss of chromium by other pathways, for example, renal excretion, and therefore, a reduced liver storage efficiency.

In contrast to the chromium storage data of this study, Onkelinx [21] and Mutti

[29], have demonstrated in rats that both the liver and kidney increase their chromium concentration after a single intravenous injection of $^{51}CrCl$. Also, bone, lung, and spleen have the capacity to retain ^{51}Cr, particularly after $^{51}CrCl_3$ infusion. Again, whether the dose and route of administration as used in these cited studies have any similarity to the implant corrosion process and eventual distribution of corrosion products is not clear. That is, organ accumulations as found in these infusion studies may be entirely non-physiological, representing merely an acute chromium overload distribution owing to the relatively high infusion rates, or may reflect the use of an inappropriate metal ion valence.

As with liver iron accumulation data, a first approximation of the minimum 316L corrosion rate can be estimated knowing the normal liver chromium concentration coupled with the increase in liver chromium concentration after seven months of implanted 316L. Quantitatively, it was found that after seven months, the $100\times$ rabbit group demonstrated liver chromium accumulations of 3.46 ng chromium/mg dry weight, or an increase over control of 1.6 ng chromium/mg dry weight. Taking the mass of the average 4-kg rabbit liver to be 150 g and assuming dry weight is approximately 33% wet weight, yields a total increase in liver chromium of 72 µg chromium or an accumulation rate of 342 ng chromium/day. Dividing this figure by the total surface area of implanted 316L (1160 cm^2), the corrosion rate calculates to 0.3 ng chromium/cm^2/day, or 1.5 ng 316L/cm^2/day. Again, it should be recalled that the liver makes up only one third or less of the total body reticuloendothelial mass and, therefore, this figure represents a minimal rate. Clearly, chromium retained at other sites, either locally or systemically, or chromium excreted through the urine or bile are not represented by this calculation and would raise the overall implant corrosion rate.

Nevertheless, even through such a crude calculation, relatively good agreement is achieved comparing in vivo corrosion rates obtained by different means: from plasma iron turnover (approximately 500 ng/cm^2/day), from the plasma iron deterministic model (40 to 60 ng/cm^2/day), from liver iron accumulations (>6.3 ng/cm^2/day), and from liver chromium accumulations (>1.5 ng/cm^2/day). The last three extrapolations are within the range of Steinemann's in vivo determination of 316L corrosion, 20 to 70 ng/cm^2/day [47] (Table 6).

Metallography revealed obvious differences in microstructure between pins and microspheres, reflecting the respective mode of manufacture. The significance of such differences in terms of in vivo corrosion kinetics could not be established within the limits of this experiment nor from the literature. The size and configurational discrepancies between pins and powders were theorized to have negligible influence on the relative surface free energy or corrosion potential. For the purposes of this experiment, therefore, the corrosion characteristics of the Steinmann pins and microspheres were assumed to be equivalent. Though potentially an oversimplification, this assumption facilitates the experimental objective of having a quantifiable surface area of implanted 316L as a corrosion product source. Obviously, a separate experiment to precisely determine the relative corrosion rates of pins versus powders is in order. Should such an experiment eventuate, the "absolute" surface areas as used here could easily be modified to include "relative" corrosion rates. Despite these reservations, however, the accuracy (or inaccuracy) of the simplifying assumption of corrosion equivalence (pins versus powders) is commensurate with inherent variation of the biological system and presumably adequate for the objectives of this investigation.

Conclusions

Methodology has been developed to accurately and reproducibly measure very small quantities (ppm-ppb) of iron and chromium in ultra-small samples (20 to 50 μl) of biological fluid or tissue. Results of plasma iron determinations in rabbits demonstrate significant alterations in iron homeostasis under conditions of high 316L implanted surface area. Plasma chromium concentrations were likewise elevated significantly, apparently as a function of implanted 316L surface area and duration of implantation. Tissue accumulations of iron and chromium were associated with organ reticuloendothelial capacity, suggesting that implant corrosion rate does *not* exceed the specific binding capacity of the blood transport compartment.

Approximate corrosion rates of implanted 316L were calculated using information from plasma trace element kinetics and storage site data and found to be within reasonable agreement with stainless steel's *in vitro* and *in vivo* measured corrosion rates.

The clinical significance of these findings relative to human populations is unclear and is beyond the scope of this paper. There is clearly a need to extend these studies to longer term animal models to investigate the precise pathophysiology of the corrosion-mediated disease processes and to examine correlations with human clinical populations.

Acknowledgments

We wish to acknowledge the support and encouragement of Dr. Carl Brighton, Chief, Department of Orthopaedic Surgery, and also, Dr. Irving Shapiro who generously volunteered his expertise and his laboratory facility for trace element analysis. This work was supported by the Veterinary Medical Scientist Training Program at the University of Pennsylvania (National Institutes of Health Grant No. GM02051) and by a direct grant (National Institutes of Health AM 25752).

References

[1] Greene, N. D. and Jones, D. A., *Journal of Materials*, Vol. 1, 1966, p. 345.
[2] Colangelo, V. J., Greene, N. D., Kettlekamp, D. B., Alexander, H., and Campbell, C. J., *Journal of Biomedical Materials Research*, Vol. 1, 1967, p. 405.
[3] Steinemann, S. G. and Perren, S. M. in *Transactions, Society for Biomaterials*, 3rd Annual Meeting, 1977, p. 115.
[4] Pourbaix, A., *Corrosion Science*, Vol. 5, 1965, p. 677.
[5] Aragon, P. J. and Hulbert, S. F., *Corrosion*, Vol. 30, 1974, p. 432.

[6] Mueller, H. J. and Greener, E. H., *Journal of Biomedical Materials Research*, Vol. 4, 1970, p. 29.

[7] Ferguson, A. B., Akahoshi, Y., Laing, P. G., and Hodge, E. S., *Journal of Bone and Joint Surgery*, Vol. 44-A, 1962, p. 323.

[8] Pourbaix, M., *Atlas of Electrochemical Equilibria in Aqueous Solutions*, National Association of Corrosion Engineers, Houston, 1974.

[9] Brussatis, G. and Nonhoff, T., *Archiv für Orthopaedische und Unfallchirurgie*, Vol. 62, 1967, p. 64.

[10] Grasser, V. H. H., *Deutsches Zahnarzteliches, Zeitschift*, Vol. 27, 1972, p. 413.

[11] Lux, F. and Zeisler, R., *Zertscrift für Analytische Chemie*, Vol. 261, 1972, p. 314.

[12] Nicole, R., *Helvetica Chirurgia Acta*, Vol. 14, Supplement III, 1947.

[13] Scales, J. T., Winter, G. D., and Shirley, H. T., *Journal of Bone and Joint Surgery*, Vol. 41-B, 1959, p. 810.

[14] Ferguson, A. B., Laing, P. G., and Hodge, E. S., *Journal of Bone and Joint Surgery*, Vol. 42-A, 1960, p. 77.

[15] Ludinghausen, M. V., Meister, P., Rossman, E., and Guckel, W., *Verhandlugen der Deutschen Gesellschaft für Pathologie*, Vol. 54, 1970, p. 414.

[16] Ludinghausen, M. V., Meister, P., and Probst, J., *Medische Welt*, Vol. 21, 1970, p. 1913.

[17] Emneus, H., *Journal of Bone and Joint Surgery*, Supplement, Vol. 44-A, 1961, p. 3.

[18] Scales, J. T., *Journal of Bone and Joint Surgery*, Vol. 53-B, 1971, p. 344.

[19] Scales, J. T., Winter, G. D., and Shirley, H. T., *British Medical Journal*, Vol. 19, 1961, p. 478.

[20] Cohen, J., *Journal of Materials*, Vol. 1, 1966, p. 354.

[21] Emneus, H. and Stenram, V., *Acta Orthopaedica Scandinavica*, Vol. 36, 1965, p. 115.

[22] Cahoon, J. R. and Paxton, H. W., see Ref 2, Vol. 2, 1968, p. 1.

[23] Ludinghausen, M. V., Meister, P., and Probst, J., *Pathologia Europaea*, Vol. 5, 1970, p. 307.

[24] Williams, D. F. and Meachum, G., *Proceedings*, 5th Annual Biomaterials Symposium, Clemson, 1973, p. 1.

[25] Smith, G. K. in *Systemic Aspects of Biocompatibility*, Vol. II, D. F. Williams, Ed., CRC Press, Boca Raton, FL, 1981, Chapter 1.

[26] Owen R. Merchim, G., and Williams, D. F., see Ref 2, Vol. 10, 1976, p. 91.

[27] Michel, R., Hofmann, J., and Zillens, J., in *Proceedings*, International Symposium on Nuclear Activation Techniques in the Life Sciences, Paper IAEA-Sm-227/10, Vienna, 1978.

[28] Merritt, K. and Brown, S. A., in *Proceedings*, 8th Annual Meeting, Society for Biomaterials, 1982, p. 73.

[29] Mutti, A. Cavatorta, A., Borghi, L., Canali, M. Giavoli, C., and Franchini, I., *Medicia de Lavoro*, Vol. 3, 1979, p. 71.

[30] Cikrt, M. and Bencko, V., *Journal of Hygiene, Epidemiology, Microbiology, and Immunology*, Vol. 23, 1979, p. 241.

[31] Onkelinx, C., *American Journal of Physiology*, Vol. 232(s), 1977, p. E-478.

[32] Brune, D., Nordberg, G., and Wester, P. O., *Science of the Total Environment*, Vol. 16, 1980, p. 13.

[33] Mutti, A., Cavatorto, A., Pedroni, C., Borghi, A., Giaroli, C., and Franchini, I., *International Archives of Occupational and Environal Health*, Vol. 43, 1979, p. 123.

[34] *Annual Book of ASTM Standards*, American Society for Testing and Materials, Vol. 46, 1983.

[35] Underwood, E. J., *Trace Elements in Human and Animal Nutrition*, 3rd ed., Academic Press, New York, 1971, p. 1.

[36] Olsen, E. D., Jatlow, P. I., Fernandez, F. J., and Kahn, H. L., *Clinical Chemistry*, Vol. 19, 1973, p. 326.

[37] Davidson, I. W. F. and Secrest, W. L., *Analytical Chemistry*, Vol. 44, 1972, p. 1808.

[38] Edwards, A. L., *An Introduction to Linear Regression and Correlation*, W. H. Freeman and Co., San Francisco, CA, 1976, p. 106.

[39] Jacobs, A. and Worwood, M., *Iron in Biochemistry and Medicine*, New York, Academic Press, New York, 1974.

[40] Tipton, I. H. and Cook, M. J., *Health Physics*, Vol. 9, 1963, p. 103.

[41] Ferguson, A. B., Akahoshi, Y., Laing, P. G., and Hodges, E. S., see Ref 7, Vol. 44-A, 1962, p. 317.

[42] Versieck, J. and Cornelius, R., *Analytica Chimica Acta*, Vol. 116, 1980, p. 217.
[43] Finch, C. A., Deubelbeiss, K., Cook, J. D., et al, *Medicine (Baltimore)*, Vol. 49, 1970, p. 17.
[44] Finch, S. C. and Finch, C. A., *Medicine (Baltimore)*, Vol. 34, 1955, p. 381.
[45] Sharney, L., Gevirtz, N. R., Wasserman, L. R., et al, *Journal*, Mt. Sinai Hospital, Vol. 32, 1965, p. 338.
[46] Najean, Y., Dresch, C., Ardaillow, N., et al, *American Journal of Physiology*, Vol. 213, 1967, p. 533.
[47] Steinemann, S., *Fortschritte der Kicfer-und Gesichts-chirurgie*, 1975, p. 19.
[48] Mueller, H. J. and Greener, E. H., see Ref 2, Vol. 4, 1970, p. 29.
[49] Colangelo, V. J., Greene, N. D., Kettlekamp, D. B., Alexander, H., Campbell, C. J., see Ref 2, Vol. 1, 1967, p. 405.
[50] Hoar, T. P. and Mears, D. C., in *Proceedings*, Royal Society, Vol. A294, 1966, p. 486.
[51] Gitlow, S. E. and Beyers, M. R., *Journal of Laboratory and Clinical Medicine*, Vol. 39, 1952, p. 337.
[52] Petrilli, F. L. and DeFlora, S., *Mutation Research*, Vol. 58, 1968, p. 167.
[53] Wapner, K. L. and Black, I., *Journal of Biomedical Materials Research*, in press, 1985.

DISCUSSION

Manfred Semlitsch[1] (*written discussion*)—Your animal tests have been carried out with microspheres of 316L, delivered by Nuclear Metals Corp. Is the micro and surface structure of these 55-μm spheres comparable to the structure of wrought 316L? What is the procedure for surface cleaning and sterilization? Is there an ASTM F-4 standard existing for the preparation and handling of metal powder? Can we therefore rely on metal powder introduction results in animals? Can we draw conclusions to the behavior of abrasion particles of implants in humans?

Gail K. Smith and Jonathan Black (*authors' closure*)—The 55-μm microspheres upon metallographic examination fell into two distinct populations. Approximately 60% of the microspheres exhibited a typical as-quenched microstructure while the remainder typified an austenitic grain structure. A few microspheres demonstrated both types of microstructure. This differs somewhat from the wrought 316L stainless steel Steinemann pins that were made up of austenite with a cold-worked history. X-ray fluorescence analysis confirmed the pins and powders to have similar composition, both of which fell within the specifications for 316L stainless steel. X-ray diffraction of the microspheres revealed a face-centered-cubic austenitic microstructure. Faint lines on the diffraction film, however, suggested the presence of a small amount of retained ferrite (body-centered-cubic). No specific tests were performed to characterize the surface of the microspheres relative to wrought 316L stainless steel. However, I would not expect the surface and microstructural variations to be more significant, for example, than the accepted differences between a stainless steel plate

[1]Sulzer Bros. Ltd., Dept. R&D, CH-8401 Winterthur, Switzerland.

relative to the associated screws having cold-work-induced martensitic transformation.

Microsphere preparation consisted of copious flushing with ethanol followed by lyophilization prior to routine sterilization (250°F for 30 min under 18 psi with a 20 min drying phase). Attempts at passivation according to ASTM protocol resulted in undesirable clumping of the powder and, therefore, the method was not utilized in this investigation. To my knowledge, there does not presently exist an ASTM standard for handling and preparing metal powders; however, ASTM is aware of the need for such a standard.

The 316L stainless steel microspheres provide the means to surgically introduce a quantifiable surface area of implant metal into experimental rabbits. Although differences in corrosion behavior between surgical-grade stainless steel and the microspheres were not established within the limits of this experiment, discussions with corrosion engineers in both industrial and academic positions would suggest, empirically, that wrought stainless steel (pins) and cast stainless steel (powders) have comparable corrosion rates *in vivo*. Moreover, because rabbits have a higher metabolic rate than man, the act of implanting stainless steel in rabbits on a per weight basis relative to man, resulted in an overestimate of the effective surface area of stainless steel that the animals were receiving. That is, in this study, instead of an implanted surface area to body weight ratio of $1 \times$, $10 \times$, and $100 \times$, the rabbits on a metabolic basis, actually received only 0.4, 4, and 40 times the human dose, respectively. Also, from the size, shape, and observed microstructure of the microspheres, one would anticipate that their corrosion potential is far less than that of highly worked wear particles observed emanating from implanted multi-component devices. Moreover, it should be noted that the 55-μm microspheres unlike wear particles are too large to be transported systemically either via bulk lymph flow or phagocytic mechanisms. Therefore, the results of this study represent, at the very least, a conservative estimate of the systemic effects to be anticipated using standard multi-component stainless steel devices of comparable composite surface area.

Polymeric Materials: *In Vivo* **and** *In Vitro*

Roger E. Marchant,[1] *Tsutomu Sugie,*[2] *Anne Hiltner,*[1] *and*
James M. Anderson[1]

Biocompatibility and an Enhanced Acute Inflammatory Phase Model

REFERENCE: Marchant, R. E., Sugie, T., Hiltner, A., and Anderson, J. M., "Bio-compatibility and an Enhanced Acute Inflammatory Phase Model," *Corrosion and Degradation of Implant Materials: Second Symposium, ASTM STP 859,* A. C. Fraker and C. D. Griffin, Eds., American Society for Testing and Materials, Philadelphia, 1985, pp. 251–266.

ABSTRACT: The evaluation of the biocompatibility of biomaterials centers on appreciating the events that occur at the tissue/implant interface. In determining the biocompatibility of biomaterials, both the host response and the material response to the host must be elucidated. The *in vivo* biocompatibility of candidate biomaterials may be investigated using the cage implant system. The cage implant system is presented, which allows the quantitative determination of the dynamic nature of cell function at the implant site. Results from the quantitative determination of cells and products of cell activation over the implantation period can be subjected to statistical analysis methods. In addition, studies utilizing a chemotactic tripeptide have been carried out to provide an enhanced or accelerated acute inflammatory phase model. These studies are important as they provide an *in vivo* situation where the effect of cells on implanted materials and the effect of biomaterials on the cellular components of the inflammatory response may be investigated.

KEY WORDS: biocompatibility, cage implant system, inflammatory response, N-formyl-Met-Leu-Phe (fMLP), poly(2-hydroxyethyl-L-glutamine) (PHEG), implant materials, biological degradation

The biocompatibility of any implanted material is a dynamic and two-way process that involves the time dependent effects of the host on the material and of the material on the host. Numerous investigators have sought to determine the biocompatibility of a wide variety of potential implant materials. The majority of these studies were based on histological observations and morphological evaluations and directed toward understanding the cellular composition and organization immediately adjacent to the implant [*1–4*].[3] In general, the biocompatibility of a given material with tissue has been described in terms of the acute

[1]Senior research associate, Department of Macromolecular Science; professor, Department of Macromolecular Science; and associate professor, Departments of Pathology and Macromolecular Science; respectively, Case Western Reserve University, Cleveland, OH 44106.

[2]Manager research chief, Unitika Ltd., Uji-shi, Kyoto, Japan.

[3]The italic numbers in brackets refer to the list of references appended to this paper.

and chronic inflammatory responses and the fibrous capsule formation that are seen over various time periods following implantation [5–7]. The degree and extent of the inflammatory reaction around a subcutaneously or intramuscularly implanted material with time is then considered to be indicative of its biocompatibility.

The inflammatory reaction is a series of complex reactions involving various types of cells whose densities, activities, and functions are controlled by various endogenous and autocoid mediators. The acute inflammatory reaction is the host's initial response to implantation. This phase, which usually lasts three to four days, presents the most intensive period of tissue/material interactions. The events that occur during acute inflammation also influence the subsequent chronic inflammatory phase of healing. Acute inflammation is characterized by the predominance of polymorphonuclear leukocytes (PMNs), principally neutrophils, that are present at the implant site in high concentration. It has been shown that mononuclear leukocytes, which include macrophages and lymphocytes, and extracellular lysosomal enzymes are also present in high concentrations during the acute phase [8].

In the work presented here, a cage implant system has been used to appreciate quantitatively the dynamic nature of cell function at the implant site. This cage system containing the biomaterial under investigation permits the formation of an exudate that can be monitored in a serial fashion without sacrificing the animal. The biomaterial is allowed to float free within the inflammatory exudate, and tissue compression against the implant, as commonly seen with subcutaneous implants, does not occur. Using the cage system, we have examined the *in vivo* behavior of crosslinked poly(2-hydroxyethyl-L-glutamine) (PHEG), a hydrophilic, nonionic poly(α-amino acid). It has been previously shown that this polymeric hydrogel undergoes *in vivo* biodegradation when implanted subcutaneously in rats [9]. In this study, the cage implant system was also utilized to develop a model that provides a maximal acute inflammatory response. The intensity and duration of the acute inflammatory reaction was maximized by the injection of the chemotactic tripeptide, N-formyl-Met-Leu-Phe (fMLP). The tripeptide is a well-known synthetic chemotactic agent for PMNs *in vitro* [10–12], and its presence within the cage provided a stimulus for the enhanced migration of leukocytes.

Methods and Materials

The cage implant system has been previously described [8]. The system involves the use of stainless steel wire mesh cages containing appropriate polymer specimens that are subcutaneously implanted in three-month-old, female, Sprague-Dawley rats. Cages without polymer specimens are used as reference controls. The cages, 3.5 cm in length and 1 cm in diameter, were fabricated from austenic Type 316 stainless steel mesh.[4] The pale yellow inflammatory

---

[4]Tyler Industrial Products, Mentor, OH 44060.

exudate that accumulated within the implanted cage was periodically aspirated under sterile procedures, using a 22-gage needle with a 1-mL Tuberculin Plastipak syringe. The exudate samples were subjected to quantitative total white cell counts performed in duplicate using a hemacytometer and then differential counts from Wright stained microslides smeared with exudate. The extracellular fraction of the exudate samples was analyzed for alkaline phosphatase. The possibility of infection was checked by culturing aliquots of exudate on brain heart infusion agar plates. Infected exudates or exudates with appreciable erythrocyte contamination were discarded.

A 3.6-mole percent crosslinked film of the biodegradable hydrogel PHEG was synthesized and prepared for implantation according to the methods described previously [8]. Sterile PHEG specimens (1.33 cm^2) were placed inside the autoclaved stainless steel cages, one specimen per cage, and implanted subcutaneously in the rat. The PHEG hydrogel samples were equilibrated in sterile phosphate buffered saline for 4 h before implantation.

In the development of an enhanced acute phase model system, a set of preliminary experiments were carried out in order to assess and optimize the chemotactic behavior of fMLP, using our cage system. To each rat a 0.1-mL injection of a chemotactic tripeptide solution was administered into both implanted cages at the cephalad end, using a 22-gage needle with a 1-mL disposable syringe. The injections were made under sterile conditions at four and seven days post-implantation time. The following solutions were prepared immediately prior to injection into the different rats: Solution A was pure dimethylsulfoxide (DMSO),[5] Solution B contained 0.13 mg of fMLP[5] per ml of DMSO, Solution C was 10 mg/mL, and Solution D was 40 mg/mL. For Solution E, fMLP was initially dissolved in DMSO and then added to a sterile solution of rat albumin[5] in phosphate buffered saline (2 mg/mL) in the ratio 2:1 (volume/volume). The final concentration of fMLP prior to injection was 7 mg/mL. For Solution F, the fMLP was initially dissolved in DMSO and mixed 1:4 (volume/volume) with a sterile solution of rat albumin in phosphate buffered saline (2 mg/mL). The solution was adjusted to pH 7.4 prior to injection. The concentration of fMLP administered was 16 mg/mL. Solution G was the same as Solution F but without added tripeptide. Inflammatory exudates were aspirated and analyzed for leukocyte concentration prior to and one day after injection.

The use of Solution F as a chemotactic agent was repeated. The chemotactic tripeptide Solution F (tripeptide control) was injected into the cages of Sprague-Dawley rats at four and seven days post-implantation time. The tripeptide solution was also injected into cages containing PHEG (PHEG + T), to provide further comparative data. The exudates aspirated from the tripeptide controls, the PHEG + T system, as well as the empty cage controls and PHEG system were analyzed for total white cell concentration, PMN and mononuclear leukocyte concentrations, and extracellular alkaline and acid phosphatase.

[5]Sigma Chemical Co., St. Louis, MO 63178.

TABLE 1—Cell counts in the exudate as a function of implantation time.

Time, days	Total White Cells, cells/μL		PMNs, cells/μL		Mononuclear Leukocytes, cells/μL	
	Control	PHEG	Control	PHEG	Control	PHEG
1	21 600 ± 900[a]	21 000 ± 900	20 100 ± 600	19 600 ± 300	1450 ± 320 (6.7)[c]	1430 ± 810 (6.8)
3	11 000 ± 2900	10 500 ± 3300	10 100 ± 3000	9 860 ± 3070	930 ± 150 (8.5)	600 ± 320 (5.7)
4	4 160 ± 520	4 150 ± 240	3 500 ± 590	3 070 ± 370	660 ± 280 (15.8)	1090 ± 250 (26.2)
5	2 890 ± 210	3 180 ± 540	2 080 ± 310	1 980 ± 570	800 ± 230 (27.8)	1200 ± 160 (37.7)
7	820 ± 180	1 100 ± 220	390 ± 110	280 ± 40	430 ± 160 (53.0)	820 ± 200 (74.9)
8	730 ± 30	910 ± 80[b]	180 ± 80	130 ± 90	550 ± 80 (75.2)	770 ± 80[b] (85.1)
11	450 ± 110	650 ± 60[b]	42 ± 14	25 ± 12	410 ± 110 (90.6)	630 ± 50[b] (96.2)
14	260 ± 50	500 ± 110[b]	12 ± 4	19 ± 6	250 ± 50 (95.3)	480 ± 110[b] (96.2)
17	250 ± 40	210 ± 20	4 ± 3	7 ± 5	250 ± 40 (98.4)	200 ± 30 (96.7)
21	120 ± 50	130 ± 60	2 ± 1	4 ± 2	120 ± 50 (98.3)	130 ± 50 (96.9)

[a]Mean value ± standard deviation.
[b]Statistically different at the 95% confidence level (p < 0.05) when compared to the mean control value. Student's t-test for unpaired samples.
[c]The percentage of mononuclear leukocytes in the exudate is given in parentheses for each time point.

Results

Response to Control and PHEG Implants

Implantation of the empty stainless steel cages provoked an acute inflammatory response that was characterized by large numbers of PMNs or granulocytes present in the exudate within the cage. The accumulation of white cells and the predominance of PMNs in the exudates over the first four days following implantation are clearly demonstrated in Table 1. When PHEG specimens were present within the cage, statistically significant differences (that is, $p < 0.05$) were noted at Days 8, 11, and 14. Differential white cell counts of these exudates showed that there were no significant differences in PMN concentrations, during the implantation period. Significant differences were noted however, in mononuclear leukocyte concentrations on Days 8, 11, and 14, indicating an elevated chronic response.

Extracellular alkaline phosphatase activity for the two types of implant as a function of implantation time are given in Table 2. The results indicated large differences between the PHEG system and controls during the first week of implantation, particularly at Days 4, 5, and 7 post-implantation. Alkaline phosphatase is located in the specific granules of PMNs and is therefore of particular significance with respect to the acute phase of inflammation. In an attempt to clarify the relationship between PMN cell number, extracellular alkaline phosphatase activity, and implant type, the enzyme activity per PMN was calculated for the two systems and plotted against the first five days of implantation time; when the PMN is the predominant cell. The results are illustrated in Fig. 1. The increased activity/PMN ratio, for the PHEG system, relative to the control system, suggested that the presence of PHEG inside the cage implant enhances

TABLE 2—*Extracellular alkaline phosphatase activity in the exudate as a function of implantation time (from Ref 8).*

Time, days	Control, U^a/dL	PHEG, U/dL
1	57.9 ± 3.2^b	53.9 ± 4.1
3	47.0 ± 11.5	55.8 ± 6.5
4	20.9 ± 3.2	34.2 ± 2.4^c
5	20.2 ± 2.0	33.8 ± 2.1^c
7	14.7 ± 1.9	21.7 ± 2.5^c
8	12.5 ± 1.9	13.6 ± 1.9
11	12.1 ± 1.3	10.7 ± 0.5
14	11.0 ± 1.2	9.9 ± 0.9
17	9.5 ± 1.8	9.2 ± 1.4
21	8.4 ± 2.1	9.3 ± 1.2

NOTE—Normal rat serum: 24.5 ± 5.6 U/dL.
^{a}U = units, see Ref 8.
^{b}Mean value $\pm$ standard deviation.
^{c}Statistically different at the 95% confidence level ($p < 0.05$) when compared to the mean control value. Student's t-test for unpaired samples.

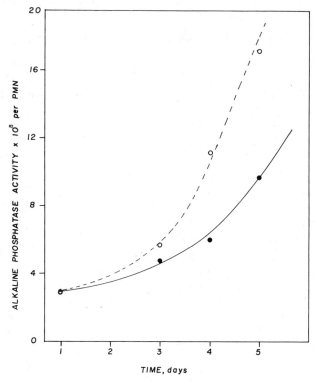

FIG. 1—*The variation in extracellular alkaline phosphatase activity per PMN in the exudate with implantation time. Control (closed circles and solid line) and PHEG (open circles and dashed line).*

TABLE 3—*The variation in white cell counts in the exudate with time after injection of potential chemotactic solutions.[a]*

Implantation Time, days	Total White Cell Counts,[b] cells/μL						
	A	B	C	D	E	F	G
4	3900	4630	3530	3700	4240	4160	4000
5	4050	5480	5760	7580	7550	10 900 ± 2000	3330 ± 500
7	505	985	1280	1710	2150	1 920 ± 300	1020 ± 220
8	645	1095	1610	950	3080	4 630 ± 1600	710 ± 220
11	480	460	500	600	820	520 ± 130	410 ± 70

[a]0.1 mL injection of the respective solutions were made at four and seven days post-implantation time. Exudates were aspirated immediately prior to and one day after injection. An additional exudate withdrawal was made at 11 days post-implantation time.

[b]A = DMSO, B = 0.13 mg/mL (fMLP/DMSO), C = 10 mg/mL (fMLP/DMSO), D = 40 mg/mL (fMLP/DMSO); 1 mL of E contained 7 mg fMLP, and 2 mg rat albumin in DMSO (0.67 mL) and PBS (0.33 mL); 1 mL of F contained 16 mg fMLP, and 2 mg rat albumin in DMSO (0.2 mL) and PBS (0.8 mL) at pH = 7.4; and 1 mL of G was the same as F but without any added fMLP.

the release of alkaline phosphatase. After the fifth day following implantation, the activity/PMN ratio increased markedly and suggested that the measured extracellular alkaline phosphatase activity was dominated by the contribution from cytolytic release.

Response to fMLP

The results of the preliminary studies to develop an enhanced acute phase model are shown in Table 3. The table shows the variation in total white cell counts in the exudate with time for the range of tripeptide concentrations that were injected into the cages at Days 4 and 7 post-implantation time. The table shows that, for Solutions A to D, the cellular response observed one day after each injection depended on the fMLP concentration. A much higher concentration was required to generate an observable cellular response in this *in vivo* model system compared to previous *in vitro* studies [11,12]. It was also noted that the injection at Day 7 stimulated a milder cellular response compared to the response following the Day 4 injection. This was possibly because of a concomitant in-

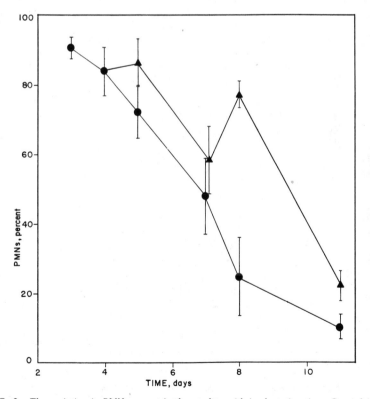

FIG. 2—*The variation in PMN percent in the exudate with implantation time. Control (closed circles and solid line) and Solution F (closed triangles and solid line).*

crease in some inflammatory feedback modulators. The presence of albumin in Solutions E and F appeared to enhance the cellular response, although the injection of Solution F without fMLP gave no observable effect on cellular migration. The increase in leukocyte concentration observed at Days 5 and 8 for Solution F, shown in Table 3, was therefore attributed to the presence of the chemotactic tripeptide, fMLP. Figure 2, which compares percent PMNs in the exudate for Solution F and empty cage controls, shows that the predominant cell type that responded to the chemotactic stimulus was the PMN.

Response to Implants with Added Tripeptide

Solution F (tripeptide control) was used to provide a maximal acute phase model system that effectively increased the intensity and duration of the PMN predominant acute phase from the fifth to the eleventh day. Figure 3 illustrates the total white cell count results, from Day 3, for the empty cage control, tripeptide control, and PHEG + T systems. Cell counts were significantly elevated from Day 5 to Day 8 for the tripeptide control system and from Day 5 to Day

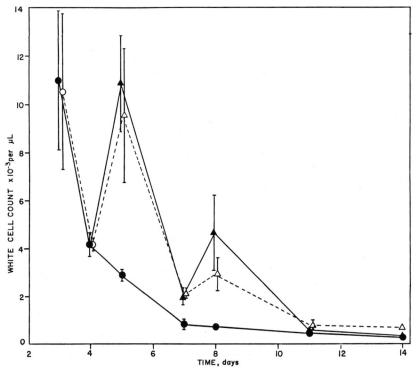

FIG. 3—*The variation in white cell concentration in the exudate with implantation time. Control (closed circles and solid line), tripeptide control (closed triangles and solid line), PHEG + T (open triangles and dashed line), and PHEG at Days 3 and 4 (open circles and dashed line).*

14 for the PHEG + T system, when statistically compared against the empty cage controls. While the cell count results for the tripeptide control and PHEG + T systems were similar from Days 5 to 8, a higher white cell concentration in the PHEG + T system was maintained for a longer period. Figure 4 illustrates the dramatic rise in PMN concentration that was observed following the injection of fMLP. The PMN response was slightly higher for the tripeptide controls than for the PHEG + T system, although statistically significant differences between them were not observed. The mononuclear leukocyte response, following injection was much less apparent. Figure 5, which shows the variation in mononuclear leukocyte concentration with time for the PHEG + T, tripeptide control, and empty cage control systems, suggests that mononuclear leukocytes responded to the chemotactic stimulus, although the response was relatively small compared to the PMN response. The PHEG + T system maintained the highest mononuclear leukocyte concentration throughout the three-week implantation period and only by Day 14 was significantly higher than either the tripeptide or empty cage controls.

The cell type principally responsible for the increased white cell counts was the PMN (Fig. 4). The marked elevation in PMN concentration was paralleled

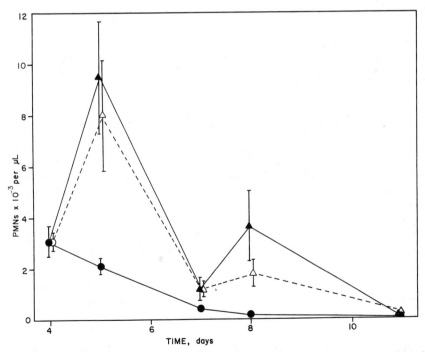

FIG. 4—*The variation in PMN concentration in the exudate with implantation time. Control (closed circles and solid line), tripeptide control (solid triangles and solid line), PHEG + T (open triangles and dashed line), and PHEG at Day 4 (open circle).*

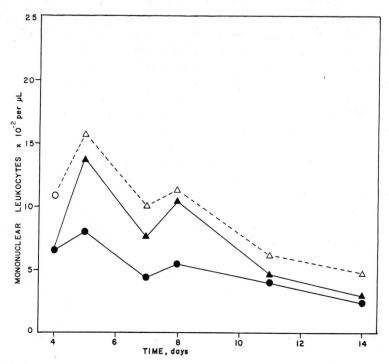

FIG. 5—*The variation in mononuclear leukocyte concentration in the exudate with implantation time. Control (closed circles and solid line), tripeptide control (closed triangles and solid line), PHEG + T (open triangles and dashed line), and PHEG at Day 4 (open circle). For clarity, standard errors are not shown.*

by the results for extracellular alkaline phosphatase activity, shown in Fig. 6. The activity was significantly elevated from Day 5 to Day 11 for the tripeptide control and from Day 4 to 8 for the PHEG + T system compared to the empty cage controls. The slightly higher activity for the tripeptide control compared to the PHEG + T system was probably a reflection of the slightly higher PMN concentration, since the activity/PMN ratio appeared to be similar for each system.

While the results of the cell counts and extracellular alkaline phosphatase activity with time for the tripeptide control and PHEG + T systems were similar, analysis of the extracellular acid phosphatase activity, shown in Table 4, revealed some differences. Compared to empty cage controls, significant differences in activity were noted throughout the three-week implantation period for the PHEG + T system, while significant differences for the tripeptide control were noted at only two points, Days 8 and 11. This would suggest that the measured acid phosphatase activity was a function of both PMNs and macrophages and that the differences between the tripeptide control and PHEG + T system may be attributed to macrophage activity.

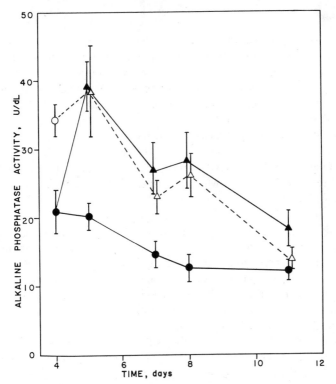

FIG. 6—*The variation in extracellular alkaline phosphatase activity in the exudate with implantation time. Control (closed circles and solid line), tripeptide control (closed triangles and solid line), PHEG + T (open triangles and dashed line), and PHEG at Day 4 (open circle).*

TABLE 4—*Extracellular acid phosphatase activity in the exudate as a function of implantation.*

Time, days	Control, SU[a]/dL	Tripeptide Control, SU/dL	PHEG + T, SU/dL
1	62.4 ± 17.9[b]	...	...
3	75.4 ± 11.7	...	109.5 ± 9.2[c]
4	86.8 ± 9.5	...	100.9 ± 7.0[c]
5	51.5 ± 4.7	65.8 ± 17.7	93.5 ± 21.2[c]
7	48.0 ± 5.7	48.6 ± 8.1	75.8 ± 8.8[c]
8	38.4 ± 5.7	71.0 ± 12.6[c]	73.3 ± 7.1[c]
11	21.5 ± 5.1	49.7 ± 7.5[c]	51.6 ± 15.7[c]
14	24.2 ± 5.0	35.0 ± 5.6	48.0 ± 12.6[c]
17	23.0 ± 4.6	23.5 ± 8.6	38.8 ± 3.7[c]
21	21.7 ± 5.0	22.9 ± 8.7	32.9 ± 5.6[c]

NOTE—Normal rat serum: 80.2 ± 4.9 SU/dL.
[a]SU = sigma units.
[b]Mean value ± standard deviation.
[c]Statistically different at the 95% confidence level ($p < 0.05$) when compared to the mean control value. Student's t-test for unpaired samples.

Discussion

The implantation of any biomaterial initiates a would healing mechanism that is initially characterized by the inflammatory response and followed by the reparative and remodeling phases. The inflammatory response involves a series of interdependent events that begin with hemodynamic changes, followed by alterations in vascular permeability and proteinaceous and leukocytic exudation. The intensity and duration of the response is controlled by various endogenous and autocoid mediators, and determined by the size and nature of the implanted biomaterial, the site of implantation, and the reactive capability of the host. Inflammation represents the host's principal effort to eliminate the foreign material, the reciprocal effects of which will largely determine the eventual success or failure of the biomaterial.

Implantation of the stainless steel cages provoked an acute inflammatory response, which was marked by high white cell concentrations, particularly neutrophils (Table 1). There appeared to be no significant difference in the inflammatory cell concentrations between control and PHEG exudates up to Day 8. However, examination of the respective exudates for extracellular alkaline phosphatase revealed significant differences for the PHEG system at Days 4, 5, and 7. The extracellular release of lysosomal enzymes from viable neutrophils is thought to occur through an exocytosis mechanism following cell-implant surface adhesion and cellular activation [8]. Thus, it appears that the PMNs in the PHEG exudate were experiencing a greater stimulus to release their granular constituents. This effect is illustrated by the elevated enzyme activity/PMN ratio for the PHEG system shown in Fig. 1. One may conclude that while the presence of PHEG within the cage probably does not affect the PMN migration from the vasculature, it does have a perturbing effect on leukocyte activation, which through the action of neutrophil products, may also account for the elevation in mononuclear leukocytes observed at later time points (Table 1).

If the presence of PHEG, within the cage implant, causes a perturbation of the inflammatory process, to what extent can a material within the cage ultimately modify the acute inflammatory reaction? The enhanced acute phase model effectively provides this positive control. With the possible exception of a proliferating bacterial infection within the cage, the results for this model system, using fMLP, provide what may be considered a maximal acute inflammatory phase response. The injection of the chemotactic tripeptide enhanced the intensity and duration of the acute inflammatory response by stimulating the migration of PMNs and to a lesser extent mononuclear leukocytes into the cage implant. Exudation into the implanted cages is the host's initial mechanism for filling the artificially created void, which is eventually replaced by connective tissue. Clearly, the inflammatory response to the implanted cage is much greater than one would observe following the simple subcutaneous implantation of a biomaterial film. Thus, the use of the cage implant effectively enhances and prolongs the biological interactions that are experienced by an implanted biomaterial. The use of the enhanced acute phase model takes this process a step further. By maximizing the concentration of PMNs present at the implant site, the whole inflammatory process is prolonged. In this study, the enhanced acute phase model was used to provide a positive control for acute inflammation.

Many of the exudate analyses for the tripeptide control and PHEG + T systems were similar, particularly for those related to the acute phase such as PMN concentrations and alkaline phosphatase activity. These similarities suggest the dominating effect of fMLP within the cage during the acute phase. However, both mononuclear leukocyte and extracellular acid phosphatase data showed differences (Fig. 5 and Table 4). Macrophages followed by PMNs possess the greatest intracellular amounts of acid phosphatases, although their relative amounts of extracellular release may differ. Thus, it remains difficult to associate the measured extracellular activity of acid phosphatase with any one particular leukocyte. However, if the extracellular acid phosphatase activity was solely a reflection of PMN release, one would expect the results for the tripeptide control to be greater than empty cage control results and to be approximately the same as the PHEG + T results. Table 4 shows that this was not entirely the case. The results indicate that the presence of PHEG in the cage increases the release of acid phosphatase from leukocytes, and that the differences between the tripeptide controls and the PHEG + T system may be attributed to macrophage activity. Therefore, the extracellular release of acid phosphatase appears to involve both PMNs and macrophages. The contribution from macrophages undoubtedly becomes the more dominant at later time points. The results appear to provide further evidence that PHEG elevates the chronic phase of inflammation and has a strong effect on acid phosphatase release extracellularly.

The enhanced acute phase model system may not only be utilized to provide a positive control, as in this study, but it may also be used to investigate the effects of a maximized acute inflammation on biomaterial properties. Any biomaterial that is suspected of deteriorating *in vivo* with time may be subjected to this enhanced acute inflammatory phase model, in order to possibly accelerate the material's deterioration. Thus, the dynamic two-way relationship between host and biomaterial, that determines biocompatibility, may be quantitatively investigated using the cage implant system and innovative modifications of the system.

Acknowledgments

The authors wish to thank Rimma Slobodkin for her assistance with the cell counts, Fran Cverna for her assistance with the enzyme analyses, and Bonnie Berry for her assistance with the preparation of the manuscript. This work was supported by the National Institutes of Health (NIH) under Grant No. HL-25239 and in part by a NIH Research Career Development Award, No. HL-00779, to J. M. Anderson.

References

[1] Gourlay, S. J., Rice, R. M., Hegyeli, A. F., Wade, C. W. R., Dillon, J. G., Jaffe, H., and Kulkarni, R. K., *Journal of Biomedical Materials Research*, Vol. 12, 1978, pp. 219–232.

[2] Turner, J. E., Lawrence, W. H., and Autian, J., *Journal of Biomedical Materials Research*, Vol. 7, 1973, pp. 39–58.

[3] Autian, J., *Artificial Organs*, Vol. 1, 1977, pp. 53–56.

[4] Marion, L., Haugen, E., and Mjor, I. A., *Journal of Biomedical Materials Research*, Vol. 8, 1974, pp. 199–211.

[5] Coleman, D. L., King, R. N., and Andrade, J. D., *Journal of Biomedical Materials Research*, Vol. 8, 1974, pp. 199–211.

[6] Coleman, D. L., King, R. N. and Andrade, J. D., *Journal of Biomedical Materials Research Symposium*, Vol. 5, No. 1, 1974, pp. 65–76.

[7] Rigdon, R. H., Tissue Reaction to Foreign Materials, *Critical Reviews in Food Science and Nutrition*, Vol. 7, 1975, pp. 435–476.

[8] Marchant, R. E., Hiltner, A., Hamlin, C., Rabinovitch, A., Slobodkin, R., and Anderson, J. M., *Journal of Biomedical Materials Research*, Vol. 17, 1983, pp. 301–325.

[9] Dickinson, H. R., Hiltner, A., Gibbons, D. F., and Anderson, J. M., *Journal of Biomedical Materials Research*, Vol. 15, 1981, pp. 577–589.

[10] Schiffmann, E., Corcoran, B. A., and Wahl, S. M., *Proceedings, National Academy of Sciences*, Vol. 72, 1975, pp. 1059–1062.

[11] Showell, H. J., Freer, R. J., Zigmond, S. H., Schiffmann, S., Aswankumar, S., Corcoran, B., and Becker, E. L., *Journal of Experimental Medicine*, Vol. 143, 1976, pp. 1154–1169.

[12] Becker, E. L., *American Journal of Pathology*, Vol. 85, 1976, pp. 385–394.

DISCUSSION

David Parins[1] *(written discussion)*—Is it possible that the second injection of chemotactic agent is modifying and capsule and altering the cellular response?

R. E. Marchant, T. Sugie, A. Hiltner, and J. M. Anderson (authors' closure)— We have no evidence that would support this suggestion.

Leukocytes are known to be refractory to repeated exposures of chemotactic peptides. See Refs *10* and *11*.

Ann C. VanOrden[2] *(written discussion)*—Can you adequately model the long-term response of an implant designed for a 20-year lifetime with an accelerated test?

R. E. Marchant, T. Sugie, A. Hiltner, and J. M. Anderson (authors' closure)— In polymer science and particularly in the rubber industry, accelerated chemical degradation methods using agents such as ultraviolet light, heat, or ozone have been successfully used to predict approximate material lifetimes or more precisely, the susceptibility of a material to a particular degradative agent. In biomaterials research, however, a major problem is really knowing which potential degradative agent to amplify for an accelerated test. Indeed, the biologic milieu in which implants reside is highly complex and although the same implant may

[1]Cardiac Pacemakers, St. Paul, MN 55112.
[2]National Bureau of Standards, Washington, DC 20234.

be implanted in the same anatomic position for the same period of time, the biological interactions with this implant may be highly variable.

With our enhanced inflammatory model several components of acute inflammation were enhanced. Thus, if a material was found to be susceptible to one or more components of this enhanced inflammation, this would suggest excluding the material as a possible candidate for long-term applications. As was suggested, it would not of course predict how many days or years the material would maintain its integrity.

While accelerated testing methods may adequately model long-term environmental effects on a material, we do not believe sufficient information exists to form the basis for accelerated tests where cellular and tissue interactions with the material may be important.

E. Horowitz[3] (*written discussion*)—Why can't you make the cage of polymeric materials to circumvent the use of stainless steel that may be introducing complicating factors? In the ideal case, the polymeric cage would be made of the same material as the polymer test sample.

R. E. Marchant, T. Sugie, A. Hiltner, and J. M. Anderson (*authors' closure*)—We have elected not to construct the cage from the same polymeric material as the polymer test sample for several reasons. The first involves appropriate controls for comparison between different materials. If a specific polymer cage were to be used in each instance, we would not have a common control to which to compare all materials tested in the cage. Secondly, constructing the cage from the same polymeric material would preclude the investigation of small amounts of new polymeric materials. Obviously, if a polymer were to be used for the cage construction it would have to meet certain physical and mechanical requirements as well as fabrication requirements. These requirements might not be readily attained in polymers that would be of interest because of their surface properties.

We have investigated the cages and the connective tissue capsule surrounding the cages in both control and polymer studies and find no evidence by light microscopy of variations in the capsule regarding its healing response or on the surface of the stainless steel mesh. This, of course, does not preclude the possibility that corrosion of the stainless steel is taking place, but if it is, we find no evidence that it has altered the inflammatory response or healing response. We have recognized the possibility that stainless steel may undergo corrosion, even for the short implant times employed in our studies, and are currently investigating titanium and carbon-coated titanium as substitutes for the stainless steel mesh.

Jonathan Black[4] (*written discussion*)—Perhaps the author is too modest in his defense of the "cage" implant system. The use of a metal cage might be

[3]John Hopkins University, Baltimore, MD 21218.
[4]University of Pennsylvania, Phila., PA 19104.

seen, rather than as a deficit, as a virtue in that it provides a "standard inflammatory environment" in which the host/material response of polymers may be examined without the complication of polymer-tissue mechanical interactions.

R. E. Marchant, T. Sugie, A. Hiltner, and J. M. Anderson (authors' closure)— We appreciate Dr. Black's comments. We have continually pointed out that implants are initially subjected to an enhanced inflammatory reaction initiated by the surgical procedure itself. This effect is rarely considered by individuals using *in vitro* model systems to investigate the biocompatibility of candidate materials. We agree with Dr. Black that the cage system provides a "standard inflammatory environment" in which the host/material response of polymers may be examined without the complication of polymer-tissue mechanical interactions.

Ron Yapp[5] (written discussion)—Do I understand then that the purpose of the cage is to limit the relative motion between the polymer implant and surrounding tissue?

R. E. Marchant, T. Sugie, A. Hiltner, and J. M. Anderson (authors' closure)— The cage serves two major purposes. The first is to limit the relative motion between the polymeric implant and the surrounding tissue. Secondly, the cage prevents the surrounding tissue from collapsing onto the test specimen and thus permits the inflammatory exudate that forms within the cage and surrounds the test specimen to be withdrawn and analyzed.

[5]Carbomedics, Austin, TX 78752.

Alfred B. Swanson,[1] *B. Kent Maupin,*[2] *Robert M. Nalbandian,*[3] *and G. deGroot Swanson*[4]

Host Reaction to Silicone Implants: A Long-Term Clinical and Histopathological Study

REFERENCE: Swanson, A. B., Maupin, B. K., Nalbandian, R. M., and deGroot Swanson, G., "Host Reaction to Silicone Implants: A Long-Term Clinical and Histopathological Study," *Corrosion and Degradation of Implant Materials: Second Symposium, ASTM STP 859,* A. C. Fraker and C. D. Griffin, Eds., American Society for Testing and Materials, Philadelphia, 1985, pp. 267–277.

ABSTRACT: Host tolerance to implant materials has been evaluated since the initiation of a research project for use of silicone for joint replacement since 1962. Both the local and systemic tolerance to implants was analyzed in biomechanical and biological studies in laboratory animals and humans. Both soft tissues and bone reactions have been evaluated by pathological and radiological studies. The early studies included implant procedures on 25 dogs. Three dogs were sacrificed after 10 years and complete autopsies were performed to evaluate the local and systemic biocompatibility. End-bearing amputation stump implants were inserted in two of these dogs and the third animal had implant capping of ribs at the costovertebral junction and a silicone active tendon in a hind leg. A smooth-surfaced fibrous capsule surrounded the implants with no evidence of bone absorption or bursa formation; small silicone wear particles were well tolerated by giant cells. A complete organ and reticuloendothelial system review failed to show evidence of metastases of silicone particles. A similar, excellent biocompatibility of silicone implants was demonstrated in the autopsy of a rheumatoid patient who had silicone implants in both hands, radial heads, and feet starting 12 years before death. A few, small, well-tolerated silicone particles were seen in giant cells in the synovium with minimal inflammatory cells and no focal necrosis. A small amount of well-tolerated silicone particles were present in giant cells in an axillary node. A thorough organ study failed to show distant silicone particle metastases. These studies showed the good biocompatibility of silicone implants.

KEY WORDS: host tolerance, implant materials, silicone, wear particles, foreign body giant cell, fibrous capsule, biological degradation, fatigue (materials)

[1]Director of the Grand Rapids Hospital's Orthopaedic Surgery Training Program, Director of Hand Surgery Fellowship and Orthopaedic Research, Blodgett Memorial Medical Center, Grand Rapids, MI 49506; and professor of Surgery, Michigan State University, Lansing, MI.
[2]Orthopaedic surgeon, Blodgett Memorial Medical Center, Grand Rapids, MI 49506.
[3]Pathologist, formerly of King Fahad Hospital at Al Baha, Al Baha, Saudi Arabia.
[4]Plastic surgeon and coordinator of Orthopaedic Research, Blodgett Memorial Medical Center, Grand Rapids, MI 49506; and assistant clinical professor of Surgery, Michigan State University, Lansing, MI.

The development and use of medical-grade, silicone elastomer as an implant material has allowed significant advances to be made in reconstructive surgery. Since 1962, Dr. A. B. Swanson has been carrying out a research project for development of flexible silicone implants for reconstruction of the small joints in the extremities. These investigations have included design, development, characterization of physical and biochemical properties, biological responses in animal and human recipients, evolution of surgical techniques, and post-surgical management regimens [1].[5] The study of the host response to implanted silicone has been an ongoing project for two decades in the Orthopaedic Research Department of Blodgett Memorial Medical Center.

Procedure

The autopsy findings in three dogs in which silicone implants had been maintained for at least 10 years and in one female patient who had multiple silicone implants 12 years prior to her death are reported here. The characteristics of the host response to silicone will be described and compared to those observed with other implant materials.

Since 1963, laboratory animal studies have been carried out in the Orthopaedic Research Department of Blodgett Memorial Medical Center, Grand Rapids, MI. Early experiments involved the insertion of silicone elastomer amputation stump implants into both front and one hind leg of the dogs [1]. One hind leg of each dog was left intact for proprioception. These implants are mushroom shaped and consist of a cylindrical stem that is inserted into the marrow canal of the amputated tibia or radius, and a broad, flattened, convex head that forms a cushion between the end of the bone and the overlying soft tissue of the amputation site.

Another dog had two different surgical procedures. In 1967, a tendon in the right lower leg was replaced by an active tendon implant consisting of a silicone elastomer tube, anchoring devices at either end, and a Dacron thread, which passed through the tube. Two years later, silicone elastomer ulnar head (bone cap) implants were inserted into osteotomies near the neck of Ribs 9 and 11 on the right side.

The dogs were followed with regular veterinary supervision. Progress notes were recorded on the systemic condition of the dogs, and on the use and local appearance of the operated limbs. Radiograms of the implant sites were taken at regular intervals. Three dogs were maintained for 10 years with the implants *in situ* to observe the long-term host response to the implant material. Two of the dogs had the amputation stump implants and the third dog had a tendon graft implant and ulnar head implants in the ribs.

The aims of these studies included improvement of the end-bearing characteristics of amputated limbs; a study of the reaction of the implant material to stress *in vivo*; and a study of the host reaction to the implant material, both locally and systemically under conditions of weight bearing and activity.

---

[5]The italic numbers in brackets refer to the list of references appended to this paper.

The necropsy examination of the first two dogs was performed by a veterinary laboratory specializing in autopsies of scientific animals, and their histo-pathological examination was supplemented by further review and examination by the Pathology Department of Michigan State University. The necropsy of the third dog was performed by Michael L. Chandler, Ph.D., Group Leader of Medical Products Research, Dow Corning Corporation, Midland, MI.

Representative sections from the following organs and tissues were fixed in 10% neutral, buffered formalin (eyes and gonads fixed in Bouin's solution): adrenal glands, aorta (thoracic), bone marrow (sternum), brain (cerebrum, cerebellum, and pons), caecum, colon, esophagus, eyes, gallbladder, gonads, heart, implant sites and adjacent bone, kidneys, liver, lungs, lymph nodes (cervical, mesenteric), muscles, optic nerve, pancreas, peripheral nerve (sciatic), pituitary gland, prostate gland, salivary gland (submaxillary), skin, small intestine (duodenum, jejunum, ileum), spinal cord, spleen, stomach (cardia, fundus, pylorus), trachea, thyroid gland, tongue, and urinary bladder.

Tissues were processed by standard techniques, cutting sections at 4 to 6 μm, and staining with hematoxylin and eosin, and other special stains for light microscopy.

Clinical Case

A 58-year-old female was first seen in 1968 with a 10-year history of rheumatoid arthritis. Initially, only her hands were involved. The disease progressed over the next five years and affected both knees, feet, ankles, and elbows, in addition to increased hand disabilities.

From June 1968 until December 1975 she had reconstructive procedures on six occasions. On 29 Feb. 1980, she died of a myocardial infarct. This patient had 14 silicone implants in her hands, feet, and elbows as well as metal, polyethylene, and methylmethacrylate materials in her knees. A complete autopsy was done, which included special studies of her hands.

Results

The findings on the laboratory animals and the human autopsy will be presented separately.

Animal Findings

Clinical Course

The three dogs survived to an old age, in good health, and without clinical evidence of intolerance to the implants. There was no case of inflammation due to tissue intolerance to the silicone elastomer implants.

There were no implant-related complications in the dog with the active tendon and ulnar head implants.

In the six dog legs with the amputation stump implants there were recurring complications, the most common being superficial or deep infection with either acute or chronic episodes occurring immediately after operation or with a latency period of weeks, months, or years; this usually followed a laceration or abrasion of the stump skin during weight bearing. These complications did not always lead to removal of the implants. Serial X-rays taken over the 10 year period showed a more intimate growth of bone to the intramedullary stem, and evidence of good bone remodeling and thickening, without evidence of bone resorption.

Pathology

Implant Site—Rib Implant—Both silicone elastomer rib capping implants were held in proper position by sheaths of connective tissue that appeared to be smoothly contiguous with the normal tissue surrounding each rib. The incised capsule was glistening and the implant stem rotated freely within the rib (Fig. 1).

Histologically, the implant capsule covering the implant head was composed of mature, well stratified connective tissue, 300 to 400 μm thick. The inner layer was strongly hyalinized with only sparse fibrocytes present. Around the implant stem within the rib, the capsule was likewise composed of mature, connective tissue, somewhat more fibrous in its deeper layers than it was above the capsule. Fibers giving anchorage to the surrounding bone arose from the capsule. While most of the capsule interface was with the cancellous bone, in

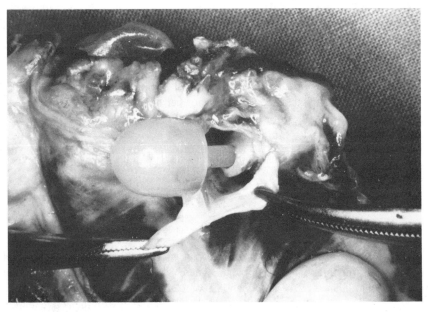

FIG. 1—*Photograph of rib cap implant at removal after 10 years. Note smooth capsule that tends to blend with periosteum of rib.*

several areas the capsule was contiguous with marrow spaces containing normal cellular populations. Careful study of the implant capsule and surrounding tissues by darkfield-phase contrast microscopy failed to reveal any strongly refringent particles indicative of "wear debris" from the silicone elastomer implant.

Implant Site—Tendon Device—The excised anterior tibial tendon had regrown along the course of, and partly surrounded, the implant. From the relative tension, it appeared that the regrown tendon was the actual load-bearing component. The implant was firmly attached to muscle and bone and had complete freedom for longitudinal motion.

Histologically, tissue ingrowth had occurred within the silicone elastomer tube near its proximal end. A cross-section of the Dacron suture, interpenetrated by very mature, hyalinized, fibrous tissue could be seen surrounded by a gap, representative of the silicone elastomer tube; the outer concentric, thin fibrous capsule also consisted of mature, fibrous tissue, 50 to 100 μm thick. A large, isolated pocket of Dacron fibers was seen surrounded by an immature, fibrous capsule with modest inflammatory cell infiltration. In ordinary light, the Dacron fibers appeared clear/pale yellow-to-brown, depending on their thickness; these fibers were moderately refractile with the use of darkfield-phase contrast viewing. The few scattered silicone elastomer particles found in the capsules were highly refractile.

Implant Site—Amputation Stump Implants—The amputation stump implants were covered by a thickened, loose skin without evidence of bursa formation.

FIG. 2—*Photomicrograph of capsule around amputation stump implant. Scattered particles of silicone can be seen. In some areas, a mild inflammatory response is noted* (center right).

On gross examination, a smooth-surfaced non-adherent pseudo-capsule surrounded the implants and was continuous with the periosteum at the distal end of the radius. This capsule was made of stratified longitudinal layers of connective tissue and was lined by a single layer of mesothelial cells. There were scattered foci of lymphocytes, macrophages, and multinucleated giant cells within the connective tissue stroma. The macrophages and giant cells were situated in the immediate vicinity of fragments of silicone elastomer that appeared as colorless to yellow, globular, amorphous particles varying from subcellular size to 100 μm or more (Fig. 2). Some of the smaller particles could be identified within the cytoplasm of macrophages and giant cells (Fig. 3). There was no necrosis. A similar mild, macrophage giant cell response to fragments of silicone elastomer was noted in the medullary cavity adjacent to the implant.

Reticuloendothelial System—No silicone elastomer particles or macrophage giant cell reaction could be found in a thorough examination of the regional and mediastinal lymph nodes, the liver, or the lungs.

Distant Organs—Complete evaluation of the organs of the central nervous, cardiovascular, endocrine, and genitourinary systems failed to reveal any traces of silicone particles.

Human Autopsy

The examination of the hands showed the implants to be completely contained within a fibrous capsule. The internal surfaces of the capsule surrounding the

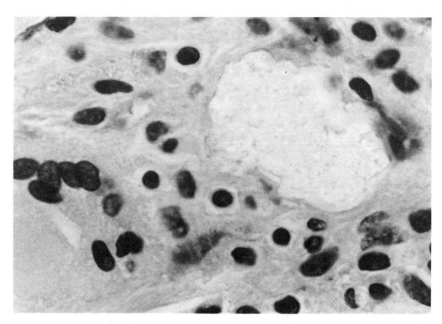

FIG. 3—*High-power photomicrograph of intracellular silicone particle within a foreign body giant cell.*

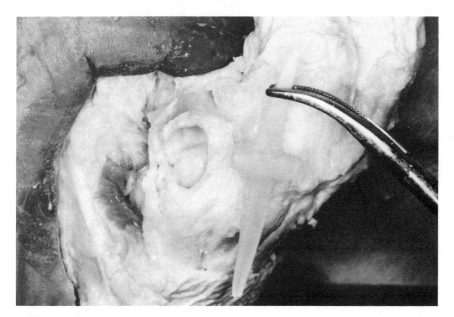

FIG. 4—*Photograph of implant from the metacarpophalangeal joint of a patient after 10 years. Note the smooth capsule that had surrounded the implant.*

finger implants were extremely smooth, shiny, and white with no evidence of inflammatory reaction. The metacarpophalangeal implants were intact (Fig. 4). The proximal interphalangeal joint implants of the third and fourth digits of the right hand were torn at their midsections. A complete duplicate set of tissue slides and protocol was evaluated by three pathologists at two different institutions. The interesting findings relating to this study were noted at the level of the digits and a single enlarged right axillary lymph node. Under microscopic examination, a mild synovitis was found at the level of the third and fourth proximal interphalangeal joints only. A few areas of macrophage response to silicone particles could be identified with some fragments located in giant cells without necrosis (Fig. 5).

A pronounced foreign body reaction to a Dacron suture could also be identified. The single enlarged axillary lymph node showed a benign foreign body giant cell reaction to silicone particles, without disruption of the overall lymph node architecture, nor any evidence of acute inflammation, microabscesses, or malignancy. A thorough organ study showed no evidence of systemic dissemination of silicone particles in organ tissues.

Discussion

The results of these long-term histopathological studies agree with published studies emphasizing the biocompatibility of silicone elastomer. Previous reports have suggested the possibility that symptoms arising from local synovitic reactions in finger joints and regional lymph nodes were due to silicone elastomer

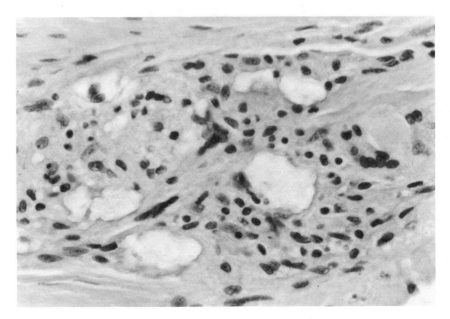

FIG. 5—*High-power photomicrograph of a focus of foreign body reaction to silicone in our patient's synovium.*

debris [2–5]. We feel that particles could aggravate an inflammatory reaction without necessarily initiating it as there is no clear separation of occurrence from local recurrent rheumatic synovitis. A mild giant-cell response to silicone particles was found in a single enlarged axillary node in our human study; no such lymphadenopathy was observed in the three dogs. In more than 5000 implant arthroplasties of the upper extremity, we have seen only three cases of particulate silicone in regional lymph nodes [6–8]. It is reassuring to note that there was no systemic dissemination of particulate silicone in either the dogs or the human studies.

It should be noted that the elastomer implants used in this study were made of Silicone No. 372, featuring a much lower tear propagation strength (75 ppi) than the newer medical-grade high-performance silicone elastomer (400% greater tear resistance), which has been used for implant fabrication since 1974.

This was a particular disadvantage in this study as the end bearing stump implants were subjected to continual weight bearing and unusual stress. However, our previous studies showed no correlation between lipid uptake and implant failure [9].

In an attempt to protect the silicone hinge-type implants from cutting by sharp bone ends, as can occur in severe rheumatoid arthritis, and to decrease the amount of wear debris, we have developed the use of grommets to line the cut bone. Several materials have been studied, including polyethylene, Proplast, pyrolytic carbon, stainless-steel screen, cobalt-chromium, and titanium. Significant wear

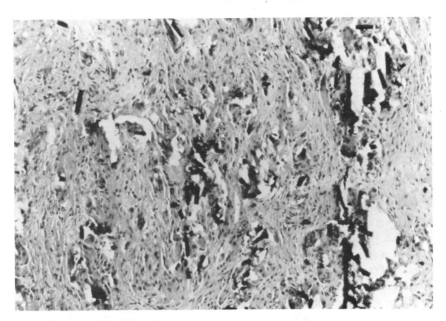

FIG. 6—*Photomicrograph of intense tissue reaction to the fragmented carbon particles.*

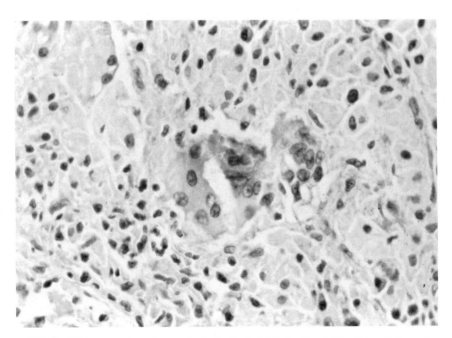

FIG. 7—*High-power photomicrograph of a foreign body giant cell containing intracellular methylmethacrylate.*

debris and fragmentation occurred with the polyethylene, Proplast, and pyrolytic carbon grommets. The stainless-steel screen was prone to fragmentation and laceration of the silicone. The titanium grommet has provided excellent implant protection and favorable bone remodeling. Our experimental work and clinical experiences have demonstrated that particulate wear debris can be generated from the surfaces of any implanted joint materials, in greater or lesser amounts [10–14]. The tissue response is determined by the physical form and physiochemical properties of the implant material, and also depends on the rate of wear and the duration of implant debris presence in the tissue (Figs. 6 and 7). It would not be surprising if some small fragments of silicone elastomer could embolize to distant sites via lymphatic drainage systems without untoward effects, as this has been noted with other implant materials such as high-density polyethylene following total hip replacement [15].

Clinically successful implant materials must exhibit an acceptable level of biological compatibility. They should not be toxic locally, nor stimulate significant inflammatory nor foreign body reactions, nor be noxious systemically to tissues or organs by way of biodegradation products (such as wear debris or reactive products), nor induce hypersensitivity or carcinogenic changes.

The long-term histopathological review supports the view that medical-grade silicone elastomer is an inert material that is well tolerated by the host tissues. It produced no local or systemic toxic effects. There were no embolic particles of silicone elastomer in distant tissues and organs. Silicone elastomer wear particles can be found free or within giant cells, and can participate in a reactive synovitis. This study confirms the results of other investigators that silicone elastomer is no more injurious to tissues than other, generally accepted implant materials.

References

[1] Swanson, A. B., Flexible Implant Arthroplasty in the Hand and Extremities, The C. V. Mosby Co., St. Louis, 1973.
[2] Aptekar, R. G., Davie, J. M., and Cattell, H. S., Clinical Orthopaedics and Related Research, Vol. 98, 1974, pp. 231–232.
[3] Christie, A. J., Weinburger, K. A., and Dietrich, M., Journal, American Medical Association, Vol. 237, No. 14, 1977, pp. 1463–1464.
[4] Gordon, M. and Bullough, P. C., Journal of Bone and Joint Surgery, Vol. 64A, 1982, pp. 574–580.
[5] Kircher, T., Human Pathology, Vol. 11, 1980, pp. 240–244.
[6] Swanson, A. B., Letter to Editor, Journal, American Medical Association, Vol. 238, 1977, p. 939.
[7] Nalbandian, R. M., Letter to Editor, Journal, American Medical Association, Vol. 238, 1977, p. 939.
[8] Nalbandian, R. M., Letter to Editor, Journal of Bone and Joint Surgery, Vol. 65A, 1983, pp. 280–281.
[9] Swanson, A. B., Meester, W. D., deGroot Swanson, G., Rangaswamy, L., and Schut, G. E. D., Orthopaedic Clinics of North America, Vol. 4, 1973, pp. 1097–1112.
[10] Buchorn, G. H. and Willert, H. G. in Biocompatibility of Orthopedic Implants, F. Williams, Ed., Vol. 1, CRC Press, Inc., Boca Raton, FL, 1983, pp. 249–270.
[11] Jenkins, D. H. R., Journal of Bone and Joint Surgery, Vol. 60B, 1978, pp. 520–522.

[*12*] Mirra, J. M., Marder, R. A., and Amstutz, H. C., *Clinical Orthopaedics and Related Research*, Vol. 170, 1982, pp. 175–183.

[*13*] Salvati, E. A., Bullough, P., and Wilson, P. D., *Clinical Orthopaedics and Related Research*, Vol. 111, 1975, pp. 212–227.

[*14*] Willert, H. G. and Semlitsch, M., *Journal of Biomedical Materials Research*, Vol. 11, 1977, pp. 157–164.

[*15*] Charnley, J., *Low Friction Arthroplasty of the Hip*, Springer-Verlag, Berlin-Heidelberg-New York, 1979, pp. 330–331.

DISCUSSION

Manfred Semlitsch[1] *(written discussion)*—Have the silicone elastomer implants of the lady been investigated for rate of degradation and decrease of flexibility after more than 10 years implantation? Discoloration is the first sign of degradation, leading to fracture of the joint implants made of silicone elastomer after three to five years.

A. B. Swanson, B. K. Maupin, R. M. Nalbandian, and G. deGroot Swanson (authors' closure)—The preceding question and comment on biodurability of silicone elastomer implants are not directly related to the topic of the article presented, which is host tolerance. The silicone elastomer implants retrieved from the patient after 12 years of implantation were not studied for biodegradation because they were made of the original elastomer material that has not been used for implant fabrication since 1974. Furthermore, no standard test methods are available for irregular microspecimens. A previous study of the original elastomer material was done in dogs using ASTM methods showing no evidence of degradation after two years of implantation.[2] Mr. Semlitsch's statement is incorrect, both from 20 years of clinical experience and from the scientific data currently available. A detailed study on biodurability of high-performance silicone elastomer is presented by Frisch in this current ASTM Symposium[3] and has also been previously reported.[4] Silicone elastomer presents color variations naturally. This typically becomes more pronounced after implantation because of moisture and lipid absorption.[5] However, detailed previous studies have shown that there is no correlation between the rate of lipid absorption, implant fracture, and duration of implantation.

[1]Sulzer Bros. Ltd., Department R & D, CH-8401 Winterthur, Switzerland.
[2]Swanson, J. W. and Lebeau, J. E., *Journal of Biomedical Materials Research*, Vol. 98, 1974, pp. 357–367.
[3]Frisch, E. E. and Langley, N. R., "Biodurability Evaluation of Medical-Grade High-Performance Silicone Elastomer," in this volume.
[4]Swanson, A. B., deGroot Swanson, G., and Frisch, E. E. in *Biomaterials in Reconstructive Surgery*, L. R. Rubin, Ed., The C. V. Mosby Co., St. Louis, 1983, pp. 595–623.
[5]Swanson, A. B., Meester, W. D., deGroot Swanson, G., Rangaswamy, L., and Schut, G. E. D., *Orthopaedic Clinics of North America*, Vol. 4, 1973 Extremities, pp. 1097–1112.

Eldon E. Frisch[1] and Neal R. Langley[1]

Biodurability Evaluation of Medical-Grade High-Performance Silicone Elastomer

REFERENCE: Frisch, E. E. and Langley, N. R., "**Biodurability Evaluation of Medical-Grade High-Performance Silicone Elastomer,**" *Corrosion and Degradation of Implant Materials: Second Symposium. ASTM STP 859*, A. C. Fraker and C. D. Griffin, Eds., American Society for Testing and Materials, Philadelphia, 1985, pp. 278–293.

ABSTRACT: A new silicone elastomer provides maximum resistance to flexural fatigue, crack growth, and the high durability required in an elastomer used as a material of construction for the flexible-hinge finger and toe joint implants designed by A. B. Swanson, M. D. The high-performance characteristics of this new elastomer are derived from its excellent physical properties, particularly its exceptionally high tear propagation strength (ASTM Test for Tear Resistance—Rubber Property (D 624-73), Die B) and high resistance to crack propagation (ASTM Test for Rubber Deterioration—Crack Growth (D 813-59)). *In vitro* testing indicated its use in finger joint implants would provide excellent flexural durability. Biocompatibility evaluations to qualify this new material for clinical use included chronic biodurability studies. Test specimens were implanted subcutaneously in dogs and retrieved at various time intervals for periods up to two years. Physical properties of test and control specimens were compared. The properties measured included tensile strength, tensile stress, ultimate elongation, tear strength (Die B), and flaw propagation characteristics. All testing was done by ASTM standard methods except for flaw propagation measurements where implantability limitations in specimen size mandated use of a smaller specimen than required in standard test methods. Thus, a flaw propagation test method was specifically designed for this experiment.

The findings indicated that this new high-performance silicone elastomer had excellent biodurability. The performance characteristics evaluated in test and control specimens were stable over the two-year implant time. Lipid absorption stabilized at approximately 1.5%, a level reached relatively early after implantation and remaining constant thereafter, apparently representing the lipid saturation level when this elastomer is implanted subcutaneously in the dog.

KEY WORDS: silicone elastomer, high-performance elastomer, flexible-hinge finger joint, flexible-hinge toe joint, tear propagation strength, crack propagation resistance, flexural durability, physical properties, flaw propagation resistance, biocompatibility, biodurability, lipid absorption, implant durability, *in vitro* testing, clinical testing, subcutaneous implantation, implant materials

Silicones are a generic class of materials with polymer chains of alternating silicon and oxygen atoms. Two organic ligands are bonded to essentially all silicon atoms (polydiorganosiloxane). The silicone generic class includes sub-

[1]Associate scientist, Health Care Products, and associate research scientist, respectively, Dow Corning Corporation, Midland, MI 48640.

stances that vary widely in molecular structure and composition, including formulated products that may contain mixtures of silicones and non-silicone ingredients such as fillers or other additives. Material types include fluids, antifoams, adhesives, elastomers, resins, and a wide variety of formulated compounds and greases. The stability inherent in most silicones is derived from the high energy of the covalent structural bonds. Silicones have become widely used in selected, demanding materials applications of our modern society.

Toxicological and biocompatibility studies initiated in the 1940's prior to commercial distribution of these new materials for other than applications related to military equipment and machinery, indicated that certain polydimethylsiloxanes and mixed poly(dimethyl and phenylmethyl)-siloxanes as a class were very low in toxicity and had a high degree of biocompatibility. When the findings regarding this inherent excellent biocompatibility were published [1],[2] an interest was generated in using silicones in health care applications, primarily in silicone elastomers for use of materials of construction for medical devices and artificial internal organs.

Thermosetting silicone elastomers typically contain high molecular weight polydimethylsiloxane (molecular weight greater than 300 000) compounded with high surface area fumed silica (approximately 400 m^2/g). Fumed silica (pure SiO_2) is the only material known that adequately reinforces silicone elastomer to provide strong, highly elastic materials. Polymer and silica are thoroughly mixed. This is done with high energy mixing equipment to produce a homogenous blend. The silica becomes coated and impregnated with silicone polymer in the mixing.

Fabrication includes a hot vulcanization step that cross-links polydimethylsiloxane chains into a polymeric, elastomeric, chemically bonded, network matrix. In one common type of cross-linking, silicon-hydrogen ligands (present as trace amounts of methylhydrogensiloxy units copolymerized into polymer that is otherwise polydimethylsiloxane) react with silicon-vinyl ligands (present as trace amounts of methylvinylsiloxy units copolymerized into a second polymer that is otherwise polydimethylsiloxane). The two polymers are intimately blended with catalyst and other ingredients as the elastomer formulation is prepared. Typical catalysts include trace quantities of rare metals such as platnium. When heated, cross-linking (vulcanization) occurs. The cross-links are dimethylene radicals covalently bonded between silicon atoms of separate polymer chains. Organic peroxides may also be used as vulcanization catalysts.

Holter's [2] successful development of the silicone elastomer hydrocephalus shunt in 1955 heralded the era of implant reconstructive surgery. A wide variety of silicone elastomer implants have subsequently been developed to facilitate or improve reconstructive surgery. By the mid 1960's, implants made from silicone elastomer were used in most surgical specialties including plastic [3–6], orthopaedic [7–11], gynecologic [12], ophthalmic [13–15], and others. Three grades of heat-vulcanized medical-grade silicone elastomer had been developed (firm,

[2]The italic numbers in brackets refer to the list of references appended to this paper.

medium, and soft) by the mid 1960's. Hardness variations were accomplished primarily by the amount of silica contained in the formulations.

Of the many implants developed during this era, the silicone elastomer finger joint implant designed by Swanson [7–11] represented an important advancement in the armamentarium available to hand surgeons. This implant has two intramedullary stems extending both distal and proximal from a centrally located flexible hinge-spacer. Swanson has also designed similar double-stemmed flexible hinged implants for reconstruction of wrist and toe joints, and a wide variety of single-stemmed joint spacer implants for reconstruction of the trapeziometacarpal joint, the proximal head of the radius, the base of the phalanx in the great toe, and a variety of other small joints in both upper and lower extremities.

Initially, flexible bone and joint implants were made from medium-hardness medical-grade silicone elastomer. Its safety as an implant material had become well-established by studies done in test animals and by history of extensive clinical use. In a biodurability study [16], test specimens were implanted subcutaneously in dogs and evaluated at various time intervals for up to 104 months. It was found that there were essentially no differences in the physical properties of test and control specimens, that no materials were released from the specimens to tissue or systemic circulation, biodegradation was not detected, and lipid absorption was about 1.5% by weight.

Swanson [17] evaluated the durability of silicone elastomer in finger joint implants retrieved from clinical cases after various periods of time, and found no evidence of biodegradation and no correlation between implant performance and lipid context.

Weightman [18] constructed a hypothetical model to suggest that lipid absorption could weaken and possibly degrade silicone elastomer, but neither the findings from his laboratory studies, nor the studies of others, supported his hypothesis.

High-Performance Silicone Elastomer

In the early 1970's, technology was developed in Dow Corning's silicone elastomer research laboratories that permitted development of a silicone elastomer with high-performance physical properties. This new technology provided elastomers with a combination of low modulus and excellent physical properties, particularly high tear propagation strength (ASTM Test for Rubber Property—Tear Resistance (D 624-73), Die B) and very low flexural fatigue crack growth rates (ASTM Test for Rubber Deterioration—Crack Growth (D 813-59), characteristics deemed highly important to the durability of flexible-hinge implants. The mode of fracture failure in flexible-hinge implants in clinical use was identified as propagation of flaws made in implants by articulation against bone. Thus, it was reasonable to believe that an elastomer with low-modulus, high-tear-propagation strength, high resistance to flexural fatigue crack growth, and

TABLE 1—*Typical physical properties of medical-grade high-performance and medium-hardness silicone elastomer.*

Property	Method	High Performance	Medium Hardness
Tensile strength	ASTM D 412–75	1.0343×10^7 Pa	8.274×10^6 Pa
Ultimate elongation	ASTM D 412–75	700%	450%
Modulus at 100% elongation	ASTM D 412–75	2.0685×10^6 Pa	1.3790×10^6 Pa
Tear initiation strength, Die C	ASTM D 624–73	5.3569×10^3 kg/m	varies widely
Tear propagation strength, Die B	ASTM D 624–73	5.3569×10^3 kg/m	1.3392×10^3 kg/m
Crack growth, 10^6 cycles	ASTM D 813–59	2.7 mm (0.108 in.)	1459 mm (57.3 in.)
Durometer, Shore A	ASTM D 2240–75[a]	52	extrapolated 50
Specific gravity	ASTM D 792–66[b]	1.15	1.14

[a] ASTM Test for Rubber Property—Durometer Hardness (D 2240–75).
[b] ASTM Test Methods for Specific Gravity and Density of Plastics by Displacement (D 792–66).

an excellent profile of physical properties would improve clinical flexural durability of flexible-hinge implants.

This technology was used to prepare a new medical-grade "high-performance" silicone elastomer. All ingredients in this new material were specifically selected based on considerations of safety when contained in a material of construction for implants. The final elastomer contained no types of chemical bonds not also found in conventional medical-grade silicone elastomer.

A comparison of the physical properties of high-performance medical-grade silicone elastomer and conventional medium-hardness medical-grade silicone elastomer are shown in Table 1. The four-fold increase in tear propagation strength (ASTM D 624-73, Die B), 1.339 kg/m versus 5.357 kg/m (75 lb/in. versus 300 lb/in.), resulted in an increase in crack growth resistance (ASTM D 813-59) by a factor of approximately 600. The growth of the initial 2.0 mm (0.080 in.) long cut in specimens prepared from medium-hardness medical-grade silicone elastomer, when tested by ASTM D 813-59, typically exceeded 12.7 mm (0.5 in.) after only 7333 cycles. The original material had an extrapolated average crack growth rate equal to approximately 1.46 m (57.3 in.)/10^6 cycles. By comparison, the length of the original 2.0 mm (0.080 in.) cut in specimens prepared from medical-grade high-performance silicone elastomer, tested by ASTM D 813-59 consistently remained less than 12.7 mm (0.5 in.) in all specimens after 10^6 cycles. With medical-grade high-performance silicone elastomer, the average length of the initial 2.0 mm (0.080 in.) cut was 4.7 mm (0.188 in.) at 10^6 cycles, an average growth of 2.7 mm (0.1 in.)/10^6 cycles.

Testing by ASTM D 813-59 was implemented as part of the batch-to-batch quality assurance program for medical-grade high-performance silicone elastomer with a requirement that the initial 2.0 mm (0.080 in.) cut must remain less than 12.7 mm (0.5 in.) after 10^6 cycles in all 12 specimens tested simultaneously in order to assure its flexural durability.

The flex life of Swanson-designed finger joints prepared from conventional medical-grade medium-hardness silicone elastomer and medical-grade high-performance silicone elastomer were compared. Flexural durability testing was done in a machine that flexed the implants through a 90° arc (0° to 90° and back to 0°) at a rate of 16.67/s (1000/min.). When unflawed implants were tested, the number of flexes required to initiate a flaw in both materials was essentially equal, exceeding 150 000 000 cycles. However, with implants flawed prior to testing by a 1.57 mm (0.062 in.)-long through-and-through cut made in the center of the hinge perpendicular to the long axis, medical-grade high-performance silicone elastomer flexed more than 100 times longer prior to failure than conventional medical-grade medium-hardness silicone elastomer. Flawed implants made from the conventional material typically had separated completely at 90 000 cycles, while the cut in the implants made from high-performance elastomer demonstrated some cut growth at 9 000 000 cycles, but the implants had not separated. Thus, the *in vitro* studies indicated the flexural durability of flexible-hinge finger joints could be substantially increased if fabricated from this new material.

Biocompatibility Evaluations

To qualify the high-performance elastomer for clinical use with humans, biocompatibility evaluations were conducted. The assessments included acute and subacute tissue reaction in rabbits and mice as described in USP XIX for Class VI plastics. Subchronic tests were also done by placing samples intramuscularly and subcutaneously in rabbits for 7, 30, and 90 days.

Histopathological evaluations of the tissues were then carried out. No adverse reactions were found in these acute, subacute, and subchronic tests. The material was also evaluated by direct contact cell culture testing done with WI-38 human embryonic lung cells [19,20]. High-performance silicone elastomer elicited no cytopathic reaction.

Chronic biocompatibility was evaluated in a two-year or lifetime study in albino rats. Four groups of 100 animals each (50 males and 50 females) were used. One group each received identically shaped implant specimens of medical-grade high-performance silicone elastomer, conventional medical-grade medium-hardness silicone elastomer, or USP polyethylene, respectively, implanted intramuscularly, subcutaneously, interperitoneally, and in bone marrow. The fourth group received sham surgery only. All animals that died before two years were appropriately assessed. At the end of two years, all surviving animals were sacrificed. All tumors were evaluated and classified. This study showed that there were no detectable differences in the biocompatibility of the three implant substances studied and that the reactions in all of the implant groups were essentially the same as in the sham surgery group.

Biodurability Evaluation

Assessment of biodurability is an important consideration with any material proposed for use as a material of construction for permanent implants, and is equally important and probably closely related to chronic biocompatibility characteristics.

Biodurability of medical-grade high-performance silicone elastomer was evaluated by subcutaneous implantation of test specimens in dogs for periods of up to 104 months. Evaluations done on test and control specimens included measurements of tensile strength, ultimate elongation, and tensile stress (modulus) at 200% elongation (all by ASTM Tests for Rubber Properties for Tension, D 412-75), and tear propagation strength (ASTM D 624-73, Die B). Other studies included testing for fatigue flaw propagation, weight changes, and chemical analysis of solvent extracted materials. The study started with 20 sets of test specimens, all prepared from the same lot of elastomer, each set consisting of five tensile strength specimens, five Die B tear strength specimens, and ten flaw propagation specimens. Thus, all reported physical property values are based on an average of five measurements, except for flaw propagation values, which are based on an average of ten measurements. All physical property testing was done by ASTM methods except for the fatigue flaw propagation test. Because of its relatively large size, it was deemed impractical to implant the test specimens prescribed in ASTM D 813-59. A flaw propagation test with a smaller specimen was specifically designed for this experiment and was not a standard test method. In flaw propagation testing, specimens were uniformly tested for fatigue flaw propagation characteristics by cyclic and progressive elongation of a small specimen with a central cut (Fig. 1). The end-point of the test was measured as a number of cycles at which separation of the specimen occurred at the location of the central cut.

All specimens were sterilized with ethylene oxide at the start of the study, and base-line values for all tests were determined. Control specimens were stored dry at 37°C, and were tested identically to test specimens. Test specimens were retrieved and tested at 2, 4, 8, 16, 32, and 104 weeks. Control specimens were evaluated at 32 and 104 weeks.

A predetermined number of control and implanted specimen sets were solvent extracted at the beginning to facilitate quantitative and qualitative observations of extractable materials and weight changes. The pre-implant extractions were done for 16 h with a refluxing 2:1 mixture of chloroform and methanol in a soxhlet extractor. This solvent mix was selected to facilitate extraction of both polar and non-polar materials such as silicone not chemically bound into the cross-linked, polymeric matrix and biological materials absorbed during the implantation period. Weight loss from these pre-implant extractions was monitored. Post-implant extractions were done for 48 h with 25 ml of the same solvent mix at room temperature rather than at reflux to avoid degrading biological materials that may have been absorbed. Weight changes were again monitored, and silicon content in the decanted solvent was determined by atomic absorption spectroscopy. Biological materials were assessed by thin-layer chromatography. Precision, accuracy, and reproducibility of the extraction techniques were not evaluated and determined, thus only the qualitative findings are considered pertinent.

Implantation and Removal

The specimens were implanted in dogs along the lower back and flank areas on both sides of the spinal column. A small skin incision parallel to the spinal column was made for each specimen. Subcutaneous fascia was bluntly dissected ventral to each incision to form a pocket for each specimen. At predetermined times specimens were removed from each dog along with fibrous tissue capsules. Specimens were weighed and physical properties measured within 7 h.

Physical Property Measurements

The physical properties as determined for nonextracted implanted and control specimens measured at the various time intervals are shown in Figs. 2 through 6. The effects of implantation on physical properties of medical-grade high-performance silicone elastomer were assessed in two ways. First, the physical property measurements obtained from specimens implanted 0, 2, 4, 8, 16, 32,

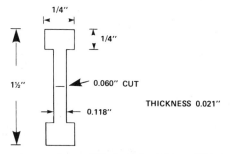

FIG. 1—*Flaw propagation test. Start at 40% elongation (254 mm (10 in.) per min) repeat through 20 cycles, increase elongation 5%, and repeat sequence until failure occurs. Test end-point is number of cycles to failure.*

and 104 weeks were monitored for changes with increasing implant time. Second, control specimens stored dry at 37°C were tested at 32 and 104 weeks for comparison to specimens implanted for the same periods of time.

Tensile Strength

Average tensile strength (ASTM D 412-75) values for test and control specimens sets involved in this study are shown in Fig. 2. Average tensile strength varied from a low of 1.0356×10^7 Pa (1502 psi) for the 32-week implanted specimen set to a high of 1.1900×10^7 Pa (1726 psi) for the 104-week control specimen set. The average tensile strength of the base-line specimen set was 1.1528×10^7 Pa (1672 psi), which may be compared to an average of 1.1542×10^7 Pa (1674 psi) for the 104-week implanted specimen set. The average tensile strength for the three non-implanted specimen sets (base-line and two controls) was 1.1425×10^7 Pa (1657 psi) compared to an average of 1.1129 Pa (1614 psi) for the six implanted specimen sets.

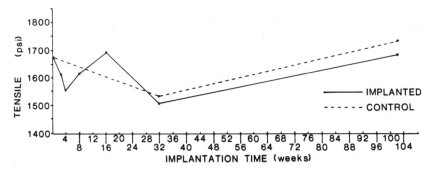

FIG. 2—*Tensile strength of high-performance silicone elastomer after implantation.*

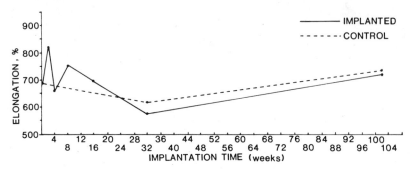

FIG. 3—*Elongation of high-performance silicone elastomer after implantation.*

Elongation

The average elongation (ASTM D 412-75) of implanted and control specimen sets evaluated in this study are shown in Fig. 3. These averages varied from a low of 576% for the 32-week implanted specimen set to a high of 812% for the two-week implanted specimen set. The average elongation of the base-line specimen set was 683%, which may be compared to an average of 716% for the 104-week implanted specimen set. The average elongation in the three non-implanted specimen sets (one base line and two control) was 676% compared to an average elongation in six implanted specimen sets of 700%.

Tensile Stress

The average tensile stress (ASTM D 412-75) at 200% elongation for the test and control specimen sets evaluated in this study are shown in Fig. 4. These

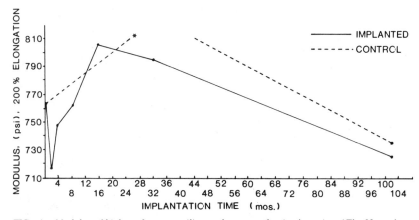

FIG. 4—*Modulus of high-performance silicone elastomer after implantation. *The 32-week control modulus value was 821.*

average tensile stress values varied from a low of 4.9437 × 10⁶ Pa (717 psi) for the two-week implanted specimen set to a high of 6.6608 × 10⁶ Pa (821 psi) for the 32-week control specimen set. The average tensile stress at 200% elongation for the base-line specimen set was 5.2609 × 10⁶ Pa (763 psi) compared to an average of 4.9989 × 10⁶ Pa (725 psi) for the specimen set implanted 104 weeks. The average tensile stress of the six implanted sets at 200% elongation was 5.2333 × 10⁶ Pa (759 psi) compared to an average of 5.3229 × 10⁶ Pa (772 psi) for the three non-implanted specimen sets (one base line and two controls).

Tear Propagation Strength

The average tear propagation strength (ASTM D 624-73, Die B) of the test and control specimen sets involved in this study are shown in Fig. 5. The highest average tear propagation strength was 5.410 × 10³ kg/m (303 lb/in.) in the 32-week control specimen set. The lowest was 5.1605 kg/m (189 lb/in.) in the two-year control set. The average for the base-line and 104-week implant specimen sets was essentially identical, 5.2855 × 10³ kg/m (296 ppi). The average tear propagation strength of the six implanted specimen sets and the three non-implanted sets (one base line and two control) was also 5.2855 × 10³ kg/m (296 lb/in.).

Flaw Propagation Resistance

The average flaw propagation resistance of test and control specimen sets involved in this study is shown in Fig. 6. The lowest average flaw propagation was 133 cycles in the 32-week control specimen set. The highest average flaw propagation resistance was 190 cycles in the four-week implanted specimen set. The average for the base-line specimen set was 178 cycles, which may be

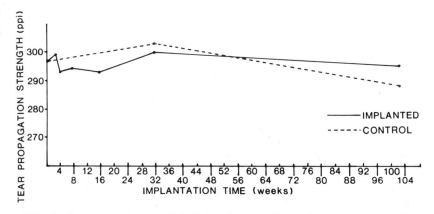

FIG. 5—*Tear propagation strength in high-performance silicone elastomer after implantation.*

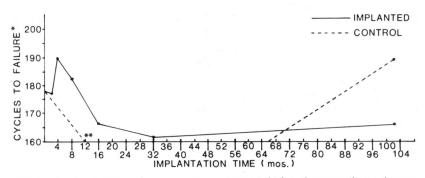

FIG. 6—*Accelerated fatigue flaw propagation resistance in high-performance silicone elastomer after implantation. *Starting at 40% elongation, 20 cycles, and increasing elongation 5% with each additional 20 cycles. **The 32-week control value was 133 cycles.*

compared to an average of 166 cycles for the specimen set implanted 104 weeks. The average for the six implanted sets was 174 cycles compared to an average of 166 cycles for the three non-implanted specimen sets (one base line and two control).

Extraction Studies and Discussion

The average weight loss from 16 h of soxhlet extraction of four base-line specimen sets using a refluxing solvent mix (chloroform:methanol = 2:1) was 2.13%. This type of extraction typically removes polydimethylsiloxane not chemically bound into the cross-linked elastomer network matrix. It may be compared to an average weight loss of 1.73% from room-temperature 48-h static extraction of four previously unextracted control specimen sets that had been stored dry, two sets for 32 weeks and two sets for 104 weeks, using 25 mL of the same solvent mix. Since these specimen sets could reasonably be expected to have the same extractable silicone content, it appeared that room-temperature extraction with static solvent was less effective and left on the average 0.4% by weight of residual extractable silicone compared to soxhlet extraction with refluxing solvent. This also probably accounted for the relatively high variation seen in weight loss from these static solvent extractions. Static solvent extraction thus did not appear to provide the accuracy and precision required for exact quantitative materials balance studies.

Atomic absorption studies of the solvent extracts from the four previously unextracted control specimen sets indicated the average weight loss attributable to extracted silicone also was 1.73%, and while this average matched well with the average total weight loss, significant variations were found between weight loss and atomic absorption determination of silicone in three of the four specimen sets indicating that the technique utilizing atomic absorption does not provide the accuracy required for precise quantitative evaluation.

In the two specimen sets that had been extracted with refluxing solvent for 16 h in a soxhlet extractor prior to implantation, the average weight gains were

1.12% after 32 weeks and 1.98% after 104 weeks with considerable specimen-to-specimen variation; the weight losses from the terminal 48 h room-temperature extraction with static solvent were 1.21% and 1.71%, respectively. These data would indicate that lipid absorption in specimens implanted after extraction is approximately 1.16% at 32 weeks implantation and 1.85% after 104 weeks implantation.

In the two specimen sets implanted without previous extraction, the weight gains were 1.06% at 32 weeks and 1.31% after 104 weeks. The weight losses from the terminal 48 h room-temperature extraction with static solvent were 2.85% and 2.95%, respectively. The average weight loss attributable to silicone as determined by atomic absorption techniques in these two specimen sets was 1.83%. High variations from specimen-to-specimen were again seen. These data would indicate that lipid absorption after 32 weeks implantation was approximately 1.04% at 32 weeks and 1.22% after 104 weeks. Furthermore, the residual silicone content in specimens implanted for either 32 or 104 weeks is essentially the same as the residual silicone in control specimens, although this experiment lacked the precision and accuracy that would be required for exact quantitative determinations.

Thin-layer chromatographic analysis of the absorbed lipids indicated they consisted of 80% to 90% triglycerides and 10% cholesterol. Smaller amounts of partial glycerides and free-fatty acids were also found. No phospholipids were found. No significant differences in lipid composition were found between specimens that had been implanted with and without prior solvent extraction.

Conclusions

None of the physical properties monitored in this study (tensile strength, ultimate elongation, tensile stress at 200% elongation, tear propagation strength, Die B, and flaw propagation resistance) changed in a consistent manner as a consequence of subcutaneous implantation in the dog for a time period of up to two years duration.

The differences in physical properties between implanted specimens and control specimens stored dry at 37°C when tested at 32 and 104 weeks were small, generally less than the differences found between test values from testing done at various time intervals. The findings from this study indicate that changes in physical properties of medical-grade high-performance silicone elastomer from two years subcutaneous implantation in the dog either do not occur or are very small and are thus not a factor in performance of implants fabricated from this material.

Extraction studies demonstrated that vulcanized and post-cured medical-grade high-performance silicone elastomer contains approximately 2% by weight of polydimethysiloxane that is not chemically bound into the cross-linked polymeric matrix, and may thus be extracted with an appropriate solvent. Lipid absorption after 32-weeks implantation was approximately 1.1% and after 104 weeks implantation approximately 1.5% based on averages of specimen weight gains while implanted and weight losses upon terminal extraction attributable to non-silicone material in specimens that had been implanted both with and without solvent extraction. The technique of room-temperature static solvent extraction used for extraction after explantation was less efficient and less predictable than the pre-implant extractions done with refluxing solvent in a soxhlet extractor. When these differences are taken into consideration, there is no apparent difference between the content of free silicone in these specimens after 104 months implantation and the content at the time of implantation. Thus, there did not appear to be a significant loss of silicone to tissues or systemic circulation, nor generation of additional free silicone.

Lipid absorption appeared to occur relatively early after implantation and probably became constant when the silicone elastomer became saturated in the specific implant environment. The absorbed lipids consisted of 80% to 90% triglycerides and 10% cholesterol. Smaller amounts of partial glycerides and free fatty acids were also found. No phospholipids were found. No significant differences in lipid composition were found between specimens that had been implanted with and without solvent extraction.

It was found by this study that medical-grade high-performance silicone elastomer had excellent biodurability during two years subcutaneous implantation in the dog. The properties of the material did not appear to change as a result of lipid absorption.

References

[1] Rowe, V. K., Spencer, H. C., and Bass, S. L., *Journal of Industrial Hygiene and Toxicology*, Vol. 30, No. 6, Nov. 1948, pp. 332–352.
[2] LaFay, H., *Readers Digest*, Vol. 57, Jan. 1957, pp. 29–32.
[3] Safian, J., *Plastic and Reconstructive Surgery*, Vol. 37, 1966, pp. 446–452.
[4] Cronin, T. D. and Gerow, F. J., "Augmentation Mammaplasty: A New "Natural Feel" Prosthesis," Excerpta Media International Congress Series No. 66. *Proceedings*, IIIrd International Congress of Plastic Surgery, Washington, DC, 1963, pp. 41–49.
[5] Cronin, T. D., *Plastic and Reconstructive Surgery*, Vol. 37, No. 5, May 1966, pp. 399–405.
[6] Snyder, G. B., Courtiss, E. H., Kaye, B. M., and Gradinger, G. P., *Plastic and Reconstructive Surgery*, Vol. 61, 1978, pp. 854–860.
[7] Swanson, A. B., "A Flexible Implant for Replacement of Arthritic or Destroyed Joints in the Hand," New York University Inter-clinical Information Bulletin, No. 6, 1966, pp. 16–19.
[8] Swanson, A. B., U.S. Patent No. 3,462,765.
[9] Swanson, A. B., U.S. Patent No. 3,875,594.
[10] Swanson, A. B., *Flexible Implant Resection Arthroplasty in the Hand and Extremities*, St. Louis, The C. V. Mosby Co., 1973.
[11] Swanson, A. B., *Journal of Bone and Joint Surgery*, Vol. 54A, 1972, pp. 435–455.
[12] Mulligan, W. J., *International Journal of Fertility*, Vol. 11, 1966, pp. 424–430.
[13] Lincoff, H. A., Baras, I., and McLean, J., *Archives of Ophthalmology*, Vol. 73, 1956, p. 160.
[14] Lincoff, H. A., Ramirez, V., Kreissig, I., Baronberg, N., and Kaufman, D., *Modern Problems of Ophthalmology*, Vol. 15, 1975, p. 188.
[15] Lincoff, H. A. and McLean, J., *American Journal of Ophthalmology*, Vol. 64, 1967, p. 877.
[16] Swanson, J. W., and LeBeau, J. E., *Journal of Biomedical Materials Research*, Vol. 8, 1974, pp. 357–367.
[17] Swanson, A. B. et al, *Orthopedic Clinics of North America*, Vol. 4, No. 4, Oct. 1973, pp. 1097–1112.

[18] Weightman, B. et al, *Journal of Biomedical Materials Research Symposium*, No. 3, John Wiley & Sons, Inc., New York, 1972, pp. 15–24.
[19] Wilsnack, R. E.: *Biomaterials Medical Devices, and Artificial Organs*, Vol. 4, Nos. 3 & 4, 1976, pp. 235–261.
[20] Wilsnack, R. E., Meyer, F. S., and Smith, J. G., *Biomaterials Medical Devices, and Artificial Organs*, Vol. 1, No. 3, 1973, pp. 543–562.

DISCUSSION

Martin W. King[1] *(written discussion)*—I have two questions:

1. When evaluating explanted silicone elastomers, what solvent system and conditions do you use to extract: (a) absorbed biological species, and (b) non cross-linked silicone oligomer and polymer.

2. During your presentation, your indicated that silicone implants degrade and fragment into particles *in vivo* only when exposed to abrasive mechanical action, and not due to biochemical degradation. We have previously reported the presence of numerous discrete particles of silicon-rich material in the thick outer tissue capsules of two silicone testicular prostheses after 72 and 99 months residency, respectively.[2] We do not believe that abrasive mechanical stresses were responsible for this phenomenon. Would you please comment on our finding in the light of your experience.

E. E. Frisch and N. R. Langley (author's closure)—The extractions were done using a 2:1 by volume mix of chloroform and methyl alcohol to extract both free silicone and biologicals. Prior to implantation, the extractions were done under conditions of reflux in a soxhlet extractor. In the explanted samples, the extractions were done at room temperature under static conditions. Soxhlet extraction of explanted samples was not done because it was found that the temperature of reflux would decompose and chemically alter materials of biological origin.

Concerning release of silicone from implanted devices, consideration must be given to the type of material used in the implant. There are many different types of silicones. Medical-grade high-performance silicone elastomer is a high consistency, solid elastomer that typically contains less than 2% extractable silicone when properly processed. Most testicular and mammary implants have a thin outer cover of similar elastomer, but the bulk of the implant is composed of a soft, low-modulus silicone gel that typically contains a significant amount of extractable silicone. The outer cover acts as a semi-permeable membrane and free silicone from the gel that dissolves in the membrane can pass through it by osmosis. Thus, with testicular and mammary implants silicone external to the implant is a predictable occurrence. However, this does not occur with implants

[1]University of Manitoba, Winnipeg, Manitoba, Canada, R3T 2N2.
[2]Genest, H. et al, *Journal d'Urologie*, Vol. 88, No. 6, 1982, pp. 337–343.

made from solid medical-grade silicone elastomers. The small quantity of free silicone these elastomers contain is highly soluble in the cross-linked elastomer matrix, and highly insoluble in aqueous-based materials, thus the free silicone remains dissolved in the mass of the elastomer.

Manfred Semlitsch[3] (*written discussion*)—Which type of fluid (serum or Ringer's solution) is surrounding the silicone-elastomer implants in the fatigue testing equipment? How many cycles are carried out and at what frequency to rupture and non-fracture? Are these fatigue tests in the laboratory able to simulate the fatigue conditions in the human body?

E. E. Frisch and N. R. Langley (*authors' closure*)—All of the fatigue testing done in this study was done in air, using ASTM methods to the greatest extent possible, on specimens that had either been stored at ambient conditions or had been implanted into the subcutaneous tissue of dogs for time periods of up to two years. No standard test methods have been developed for testing elastomers in simulated implant conditions. The actual implant environment provides a more valid assessment of biodurability than any type of artificial, laboratory-simulated environment.

Laboratory and animal implant studies considered alone are seldom, if ever, adequate to provide a valid measurement of clinical and functional performance. With the SILASTIC Finger Joint Implant H. P. (Swanson Design) (a flexible hinge implant made from medical-grade high-performance silicone elastomer), field clinic studies were conducted to assess clinical and functional performance. The fracture rate from all causes was less than 5% in a group of patients with 221 metacarpophalangeal joint implants followed for an average of 55 months. Thus, the clinical finding of excellent flexural fatigue durability with this material was similar to the findings in laboratory and animal studies.

H. R. Piehler[4] (*written discussion*)—Are you comfortable using laboratory tests involving hundreds of percent strain to simulate *in vivo* performance that involves strains that are an order of magnitude smaller?

E. E. Frisch and N. R. Langley (*authors' closure*)—Laboratory and animal testing is useful to compare various materials of construction and implant designs, and to assess basic biological safety. These tests may also provide an indication of the expected clinical and functional performance, but generally clinical and functional performance can be determined only by carefully done clinical studies.

In laboratory testing it is frequently necessary to use severe test conditions in order to accelerate failure. As a general principle, any failure testing procedure should produce quantitative or qualitative information in a reasonable period of time. Performance testing done when test conditions are too mild such that a

[3]Sulzer Bros. Ltd., Department R & D, CH-8401 Winterthur, Switzerland.
[4]Department of Metallurgical Engineering & Materials Science, Carnegie-Mellon University, Pittsburgh, PA 15213.

failure rate is not included in a reasonable period of time generally have very little practical value. With many materials of construction and the implant devices made therefrom, the primary assurance that materials and implant performance have been duplicated from lot-to-lot must rely on short-term testing as done for lot-to-lot control. Thus, the most meaningful and valuable performance tests are those involving conditions adequately severe to produce quantifiable failure rates.

Martin W. King,[1] Robert Guidoin,[2] Pierre Blais,[3] Andrew Garton,[4] and Kingsley R. Gunasekera[1]

Degradation of Polyester Arterial Prostheses: A Physical or Chemical Mechanism?

REFERENCE: King, M. W., Guidoin, R., Blais, P., Garton, A., and Gunasekera, K. R., "**Degradation of Polyester Arterial Prostheses: A Physical or Chemical Mechanism?**" *Corrosion and Degradation of Implant Materials: Second Symposium, ASTM STP 859*, A. C. Fraker and C. D. Griffin, Eds., American Society for Testing and Materials, Philadelphia, 1985, pp. 294–307.

ABSTRACT: Despite earlier claims of biological stability, there is growing evidence from case histories, *in vitro* testing, and implant retrieval programs that poly(ethylene terephthalate) fibers experience some degradation when implanted in humans as arterial prostheses.

Previously, this was of minor significance when prognoses were of limited duration. Today these devices are being implanted in a wider spectrum of patients with the expectation of longer service lifetimes. There is growing urgency, therefore, to identify the rate and cause of this degradative behavior.

The authors have previously reported a progressive loss in the bursting strength of commercial polyester arterial prostheses with increasing times of implantation in humans. They have suggested that physical factors, such as the changes in the geometry of the knitted and woven structures have contributed to this loss of mechanical performance. This paper reports additional findings from an implant retrieval program. After residency periods from a few hours to 14 years, the chemical properties of the explanted devices have been analyzed. The results are compared with those from unused controls in order to determine the nature and extent of the chemical changes to the polymer during implantation. Losses in bursting strength are accompanied by losses in molecular weight and increases in carboxyl group concentration, suggesting that a chemical mechanism is also responsible for the degradative process. The kinetics of this chain scission reaction approximate a logarithmic decay model rather than that expected from autocatalytic initiation or a simple random or end-group mechanism with or without diffusion control of the initiator. It is estimated that 25% of the initial bursting strength is lost after 162 ± 23 months and 25% of the initial molecular weight is lost after 120 ± 15 months of implantation in humans.

[1]Assistant professor and research assistant, respectively, Department of Clothing & Textiles, University of Manitoba, Winnipeg, Manitoba, R3T 2N2, Canada.

[2]Professeur agrégé, Laboratoire de Chirurgie Expérimentale, Université Laval, Sainte-Foy, Québec, G1K 7P4, Canada.

[3]Senior scientific advisor, Bureau of Medical Devices, Health and Welfare (Canada), Ottawa, Ontario, K1A OL2, Canada.

[4]Research officer, Division of Chemistry, National Research Council of Canada, Ottawa, Ontario, K1A OR9, Canada.

KEY WORDS: implant materials, biological degradation, fatigue (materials), biodeterioration, bursting strength, carboxylic acids, chemical analysis, depolymerization, dimensional stability, implantation, infrared spectroscopy, molecular weight, peripheral vascular diseases, polyester fibers, poly(ethylene terephthalate), prosthetic devices

Thirty years ago when vascular surgery was in its infancy, the immediate objective of the surgeon was to prolong the life of acutely-ill patients for several months. Today with considerable advances in surgical techniques, patient management protocols, and prosthesis design, the life expectancy of patients undergoing peripheral vascular surgery is considerably longer, in many cases exceeding 20 years. At the same time, however, there has been an increasing use of lighter weight knitted and velour Dacron® polyester vascular prostheses for large and medium caliber vessel replacement and bypass. Since the Dacron material in today's commercial grafts is essentially the same poly(ethylene terephthalate) (PET) evaluated by Harrison 25 years ago [1],[5] it is timely to review the biostability of this synthetic polymer to ensure that its life expectancy exceeds that of today's patients.

Over the years, the literature has contained evidence of early and late failure of knitted and woven Dacron polyester arterial prostheses. Complications involving graft dilatation, hemorrhaging, and the formation of false aneurysms have been reported. In many of these cases, the authors have identified a weakening of the prosthesis during implantation accompanied by a deterioration of the polyester fibers or fabric structure to be the main contributing factors for such complications [2–19]. A number of alternative explanations have been suggested, including manufacturing errors, incorrect storage, sterilization and handling techniques, surgical trauma, and defective material [20]. In addition, the likelihood of mechanical fatigue and biodegradation of the polyester material by biological species appears plausible [18,19,21].

Stronger evidence supporting a slow, inherent biodegradation of Dacron polyester comes from implant retrieval studies that have analyzed larger numbers of explanted vascular prostheses from autopsies and reoperations regardless of manufacturer, type of fabric construction, sterilization procedure, surgeon, or postoperative complications [18,22,23]. All three of these studies found that PET prostheses experienced some loss in strength after residency periods of six months or longer. However, the amount of the loss could not be measured precisely because of the lack of reliable unused control samples.

The reason, or reasons, for this loss in strength of implanted polyester devices is not fully understood. A number of alternative explanations have been proposed, including chemical mechanisms such as salts or tissue electrolytes [18,24], acids [25], and enzymes [21,24–26], and physical mechanisms such as mechanical fatigue [17–19,21–23,27,28]. We have previously reported that this loss in strength is accompanied by a dilatation, permanent deformation, or creep of the

---

[5]The italic numbers in brackets refer to the list of references appended to this paper.

fabric structure as measured by losses in stitch density for knits and losses in fabric count for wovens [23]. In addition, we found that the extent of this dilatation varied with different types of knitted constructions, but did not correlate with the amount of strength loss. From this, we hypothesized that mechanical fatigue is only one of several mechanisms that contribute to the loss of strength during implantation. This hypothesis has been supported by visual evidence. For example, the fibers from damaged PET prostheses removed from the site of false aneuryms in two different patients, both after 84 months *in vivo*, can be seen in Figs. 1 and 2. Figure 1 shows a fractured fiber containing many microfibrils. This fibrillated morphology is typical of a mechanical failure mechanism due to cyclic bending, tensional, and torsional stresses [29]. In comparison, Fig. 2 shows a different failure mechanism. The transverse cracks and embrittled appearance suggests that these fibers have been exposed to chemical degradation.

The objective of this study was to establish whether, in addition to physical mechanisms, vascular prostheses constructed from PET fibers experience significant chemical degradation during implantation in humans. Should the study's findings be positive, it was anticipated that the kinetics of the chemical reaction might be identified, thus demonstrating whether the rate of strength loss of the prosthesis can be associated with the rate of chemical degradation of the polymer.

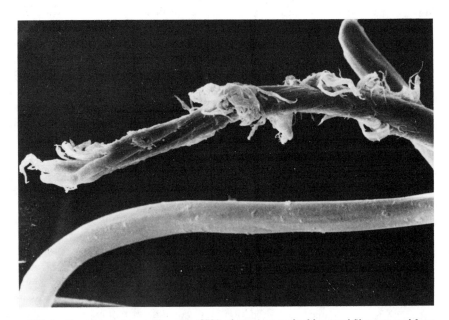

FIG. 1—*Scanning electron microscope (SEM) photomicrograph of fractured fiber removed from retrieved and cleaned vascular prosthesis after 84 months implantation showing fibrillated morphology possibly due to mechanical fatigue (original magnification ×700).*

FIG. 2—*SEM photomicrograph of fractured fibers removed from retrieved and cleaned vascular prosthesis after 84 months implantation showing embrittled appearance possibly due to chemical degradation (original magnification ×1250).*

Procedure

Explanted samples of polyester vascular prostheses were obtained from autopsies and reoperations using the retrieval protocol described previously [23]. Using a purposive random sampling technique, a series of 19 prostheses was selected to ensure representation throughout a range of residency periods extending from a few hours to over 14 years (170 months). Five different models of knitted and velour prostheses from three U.S. manufacturers were included in the sample. In view of the fact that the prostheses had been implanted in different patients by a number of different surgeons, the one feature common to all the samples was that they contained Type 56 and Type 62 semidull Dacron poly(ethylene terephthalate) (PET) polyester yarns.

Unused control samples for comparison purposes were obtained from two sources. In 13 cases, equivalent commercial samples were obtained from the manufacturers and, in the six remaining cases, unused samples had been removed from the implanted specimen in the operating room at the time of surgery and stored at room temperature in dark, moisture-free conditions. Because it eliminates batch-to-batch variability and changes due to storage, handling, and sterilization procedures, we believe this latter approach provides more reliable controls.

Following storage in formalin, the retrieved samples were opened longitudinally and cleaned. Removal of adhering tissue was achieved by boiling in 5% sodium carbonate, treating with a mild oxidative bleach at room temperature, exhaustive washing with distilled water, and air drying. This procedure was found to produce no significant changes in the properties under investigation.

The following four properties were measured on the retrieved and cleaned samples and their respective controls as long as there was a sufficient size of sample to complete a statistically reliable measurement.

Bursting Strength

Bursting strength measurements were carried out using a probe puncture test developed by Meadox Medicals and described in the current AAMI[6] and ISO[7] draft standards. Specimens were flattened to remove residual crimp prior to being clamped over the 1.13-cm-diameter circular hole in the clamping plates. A probe with a circular cross section of 0.79-cm diameter and a hemispherical head was traversed at a rate of 12.5 cm/min through the specimen using a compression cage mounted on an Instron tension tester. The maximum loads for five repeated tests were averaged and divided by the area under test (3.80 cm^2) to give the mean bursting strength of each sample. These values have been found to correlate well ($0.94 \leqslant r \leqslant 0.97$) with the alternative method of using a diaphragm bursting test[6,7] [23], when testing single jersey and tricot warp knit fabrics [30].

Stitch Density

The amount of dilatation, permanent deformation, or creep of the retrieved prostheses was determined by measuring the loss in stitch density during implantation. The number of wales/cm and courses/cm [22] were counted in both directions of each retrieved and control sample after flattening the specimen between glass slides to remove the crimp. Five repeated measurements were made along 3-cm distances using an optical microscope at ×20 magnification and the mean values calculated. The stitch density was calculated by multiplying the wales/cm by the courses/cm [23]. Any in vivo loss of stitch density represents the extent to which the structure has been mechanically deformed through distortion of the knitted loops and stretching of the fibers.

Molecular Weight

Changes in molecular weight during implantation were determined by measuring the intrinsic viscosity, [η], of the polymer in dilute solutions. The ASTM

[6]"Standard for Vascular Graft Prostheses (Draft)," Section 4.3.1.2(2) b, Association for the Advancement of Medical Instrumentation, Arlington, VA, May 1984, pp. 19–22.
[7]"Standard for Cardiovascular Implants—Synthetic Tubular Vascular Prostheses," Annex E.3, DP 7198, International Organization for Standardization, Geneva, Switzerland, Aug. 1984, pp. 22–23.

Test Method for Dilute Solution Viscosity of Polymers (D 2857-70) was followed as appropriate. Specimens of the retrieved and control samples were dried, ground, weighed, and dissolved in distilled orthochlorophenol. After centrifugation and filtration, the times of flow of a series of sequentially diluted solutions were measured using a Cannon-Ubbelohde dilution viscometer at 25.0 ± 0.1°C. The viscosity average molecular weight, M_v, of each sample was calculated using the following Mark-Houwink equation [31]

$$M_v = \sqrt[0.81]{[\eta]/1.9 \times 10^{-4}}$$

These experimental results have subsequently been confirmed using a gel permeation chromatography (GPC) technique using hexafluoroisopropanol as the solvent and micro-styragel columns in a Waters chromatograph [32]. This is considered a more desirable approach since it provides additional information about the uniformity and width of the total molecular weight distribution.

Carboxyl Group Content

Since poly(ethylene terephthalate) contains carboxylic acid end groups, a decrease in molecular weight during implantation should be accompanied by an increase in carboxyl group content. Fourier transform infra-red spectroscopy was used to measure the carboxyl group concentration indirectly. Specimens of the retrieved and control samples were exposed to sulfur tetrafluoride for 72 h to convert the carboxylic acid groups to acid fluoride groups [33]. This permitted more reliable quantitative analysis of the infra-red spectra. The specimens were prepared on KBr disks and the spectra were generated using a Nicolet 7199 FTIR instrument that measured the area under the 1820/cm acid fluoride peak [34]. The 730/cm peak was used as the reference for calibration.

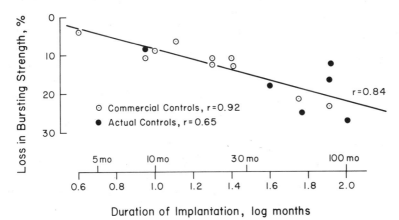

FIG. 3—*Loss in bursting strength of polyester vascular prostheses during implantation.*

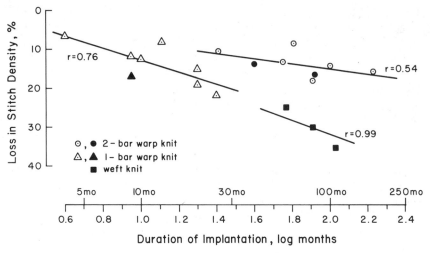

FIG. 4—*Loss in stitch density during implantation.*

Results

The percentage loss in bursting strength during implantation is presented in Fig. 3. Note that by plotting the duration of implantation on a log scale an approximately straight line relationship is obtained, which, when analyzed statistically [35], gives a Pearson's r correlation coefficient of 0.84. These data have greater precision than our previous findings [23]. They confirm the same trend as reported earlier and they can be extrapolated to predict a 25% loss in initial bursting strength after 162 ± 23 months of implantation.

Figure 4 confirms that all samples experienced some dilatation, permanent deformation, or creep during implantation. This tendancy to dilate continues throughout implantation, although the rate of change decreases with time (note log time scale). Figure 4 also confirms earlier reports that 1-bar warp knits (for example, Microvel) and weft knits (for example, Knitted Standard DeBakey) are more prone to dilatation than other models [23,36].

All samples suffered some loss in molecular weight during implantation. The extent of this loss can be seen in Fig. 5 that also shows that less scatter is achieved by using the actual controls rather than equivalent commercial samples. Additional evidence of chemical changes is to be found in Fig. 6 that shows that the carboxyl group content, as measured by the ratio of the 1820/730 peaks, increased with the duration of implantation. We believe that these two analytical techniques are closely correlated. Figure 7 demonstrates that the viscosity average molecular weight values have an inverse linear relationship with the carboxyl group data. Figure 7, however, includes one point that is outside this relationship. This point represents a retrieved sample that was stored in formalin for 18 months prior to analysis, and was not included in this study. It demonstrates that prolonged exposure to formalin can increase the carboxyl group content. This potential error has been minimized in this study by limiting the time of exposure

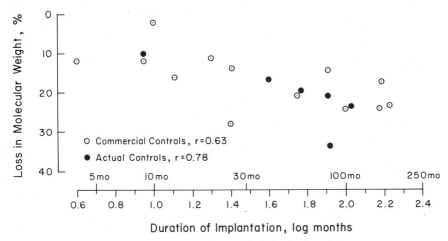

FIG. 5—*Loss in molecular weight during implantation.*

of the retrieved samples to formalin to less than two months. This duration of exposure has been found to cause no significant change in the molecular weight or the carboxyl group content of PET.

Discussion

We believe that these findings demonstrate that vascular prostheses produced from poly(ethylene terephthalate) yarns experience losses in strength during implantation due to a combination of physical and chemical factors. Because the study included a variety of different models with different textile structures manufactured by different companies and implanted by different surgeons in different patients, it is proposed that the *in vivo* changes that have been identified are primarily inherent characteristics of the PET polymer.

Our results have less variability and are therefore more reliable than those of previous retrospective studies [18,22,23,25]. We believe this is due to our partial

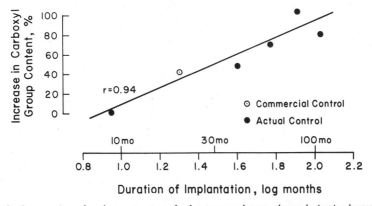

FIG. 6—*Increase in carboxyl group content of polyester vascular prostheses during implantation.*

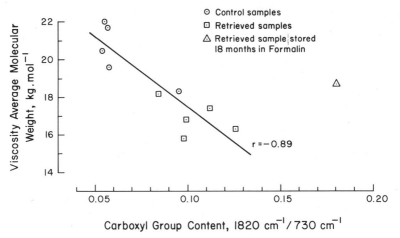

FIG. 7—*Relationship between carboxyl group content and viscosity average molecular weight.*

use of actual control samples that are not subject to batch-to-batch variations as commercially equivalent controls are. This prospective approach is an important ingredient in the success of an implant retrieval program.

The rate of *in vivo* loss of strength of the PET prostheses measured in this study is slower than that reported by other workers [21,22,25,37]. Dvorak claims a 50% loss of strength after only 23 to 32 months of implantation [21], compared to Gumargalieva et al and Rudakova et al who suggest about 100 months is required to halve the strength of a PET net [25,37]. Edwards et al found a 25% loss after only 18 months *in vivo* [22]. Two main differences exist between our work and these previous studies. All previous work has relied on animal models, either dogs or rats, whereas our data describes human clinical experience. In addition, we have measured the bursting strength of the wall of the prosthesis, whereas earlier work has measured the tensile strength of the yarns.

The rate of change in molecular weight of PET during implantation measured in this study is significantly different to that reported by Gumargalieva et al [25] and Rudakova et al [37]. These Russian authors were implanting PET subcutaneously in canine and rat models, and one was evaluating the effect of infection [25]. This points to the difficulty in attempting to predict clinical performance in humans from animal studies.

At the same time, we recognize two major limitations of the analytical approach used in this study. First, previous work in our laboratories using mass spectroscopy has demonstrated that PET absorbs a variety of biological species while implanted in the body [38]. Such species, if soluble in orthochlorophenol, would dilute the solutions for viscosity measurements, and, if they contained carboxyl groups, they would increase the level of infra-red absorption. While recognizing that these experimental errors may have affected our results, our earlier work found insufficient quantities of absorbed species to account fully for the extent of chemical changes identified in this study.

The second deficiency of clinical retrieval programs is that one is unlikely to retrieve the whole implant. Degradation of the PET produces various monomer and oligomer fragments [38] that are carried away by the blood stream. Consequently, the implant may become increasingly lighter in weight as the degraded surface passes into solution, yet maintain its original molecular weight and other bulk chemical properties.

Models of Depolymerization and Diffusion

One of the objectives of this study was to explain the kinetics of any *in vivo* chemical change in terms of established theories for depolymerization and diffusion.

Depolymerization can be initiated by one of three mechanisms; either initiation by random chain scission or by end group unzipping [39] or by an autocatalytic mechanism involving, for example, an acid catalyzed hydrolysis of the ester links [40]. In the first case, assuming a first order chain scission reaction and termination by disproportionation, the change in molecular weight with time is given by

$$\frac{1}{M} - \frac{1}{M_0} = k_s[x]t \tag{1}$$

In the second case, assuming a zero order end group unzipping reaction and first order termination, the change in molecular weight with time is given by

$$M - M_0 = -\frac{k_E[x]t}{\gamma} \tag{2}$$

In the third case, empirical work by Brown et al on polyester urethanes has shown that for small extents of degradation [41]

$$\frac{1}{M} - \frac{1}{M_0} = [x_0](e^{k_a t} - 1) \tag{3}$$

where

M = average molecular weight at time, t;
M_0 = initial average molecular weight;
k_s = rate coefficient for random chain scission;
k_E = rate coefficient for end group initiation;
k_a = rate coefficient for autocatalytic mechanism;
$[x]$ = concentration of initiator;
$[x_0]$ = initial concentration of initiator;
t = time; and
γ = reciprocal of average zip length.

Note that in all three cases all values are constant, except M and t, as long as one assumes that the initiator is present at the same concentration, $[x]$ or $[x_0]$, throughout the bulk of the material. While this is an acceptable assumption for an autocatalytic process, it does not necessarily occur in the first two cases if the initiator (for example, an absorbed biological species) is diffusing into the fibers of a vascular prosthesis from the surface.

The theory of diffusion is well established [42]. Assuming that the fibers are solid cylinders and that the concentration of initiator remains constant at the surface but has a non-steady-state distribution through the radius of the fiber, then at low levels of penetration, the concentration of initiator accumulated after time, t, is given by

$$[x] = [x_0] \frac{4}{a} \left[\frac{Dt}{\pi} \right]^{1/2} \tag{4}$$

where

$[x]$ = concentration of initiator at time, t;
$[x_0]$ = surface concentration of initiator;
D = diffusion coefficient;
t = time; and
a = radius of fiber.

By combining Eqs 1 and 4, the change in molecular weight with time for a diffusion-controlled, random chain scission reaction is defined by

$$\frac{1}{M} - \frac{1}{M_0} = \frac{4[x_0]k_s}{a} \left[\frac{D}{\pi} \right]^{1/2} t^{3/2} \tag{5}$$

And by combining Eqs 2 and 4, the change in molecular weight with time for a diffusion-controlled end group unzipping reaction is defined by

$$M - M_0 = \frac{-4[x_0]k_E}{a\gamma} \left[\frac{D}{\pi} \right]^{1/2} t^{3/2} \tag{6}$$

It is now possible to assess how well the experimental results for change in molecular weight fit these five alternative models. The variables listed in Table 1 were analyzed statistically and the Pearson's r correlation coefficients were calculated from the experimental data for each model [35]. It is evident from Table 1 that none of the five models give good correlation with our experimental results. In fact, the best fit ($r = 0.77$) is obtained when $M_0 - M$ is correlated with log t (Fig. 8). This relationship can be used to predict a 25% loss in molecular weight after 120 ± 15 months of implantation in humans.

From this, we conclude that none of the five kinetics/diffusion models describes accurately the chain scission reaction of PET when implanted in humans

TABLE 1—*Correlation of experimental molecular weight results with alternative depolymerization models.*

Method of Initiation	Diffusion Controlled	Equation	Variables		Pearson's r Correlation Coefficient
			y-axis	x-axis	
Random chain scission	no	1	$\dfrac{1}{M} - \dfrac{1}{M_0}$	t	0.43
End group unzipping	no	2	$M_0 - M$	t	0.65
Autocatalytic	no	3	$\ln\left[\dfrac{1}{M} - \dfrac{1}{M_0}\right]$	t	0.47
Random chain scission	yes	5	$\left[\dfrac{1}{M} - \dfrac{1}{M_0}\right]^{2/3}$	t	0.46
End group unzipping	yes	6	$[M_0 - M]^{2/3}$	t	0.65
?	?	...	$M_0 - M$	$\log t$	0.77

as vascular prostheses. One possible limitation is our assumption that the concentration of the initiator remains constant at the surface of the prosthesis. With the healing process continuing to change the biological environment of the prosthesis for six months or longer after implantation, it is likely that the initiator concentration at the fiber's surface changes with time. The development of encapsulating tissue may account for the observed logarithmic decay of molecular weight with time of implantation. If this is the case, then the rate of degradation of PET in diabetic patients who rarely develop surrounding tissue will be faster than reported here. The level of risk for diabetic patients therefore needs further study.

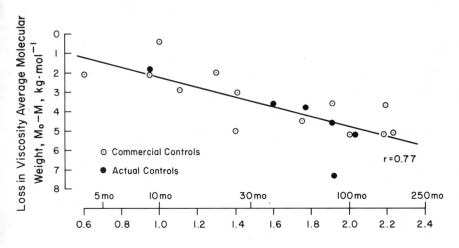

FIG. 8—*Best fit relationship found between loss in viscosity average molecular weight and duration of implantation.*

Conclusion

The loss in bursting strength of PET vascular prostheses during implantation in humans has been shown to coincide with and follow the same logarithmic decay as a loss in the molecular weight and an increase in the carboxyl group content of the polymer. This evidence confirms that poly(ethylene terephthalate) experiences a chain scission reaction *in vivo*. The kinetics of this reaction do not appear to fit simple depolymerization and diffusion models. While the rate of degradation does not put patients currently wearing PET vascular prostheses at risk, further work is required to clarify the chemical mechanism. This will lead to proposals for improving the biostability of the polymer and reducing the loss of mechanical performance to acceptable levels for future long-term implantation.

Acknowledgments

This work was supported in part by grants from the Manitoba Heart Foundation, la Fondation du Québec des Maladies du Coeur, the Department of National Health and Welfare (Canada) and the Mrs. James A. Richardson Foundation, Winnipeg. The authors are grateful to Drs. S. A. Brown, J. Couture, C. Gosselin, M. Marois, E. A. MacGregor, and D. M. Wiles for their guidance and encouragement. The collaboration of the surgeons and pathologists in the retrieval program is much appreciated. We are also indebted to B. Badour, S. Bourassa, A. Dion, J. Doig, D. Lafrenière-Gagnon, E. Nilsen, and L. Wescott for their technical assistance. The gifts of control samples from Meadox Medical (Brent Surgical, Canada), Bard Implants (Medico-Tech., Canada) and Golaski Laboratories (Medtronic, Canada) are gratefully acknowledged.

Disclaimer

The opinions expressed herein are solely those of the authors and do not necessarily represent the views of the organizations with which the authors are affiliated or those of the agencies who supported the work.

References

[1] Harrison, J. H., *American Journal of Surgery*, Vol. 95, 1958, pp. 3–15.
[2] Knox, W. G., *Annals of Surgery*, Vol. 156, 1962, pp. 827–830.
[3] Orringer, M. B., Rutherford, R. B., and Skinner, D. B., *Surgery*, Vol. 72, 1972, pp. 769–771.
[4] Cooke, P. A., Nobis, P. A., and Stoney, R. J., *Archives of Surgery*, Vol. 108, 1974, pp. 101–103.
[5] Deterling, R. A., Jr., *Archives of Surgery*, Vol. 108, 1974, p. 13.
[6] Perry, M. O., *Journal of Cardiovascular Surgery*, Vol. 16, 1975, pp. 318–321.
[7] Rais, O., Lundstrom, B., Angquist, K. A., et al, *Acta Chirurgica Scandinavica*, Vol. 142, 1976, pp. 479–482.
[8] Hayward, R. H. and Korompai, F. L., *Surgery*, Vol. 79, 1976, pp. 581–583.
[9] Blumenberg, R. M. and Gelfand, M. L., *Surgery*, Vol. 81, 1977, pp. 493–496.
[10] May, J. and Stephens, M., *Archives of Surgery*, Vol. 113, 1978, pp. 320–321.

[11] Edwards, W. S. in *Vascular Grafts*, P. N. Sawyer and M. J. Kaplitt, Eds., Appleton-Century-Crofts, New York, 1978, pp. 173–176.

[12] Sauvage, L. R., Berger, K., Wood, S. J., Smith, J. C., Davis, C. C., Hall, D. G., Mansfield, P. B., and Rittenhouse, E. A., in *Vascular Grafts*, P. N. Sawyer and M. J. Kaplitt, Eds., Appleton-Century-Crofts, New York, 1978, pp. 185–196.

[13] Yashar, J. J., Richman, M. H., Dyckman, J., Witoszka, M., Burnard, R. J., Weyman, A. K., and Yashar, J. *Surgery*, Vol. 84, 1978, pp. 659–663.

[14] Tutassaura, H., Gerein, A. N., and Sladen, J. G., *American Surgeon*, Vol. 44, 1978, pp. 262–266.

[15] Komoto, Y., Kawakami, S., and Uchida, H., *Vascular Surgery*, Vol. 12, 1978, pp. 274–279.

[16] Wesolowski, S. A., *Surgery*, Vol. 84, 1978, pp. 575–576.

[17] Koopman, M. D. E. and Brands, L. C., *Journal of Cardiovascular Surgery*, Vol. 21, 1980, pp. 159–162.

[18] Berger, K. and Sauvage, L. R., *Annals of Surgery*, Vol. 193, 1981, pp. 477–491.

[19] Godard, F., King, M. W., Guidoin, R., Marois, M., Garton, A., Blais, P., Gosselin, C., and Gunasekera, K. R., *Journal des Maladies Vasculaires*, (Paris), Vol. 6, 1981, pp. 167–171.

[20] Wright, C. B. and Hiratzka, L. F., *Archives of Surgery*, Vol. 114, 1979, pp. 665–668.

[21] Dvorak, J., *Textile Science and Research*, Vol. 16, 1975, pp. 31–66.

[22] Edwards, W. S., Snyder, R. W., Botzko, K., and Larkin, J. in *Graft Materials in Vascular Surgery*, H. Dardik, Ed., Symposia Specialists, Miami, 1978, pp. 169–182.

[23] Guidoin, R., King, M., Blais, P., Marois, M., Gosselin, C., Roy, P., Courbier, R., David, M., and Noël, H. P. in *Implant Retrieval: Material and Biological Analysis*, NBS Special Publication 601, A. Weinstein, D. Gibbons, S. Brown, and W. Ruff, Eds., National Bureau of Standards, Washington, DC, Jan. 1981, pp. 29–129.

[24] Moiseev, Y. V., Daurova, T. T., Voronkova, O. S., Gumargalieva, K. Z., and Privalova, L. G., *Journal of Polymer Science: Polymer Symposia*, Vol. 66, 1979, pp. 269–276.

[25] Gumargalieva, K. Z., Moiseev, Y. V., Daurova, T. T., and Voronkova, O. S. *Biomaterials*, Vol. 3, 1982, pp. 177–180.

[26] Salthouse, T. N., *Journal of Biomedical Materials Research*, Vol. 10, 1976, pp. 197–229.

[27] Botzko, K., Snyder, R., Larkin, J., and Edwards, W. S. in *Corrosion and Degradation of Implant Materials, ASTM STP 684*, B. C. Syrett and A. Acharya, Eds., American Society for Testing and Materials, Philadelphia, 1979, pp. 76–88.

[28] Biedermann, H. and Flora, G., *International Journal of Artificial Organs*, Vol. 5, 1982, pp. 205–206.

[29] Hearle, J. W. S., Lomas, B., and Bunsell, A. R., *Applied Polymer Symposia*, Vol. 23, 1974, pp. 147–156.

[30] Nilsen, E. J., "A Comparison of Two Bursting Strength Testers for Knitted Fabrics," M.Sc. thesis, University of Manitoba, Winnipeg, 1981.

[31] Moore, W. R. and Sanderson, D., *Polymer*, Vol. 9, 1968, pp. 153–158.

[32] Maarek, J. M., Aubin, M., Prud'homme, R. E., and Guidoin, R. in *Proceedings*, 4th Annual Meeting, Canadian Biomaterials Society, Quebec, 22–23 June 1983, pp. 30.1–30.7.

[33] Addleman, R. L. and Zichy, K., *Polymer*, Vol. 13, 1972, pp. 391–398.

[34] Heacock, J. F., *Journal of Applied Polymer Science*, Vol. 7, 1963, pp. 2319–2327.

[35] Snedecor, G. W. and Cochran, W. G. in *Statistical Methods*, 6th ed., Iowa State University Press, Ames, Iowa, 1967, pp. 172–175.

[36] Nunn, D. B., Freeman, M. H., and Hudgins, P. C. *Annals of Surgery*, Vol. 189, 1979, pp. 741–745.

[37] Rudakova, T. E., Zaikov, G. E., Voronkova, O. S., Daurova, T. T., and Degtyareva, S. M., *Journal of Polymer Science: Polymer Symposia*, Vol. 66, 1979, pp. 277–281.

[38] King, M. W., Guidoin, R. G., Gosselin, C., Godard, F., Marois, M., Gunasekera, K. R., Blais, P., and Garton, A., *European Revue of Biomedical Technology*, Vol. 4, 1982, pp. 26–33.

[39] David, C. in *Comprehensive Chemical Kinetics*, C. H. Bamford and C. F. H. Tipper, Eds., Vol. 14, Elsevier, Amsterdam 1975, pp. 9–25.

[40] Zimmerman, H. and Kim, N. T., *Polymer Engineering and Science*, Vol. 20, 1980, pp. 680–683.

[41] Brown, D. W., Lowry, R. E. and Smith, L. E., *Macromolecules*, Vol. 13, 1980, pp. 248–252.

[42] Crank, J., *The Mathematics of Diffusion*, 2nd ed., Oxford University Press, London, 1975, pp. 69–88.

Michael Szycher[1] and William A. McArthur[2]

Surface Fissuring of Polyurethanes Following *In Vivo* Exposure

REFERENCE: Szycher, M. and McArthur, W. A., **"Surface Fissuring of Polyurethanes Following *In vivo* Exposure,"** *Corrosion and Degradation of Implant Materials: Second Symposium, ASTM STP 859*, A. C. Fraker and C. D. Griffin, Eds., American Society for Testing and Materials, Philadelphia, 1985, pp. 308–321.

ABSTRACT: Compared to standard silicone elastomers, polyurethanes offer enhanced mechanical properties, improved durability, and reduced wall thickness for pacing leads. These thinner-wall polyurethane leads can be used in smaller vessels, or will allow multiple leads to be introduced into a single vein.

Recent reports have shown that pacing leads made with Pellethane 2363-80A exhibit shallow surface fissuring; some pacemaker or lead constructions have failed, particularly at areas of chronic mechanical stress. Several mechanisms have been proposed as the cause of surface fissuring: protein absorption with associated swelling, leaching of low molecular weight substances to the surface, or lipid absorption. However, there is little evidence supporting any of these mechanisms.

Our studies suggest that polyether-based polyurethanes are susceptible to *in vivo* oxidation of the polyether chain. In this chain, the most susceptible group is the -CH_2 group in the alpha position to the ether oxygen, which undergoes peroxidation, free radical dissociation, and, eventually, chain cleavage, possibly leading to a reduction in molecular weight averages. Attenuated total reflectance infrared (ATR-IR) studies at the surface of the fissured polyurethanes have shown the presence of oxidative byproducts such as hydroxyl end groups (-OH).

Based on these preliminary results, we hypothesize that progressive surface degradation is caused, in part, by stress-induced oxidation of the polyether macroglycol used in the synthesis of polyurethane elastomers. Our hypothesis has been reinforced by experimental evidence that surface cracking can be significantly reduced, if not eliminated, by using higher-durometer polyurethanes, because these polymers contain fewer polyether macroglycol chains in the molecular backbone.

KEY WORDS: polyurethanes, pacing leads, surface fissuring, stress cracking, stress-induced oxidation, implant materials, biological degradation

Compared to standard silicone rubbers, polyurethane elastomers offer several advantages including higher tensile strengths, significantly higher tear resistance, and excellent abrasion resistance. In the manufacture of cardiac pacing leads, these advantages have resulted in the introduction of polyurethane leads with significantly reduced wall thicknesses. These thinner-wall leads have resulted in

[1]Director of Biomaterials Research, Thermo Electron Corporation, Waltham, MA 02254.
[2]Pacesetter Systems, Inc., Sylmar, CA 91342.

easier surgical insertion, less traumatic introduction of multiple leads when inserted into single veins (for dual-chamber pacing), and greater elasticity for the implanted lead.

Recent reports by Parins [1],[3] Guerrant, [2], and McArthur [3] have shown that Pellethane 2363-80A commonly exhibits thin surface cracking, and some particular constructions have experienced failure because of surface fissures, particularly at areas of chronic mechanical stress and dynamic flexure points. Surface-fissuring of polyurethane leads has resulted in loss of insulation, clinically evident by inappropriate muscle stimulation or increased current drain or both that causes early depletion of the pacemaker battery.

We have undertaken an exhaustive experimental study designed to test polyurethane elastomers for biostability and performance in implanted pacemaker leads. Our preliminary results indicate that surface fissuring of the implanted polyurethane elastomers occurs principally at areas of excessive induced stresses such as the site of vein ligation where a suture is tightly tied around the lead to stabilize it to the adjacent tissue. The thin, more common surface cracking appears to be the result of the combination of residual stress from extrusion and environmentally induced stress such as dynamic flexure or bending, which promote oxidative instability.

Procedure

Two polyurethane elastomers were utilized in this study: Pellethane 2363-80A and Tecoflex EG-60D. Pellethane 2363-80A[4] is a segmented polyurethane elastomer consisting of hard segments formed by the reaction of 4,4' diphenylmethane diisocyanate (MDI) with a glycol chain extender, and soft segments formed in the reaction of the diisocyanate with polyether macroglycols [4]. Tecoflex EG-60D[5] is a segmented polyurethane elastomer consisting of hard segments formed by the reaction of 4,4' dicyclohexyl methane diisocyanate (HMDI) with a glycol chain extender, and soft segments formed in the reaction of the diisocyanate with polyether macroglycols [5].

Polyurethane leads made of Pellethane and Tecoflex were implanted in the right heart of experimental animals (dogs) for varying periods of time. Following explantation, the leads were cleaned with pepsin, a proteolytic enzyme, by digestion at 37°C with pH controlled at 1.0 for several hours to remove biologic deposits. The efficacy of the pepsin digestion was experimentally demonstrated by totally removing known biologic debris after soaking for only 30 min. The specimens were subsequently dried and readied for scanning electron microscopy (SEM).

Experimental leads were ultrastructurally compared with returned clinical leads made of Pellethane 2363-80A. These leads were explanted from patients, and the surfaces were enzymatically cleaned with pepsin prior to SEM evaluation.

---

[3]The italic numbers in brackets refer to the list of references appended to this paper.
[4]Upjohn Chemical Co., Kalamazoo, MI 49001.
[5]Thermo Electron Corp., Waltham, MA 02254.

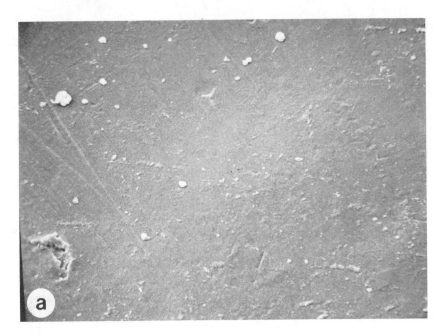

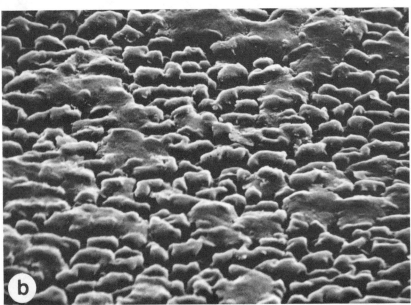

FIG. 1—*Relationship between surface cracking and stress.* (a) *Surface of molded Tecoflex EG-60D implanted subcutaneously in swine for three months. This specimen was implanted in an unstressed manner, original magnification* (× 1000). (b) *Surface of molded Tecoflex EG-60D in the same animal, for an identical period of time. This specimen was deliberately implanted in a highly stressed configuration, and shows characteristic surface fissuring, original magnification* (× 1000).

Finally, stressed and unstressed Tecoflex EG-60D specimens were implanted subcutaneously in swine for three months. After pepsin treatment, the surfaces were chemically analyzed by attenuated total reflectance infrared (ATR-IR) spectroscopy, and compared to unimplanted controls that provided base-line spectral data.

Results

Our results indicate that most surface fissuring of the polyurethanes identified in this study appears to be related to the combination of processing-induced and environmentally induced stresses. Much of the observed surface fissuring occurred at the vein ligation site, probably because of excessive pressure of a suture on the lead. Even shallow surface fissuring (not at ligation sites) seems to initiate at points of environmentally induced chronic stress (although extrusion stress probably contributes), such as the section immediately adjacent to the connector where the lead is looped in the pocket or near the electrode end where dynamic flexure is greatest in the heart.

The relationship between surface cracking and stress is dramatically illustrated in Figs. 1a and b. Figure 1a shows the surface of a molded Tecoflex EG-60D specimen implanted subcutaneously (unstressed) in a swine for three months. Figure 1b shows the surface of another molded Tecoflex EG-60D specimen

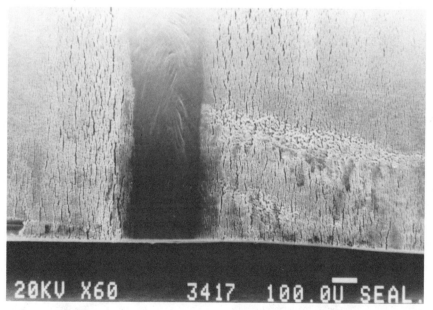

FIG. 2—*Pellethane 2363-80A lead explanted from patient after nine months at site of vein ligation showing extent and morphology of surface fissuring, original magnification (×60).*

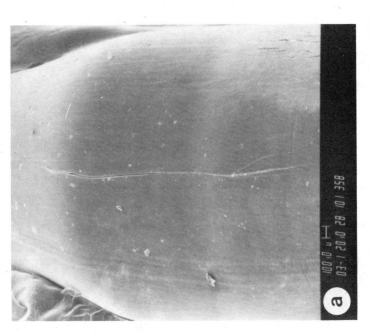

FIG. 3—*Effect of constant stress on surface fissuring.* (a) *Pellethane 2363-80A lead explanted from patient after seven months at section of tube expanded over electrode (≈60% stress) showing cracking in expanded area, none in relaxed area, original magnification (×30).* (b) *Enlargement at ×180 of cracked section of expanded Pellethane 2363-80A tube.*

implanted subcutaneously in the same animal, for an identical period of time; this sample shows unmistakable and characteristic surface fissuring, since it was deliberately implanted in a highly strained configuration.

The relationship between surface cracking and excessive stress is further illustrated in Figs. 2 and 3. These SEM micrographs of implanted leads show surface fissures confined exclusively to the ligature site (Fig. 3), or at the section of Pellethane 2363-80A polyurethane tube stretched over the electrode (Figs. 3a and b), where the polyurethane is under constant strain. The micrographs clearly depict surface cracking at the sites of stress with a rapid decrease in surface cracking with distance from the site of highest stress. Furthermore, at relaxed areas, no surface cracking is visible.

A comparison of the effects on harder (Tecoflex EG-60D) versus softer (Pellethane 2363-80A) polyurethanes is shown in Figs. 4 and 5. Leads constructed from the two materials were implanted in dogs without silicone support sleeves at the ligature sites. At varying implant times, leads were explanted and analyzed for cracking by SEM. Samples were taken from the ligature site of the lead body near the connector end, and the lead body near the electrode end. On two leads implanted for 90 days, Fig. 4b shows significantly less deep or severe cracking on the harder Tecoflex EG-60D lead at the ligature site, compared to the equivalent Pellethane 2363-80A section, shown in Fig. 5b. Similarly, Fig. 5a shows slight cracking in the lead body near the connector on the softer Pellethane 2363-80A sample, while none was seen in Fig. 4a, the corresponding section on the harder Tecoflex EG-60D lead.

No cracking was seen near the electrode in either lead at 90 days. Thus, it appeared that surface fissuring was significantly reduced in the harder Tecoflex polyurethane lead, compared to the softer Pellethane lead in the same environment. These results lead us to conjecture that if a silicone suture support sleeve had been used, the Tecoflex lead probably would have shown no surface fissuring at all.

Chemical analysis by attenuated total reflection infrared spectroscopy showed significant differences among stressed (cracked) and unstressed (not cracked) implanted specimens and the unimplanted control. These differences are shown in Figs. 6 and 7, and are summarized in Table 1. Apparent in this study was the finding of a mobile film on both the control and stressed samples. The films transferred to the IR crystal and their spectra were found to be similar and, in both, the relative amounts of polyether and polyurethane contributions were reduced. As would be expected when this film was removed by extraction from the control, the extracted control showed an increased polyether contribution. The mobile film from the stressed sample did show a small increase in the carbonyl peak that could be indicative of oxidation. The unstressed implant did not have the mobile phase, but did show a decrease in urea bands. Possible oxidation was seen in the stressed sample by the presence of -NH and -OH bands at 3500 cm^{-1}.

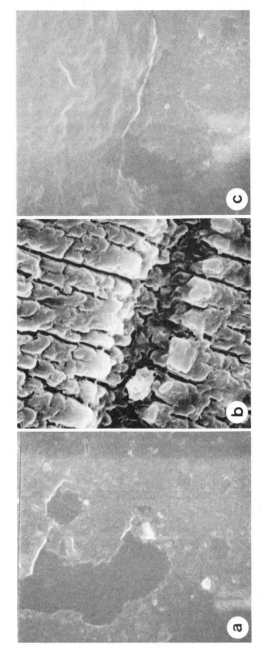

FIG. 5.—*Three-month dog implant leads, Pellethane 2363-80A. (a) Lead body section near connector end, view showing slight cracking, original magnification (×1000). (b) At site of vein ligation, view showing significant cracking, original magnification (×2000). (c) Lead body section near electrode, showing no cracking, original magnification (×5000).*

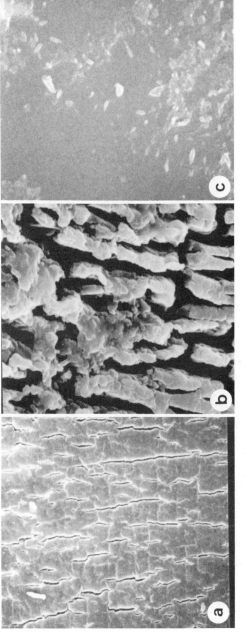

FIG. 4—*Three-month dog implant leads, Tecoflex EG-60D. (a) Lead body section near connector, showing no cracking, original magnification (×5000). (b) At site of vein ligation, view showing slight cracking, original magnification (×2000). (c) Lead body section near electrode, view showing no cracking, original magnification (×5000).*

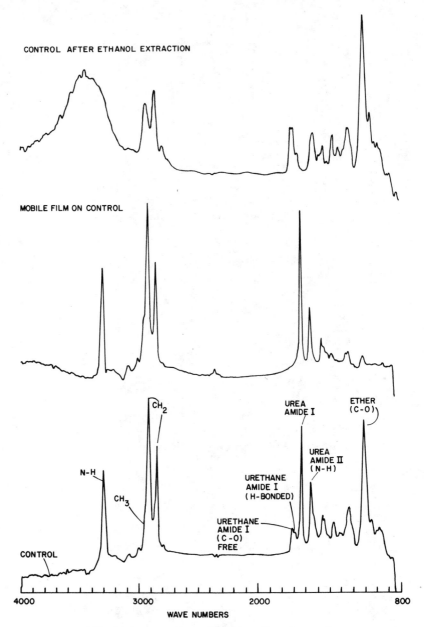

FIG. 6—*Surface analysis of unimplanted Tecoflex EG-60D.*

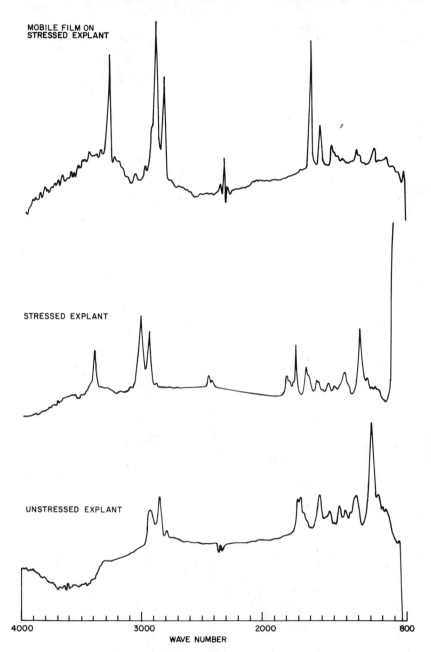

FIG. 7—*Surface analysis of cracked Tecoflex EG-60D explanted from swine after three months.*

TABLE 1—Changes in spectral features of explants compared with unimplanted control.

Mode	Sample	Urethane	Urea	Polyether	Presence of Mobile Component	Evidence of Oxidation
Unimplanted Control	Control	...	...	...	yes	no
	mobile film on control	1700 band reduced	similar	significantly reduced		no
	control after extraction with ethanol	similar to control	decreased	increased	no	...
Explants	unstressed explant	possible increase in H-bonded urethane	bands at 5310, 1640, 1560 absent	similar to control	no	no
	stressed explant	similar to control	similar to control	similar to control	yes	some indication of -OH and -NH bands at 3500 small CO band at 1700
	mobile film on stressed explant	similar to mobile film from control				

A tentative explanation suggested by these results is that polyether-based polyurethanes are susceptible to *in vivo* oxidation of the polyether chain, particularly when the polymer is under chronic stress. In the polyether chain, the most susceptible group is the -CH$_2$ group in the alpha position to the ether oxygen, which undergoes peroxidation, free radical dissociation, and eventual chain cleavage.

This analysis also indicates a net increase in soft segments at the surface of stressed implanted specimens. In the stressed specimens, the combination of mechanical stress and molecular cleavage produces a net decrease in total urethane bonds, thus increasing the relative amount and distribution of soft segments at the surface.

Discussion and Conclusions

Several mechanisms have been proposed as the cause of surface fissuring including enzymatic attack, protein and liquid adsorption with associated swelling, leaching of small molecular weight substances to the surface, and lipid absorption. There is little experimental evidence that supports any of these mechanisms.

Preliminary results from our studies indicate that *in vivo* surface fissuring occurs primarily in areas of excessively induced stress; we believe that excessively induced stresses may, in some cases, be related to specific manufacturing processes used in the production of the lead insulation and stresses induced in the implant environment. In addition, our IR analysis of the surface of cracked polyurethane specimens show the presence of hydroxyl (-OH) and end amino groups (-NH$_2$) indicative of oxidative byproducts. Amino groups could be obtained from residual protein surface layer; however, we have exhaustively washed the surface free of proteinaceous material, so we conclude that any hydroxyl or amino groups derive from the degradation of the polyurethane itself, rather than from any extraneous source.

Based on these preliminary results, we hypothesize that surface fissures may be due, at least in part, to oxidation of polyether-based polyurethanes in the -CH$_2$ group in the alpha position to the ether oxygen, which undergoes peroxidation, free radical dissociation, and eventual chain cleavage, leading to the formation of hydroxyl end groups. Our hypothesis has been reinforced by experimental evidence that surface fissuring can be significantly reduced with the use of higher durometer polyurethanes, since these polymers inherently contain fewer polyether macroglycol chains in the molecular architecture.

Pande [6] compared the long-term *in vivo* stability of a soft polyurethane (Shore 80A) to a harder one (Shore 55D). At intervals of up to one year, samples were explanted and observed under SEM. No evidence of surface fissuring was observed in the harder polyurethane. However, similar tests on the softer polyurethane, conducted for comparison purposes, did show some evidence of surface fissuring.

If long-term data substantiate these preliminary results, there are three possible ways to reduce the surface fissuring of polyurethane-based pacemaker leads: (1) utilize higher durometer polyurethanes, since they are inherently less susceptible to stress-induced oxidation; (2) lower ligature stress at the ligation site by the use of a silicone anchoring sleeve that spreads the stresses over a larger area; and finally, (3) control the manufacturing process to minimize the draw-down ratio of the tubing.

References

[1] Parins, D. J., McCoy, K. D., and Horvath, N. J., "*In Vivo* Degradation of a Polyurethane," Cardiac Pacemakers, Inc., St. Paul, MN, 1981.
[2] Guerrant, K., "Biostability of a Polyurethane," Intermedics, Inc., Freeport, TX, 1981.
[3] McArthur, W. A., "Long-Term Implant Effects on Three Polyurethane Leads in Humans," Pacesetter Systems, Inc., Sylmar, CA, April 1982.
[4] Ulrich, H. and Bonk, H. W. in *Proceedings*, 27th Annual Conference, Society of Plastics Industry, Bal Harbor, FL, 1982, p. 143.
[5] Szycher, M., Poirier, V. L., Dempsey, D. J., and Robinson, W. R., "Second Generation Biomedical-Grade Thermoplastic Polyurethane Elastomers," Society of Plastics Engineers, 1982 ANTEC, in press.
[6] Pande, G. S., "Polyurethane Insulation for Cordis Permanent Pacing Leads," Technical Memorandum 35, April 1982.

DISCUSSION

Emanuel Horowitz[1] (*written discussion*)—Have you attempted to measure the stress applied to the polyurethane materials during fabrication? What is the magnitude of the residual stress in the material at the time of implantation? Do you have any data on the relationship between the stress intensity factor and the rate (velocity) of crack propagation?

M. Szycher and W. A. McArthur (*authors' closure*)—During fabrication, we attempt to maintain the lowest possible stress. To that end, we specify an extrusion draw-down ratio of 2:1 (maximum). Under these circumstances, the residual strain in Tecoflex EG-60D tubing varies between 10 and 20%.

The magnitude of the residual stress in the material at the time of implantation depends on two factors: (1) residual stresses induced during manufacture, and (2) residual stresses induced during assembly. At implantation, total stresses are the sum of manufacture stresses plus assembly stresses.

We do not have a quantitative relationship between stress intensity and velocity of crack propagation. However, we do know that stress intensity and velocity at crack propagation are directly related. Thus, we recommend the use of Tecoflex EG-60D for pacemaker lead insulation, if residual stresses are minimized.

[1]Johns Hopkins University, Baltimore, MD 21218.

Rick Georgette[2] (*written discussion*)—Could you comment on the manufacturing variables employed with this material and their relationship to surface fissuring?

M. Szycher and W. A. McArthur (*authors' closure*)—The manufacturing variables affecting Tecoflex EG-60D are (1) extrusion draw-down ratio, (2) melt temperature, and (3) geometric relationship between die and finished tubing dimensions.

Our observations confirm that Tecoflex EG-60D tubing kept below 60% strain, do not surface fissure following *in vivo* implantation (subcutaneous).

G. I. Ogundele[3] (*written discussion*)—Comparing the ministructures of the unimplanted and explanted polyurethanes, the cracks observed in the explanted material must have initiated from the surface defects caused by manufacturing or surgical operations. Another possibility can be the presence of voids near the surface.

Did you perform detailed stress analysis around your specimen?

M. Szycher and W. A. McArthur (*authors' closure*)—We attempted to define the polyurethane surfaces by SEM and polarized light, both before and after implantation.

Photographic records were kept of representative surface characteristics, prior to implantation. From these records, we know that presence of surface voids near the surface accelerates fissuring. We hypothesize the reason for this phenomenon may be explained by realizing that voids are stress risers, and the increased stress at the imperfection accelerates fissuring.

James M. Anderson[4] (*written discussion*)—The ATR-FT-IR studies are subject to orientation of the specimen. Could this have influenced your results and were comparable polarization ATR-FT-IR studies carried out?

M. Szycher and W. A. McArthur (*authors' closure*)—Specimen orientation is indeed an important variable when analyzing surface chemical composition. We do not believe that orientation played a major role in our results, since we analyzed more than 15 specimens, both prior to and after implantation.

Polarization ATR-FT-IR studies were not carried out, since this is a new procedure and the equipment was not readily available to us.

[2]Richards Medical Co., Memphis, TN 38115.
[3]University of Calgary, Calgary, Alberta, Canada.
[4]Department of Pathology, Case Western Reserve University, Cleveland, OH 44106.

David J. Parins,[1] Kathleen D. McCoy,[2] Nicholas Horvath,[2] and
Russell W. Olson[2]

In Vivo Degradation of a
Polyurethane: Preclinical Studies

REFERENCE: Parins, D. J., McCoy, K. D., Horvath, N., and Olson, R. W., *"In Vivo
Degradation of a Polyurethane: Preclinical Studies,"* Corrosion and Degradation of
Implant Materials: Second Symposium, ASTM STP 859, A. C. Fraker and C. D. Griffin,
Eds., American Society for Testing and Materials, Philadelphia, 1985, pp. 322–339.

ABSTRACT: *In vivo* investigation of a polyether polyurethane for possible use as pacing
lead insulation material revealed surface degradation in the form of numerous cracks. The
cracking phenomenon appeared to be stress related. Cracks were observed after three
months implant time in rats and affected greater surface area with increasing implant time.

Toxicological investigation of the host animals revealed no adverse effects attributable
to the polymer cracking phenomenon.

Examination by scanning electron microscopy revealed cracking that raised concerns
about the integrity of thin-walled pacing leads. Ambient controls and *in vitro* specimens
did not develop surface cracks. Surface cracking was independent of weight averaged
molecular weight (Mw) and was observed on all tested samples, the Mw ranged from
255 000 through 636 000. No significant change in Mw after implant could be detected.

Surface degradation of this polyurethane was reproducible in controlled animal studies.

KEY WORDS: pacing leads, polyether polyurethane, scanning electron microscopy, surface cracking, failure mechanisms, stress (materials), implant materials

The advent of high-quality segmented polyether polyurethanes has provided
a polymer that possesses the superior physical properties of urethanes but with
reduced tendency for the hydrolysis and degradation exhibited by polyester
polyurethanes. Boretos [1][3] demonstrated excellent tissue response characteristics
and biostability of a polyether polyurethane in long-term subcutaneous applications. Later, the advantages of polyether polyurethanes as pacemaker lead
insulators were investigated and described by Stokes [2] and others [3–6]. The
low coefficient of friction, good mechanical strength, low moisture absorbence,
thermal setting, extrusion and injection molding capabilities should produce a
pacing lead that is thinner and stronger than conventional silicone leads. To date,
pacing leads represent the most extensive long-term blood contact application
of polyurethane polymers.

[1]Director of Research, Angiomedics, Inc., Minneapolis, MN 55441.
[2]Scanning electron microscopist, lead scientist, and microbiologist, respectively, Cardiac Pacemakers, St. Paul, MN 55164.
[3]The italic numbers in brackets refer to the list of references appended to this paper.

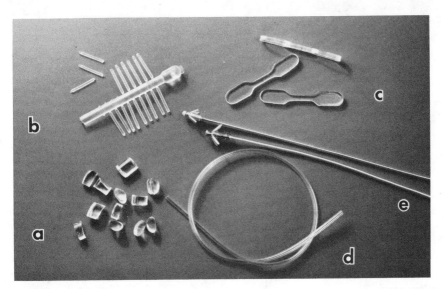

(*a*) Raw manufacturing pellets.
(*b*) Molded rod specimens 1 mm diameter by 10 mm long.
(*c*) Molded microtensile specimens 31.75 mm long by 1.57 mm thick by 1.57 mm wide in gage length.
(*d*) Commercial extruded tubing 1 mm inside diameter by 1.5 mm outside diameter by 220 mm long.
(*e*] Cardiac pacing lead; commercial and CPI prototype.

FIG. 1—*Polyurethane specimen forms tested.*

Preliminary animal implants with toxicological and tissue investigations of polyether polyurethane at Cardiac Pacemakers, Inc., (CPI) were initiated in early 1977 with a custom device implanted in 1979. The results of the preliminary examinations and the successful performance of the custom lead prompted initiation of a biomaterials evaluation of polyurethane for use in thin pacing leads.

Materials

Pellethane[4] 2363-80A was selected as a candidate lead material because it is a well-characterized segmented polyether polyurethane and is available for medical applications. This polymer was thought to possess the desired extrusion and injection molding characteristics for lead fabrication and is known to be used in commercial pacing leads.

The candidate polymer was evaluated in a variety of processed forms (Fig. 1). The choice of processed configurations was intended to facilitate implantation and to represent all anticipated fabrication methods. The final evaluation program examined actual CPI prototype leads as well as commercially available leads in a canine implant study.

[4]Registered trademark of Upjohn Chemical Co., Kalamazoo, MI 49001.

TABLE 1—*Pellets—bulk characterization.*[a]

USP Acute Systemic Toxicity (mouse)
USP Intracutaneous Toxicity (rabbit)
USP Pyrogen Test (rabbit)
30-day Intramuscular Implant (rabbit)
In Vitro Hemolysis Tests
In Vivo Hemolysis Test (rabbit)
MEM Elution Cytotoxicity (L929 cells)[b]
USP - Physiochemical Test
Dermal Sensitization (guinea pigs)
USP Mouse Safety Test (mouse)
Infrared Analysis[c]
Microscopic Examination[d]

[a]Testing performed at North American Science Associates, Inc., Northwood, OH, reported in May 1979; unless otherwise indicated.
[b]Extended elution study performed at CPI, reported Feb. 1980.
[c]Testing done by Honeywell, Inc., Hopkins, MN, reported May 1979.
[d]Performed at CPI.

Procedure

Bulk material characterization (pellets) consisted of tests summarized in Table 1.

Injection molded rod specimens suitable for hypodermic needle insertion were characterized by tests summarized in Table 2.

Injection molded microtensile specimens prepared using ASTM Test for Tensile Properties of Plastics by Use of Microtensile Specimens (D 1708-66), dimensionally modified, were implanted in Sprague Dawley albino rats, Table 3. At ½, 1, 2, 5, and 6 months, tissues around the implant site were examined histologically. To evaluate the effect of the *in vivo* environment on the polymer, physical properties, chemical analysis, and scanning electron micoscopic (SEM) examinations were made on explanted microtensile specimens. These tests were

TABLE 2—*Molded rods.*[a]

Agar Overlay Cytotoxicity (L929 cells)
MEM Elution Cytotoxicity (L929 cells)
USP Acute Systemic Toxicity (mouse)
USP Intracutaneous Toxicity (rabbit)
Dermal Sensitization (guinea pig)
Ames Test
30-day Intramuscular Implant (rabbit)
90-day Intramuscular Implant (rabbit)
120-day Subcutaneous Implant (rat)
USP Physiochemical Test
Light Microscopy[b]

[a]Tests performed at North American Science Associates, Inc., Northwood, OH, reported in March 1980, unless otherwise indicated.
[b]Performed at CPI.

TABLE 3—*First rat implant study.*

Microtensile Specimens	Host
SEM examination[a]	weight gain[e]
Tensile strength[b]	gross examination at sacrifice[g]
Elongation, %[b]	
Hardness[c]	
Light microscope[d]	local tissue histology[h]
Infrared spectrum[e]	
Molecular weight analysis[f]	

[a]Specimens were taken from saline storage solution, rinsed with deionized water, dehydrated in ascending concentrations of ethanol ending in 100% ethanol, and critical point dried. Dried specimens were mounted and sputter coated with gold or gold palladium, then examined in the SEM at 15 kV and 25° tilt.
[b]Specimens pulled at 10 mm/min crosshead speed (CPI).
[c]Performed at CPI.
[d]Performed at CPI.
[e]Performed at CPI.
[f]Performed by Jordi Associates, Millis, MA, reported in July 1980.
[g]Performed by North American Science Associates, Inc., and reported in May 1980.
[h]Performed by North American Science Associates, Inc., and reported in Nov. 1980.

also performed at the same explant intervals plus a nine-month explant. Control specimens and *in vitro* specimens (37°C, Locke Ringer's solution) were tested at each interval. As a result of this examination, an expanded test program, including microtensile and tubing specimens, was initiated to address five critical areas:

1. reproducibility of results and verification of specimen preparation methods,
2. the effects of polymer fabrication methods and residual stress,
3. the potential for systemic toxicity to the host,
4. the role of the implant site (subcutaneous, muscle, blood contact), and
5. correlation between polymer molecular weight and morphology changes.

The sample size was expanded to establish a broader statistical base and also employed an alternate independent laboratory for the implants and histological analyses. A post-cure cycle recommended by Upjohn (116°C, 16 h) was performed to minimize residual stress or other factors resulting from the polymer fabrication. Because pacing leads are usually constructed using both injection molded and extruded components, microtensile (molded) and commercial tubing (extruded) specimens were simultaneously implanted, Table 4.

To assess the potential for local or systemic toxicological response, the following factors were used to analyze the animal response over a six-month period: mortality, physical signs of toxicity, growth, food consumption, efficiency of food utilization, blood chemistry, hematology, gross examination of tissues and organs, and microscopic examination of implant sites, liver, kidneys, and lungs were examined in Wistar rats over a six-month period.

TABLE 4—*Second rat implant study (six months).*

Microtensile Specimens	Tubing	Host
Tensile strength[a]	tensile strength[a]	mortality
Elongation, %[a]	elongation, %[a]	physical indications
Light microscopy[b]	IR spectrum[b]	growth
IR examination[b]	Mw distribution[c]	food consumption
Mw distribution[c]	polarized light[d]	efficiency of food utilization
Polarized light[d]	SEM examination[e]	blood chemistry
SEM examination[e]		hemotology
		gross examination at sacrifice
		histological examination

[a]Specimens pulled at 30 mm/min crosshead speed (CPI).
[b]Performed at CPI.
[c]Performed by Jordi Associates, Millis, MA, reported June-Dec. 1981, and by Upjohn Co., D. S. Gilmore Lab, North Haven, CT, reported Dec. 1981.
[d]Performed at CPI. From *Strain Patterns*, American Laboratory, Theodore Rochow, Dec. 1980, p. 35.
[e]Same as Table 3, Footnote a.
[f]Performed at Lilly Research Laboratories, Greenfield, IL, reported Oct. 1981.

Molecular weight determination was used for detection of possible polymer backbone degradation [19,20]. Degradation that shortens the chain length should yield lowered molecular weight averages [21,22], if a significant portion of the test specimen has been affected. An examination by gel permeation chromatography (GPC) of specimens explanted at regular intervals of the study was conducted by two outside laboratories.

The possibility of interspecies differences was studied by including canine implants of microtensile and tubing specimens identical to those used in the rat protocol. Commercially available polyurethane unipolar, ventricular pacing leads and functional CPI prototype leads were implanted in canines to evaluate performance in an actual pacing application and to assess the cumulative effects of the complete manufacturing process, Table 5.

Results

The eluate from pellets extracted for 24 h in Minimal Essential Medium (MEM) at 37°C was not cytotoxic to L929 cells. However, eluate from pellets extracted for 120 h in MEM at 37°C produced a cytotoxic effect on L929 cells. The other biological testing did not reveal any significant effect (Table 1).

Histological examination of injection-molded rod specimens from intramuscular implants revealed a "cuff" of macrophages and monocytes surrounding all of the polyurethane pins in one animal of each group but not the polypropylene control pins. The agar overlay cytotoxicity test showed a 1-mm cytotoxic zone surrounding the specimen. All extracts prepared for the acute systemic and intracutaneous toxicity tests became thick and cloudy; the cottonseed-oil extract

TABLE 5—*Canine implants.*[a]

Microtensile Specimens	Tubing	CPI Prototype Lead	Commercial Pacing Leads
Elongation, %[b]	tensile properties[b]	tensile properties[b]	tensile properties[b]
Tensile strength[b]	SEM examination[b]	SEM examination[b]	SEM examination[b]
SEM examination[b]			
Visible microscopy[b]	pacing performance[b]	pacing performance[b]	pacing performance[b]

[a] All canines used in this study were random source and preconditioned for 30 days.
[b] All performed at CPI.

TABLE 6—Mechanical properties.

	Rat Study I			Rat Study II							
	Untreated - Microtensile Specimens			Microtensile Specimens				Tubing Specimens			
				Controls		In-Vivo 6 months		Controls		In-Vivo 6 months	
	Control	In Vivo 9 Months	In Vitro 6 Months	Post Cured	Untreated	Post Cured	Untreated	Post Cured	Untreated	Post Cured	Untreated
Ultimate tensile strength (kg/cm^2)	550	430	405	510	466	397	409	784	774	469	565
Standard deviation	48	36	27	90	25	7	67	52	12	108	18
Elongation, %	400	415	433	460	410	523	443	580	490	523	490
Standard deviation	19	26	23	150	106	6	49	20	16	57	10
Hardness (Shore D)	40.4	41.1	40.9	...	...	...	...	...	...	...	...
Standard deviation	0.6	0.3	0.8	...	...	...	...	...	...	...	...

Canine Study—Commercial Leads

	Implant Interval	Area	Ultimate Tensile Strength	Elongation, %
Lead 1	3 months	subcutaneous	456	600
	3 months	intravenous	284	500
Lead 2	6 months	subcutaneous	343	540
	6 months	intravenous	204	490

produced a mild response in the intracutaneous toxicity test. Other biological tests did not reveal any significant effect, Table 2.

The results of the mechanical property testing of *in vivo* and *in vitro* microtensile specimens, which included hardness, tensile strength, and elongation, are shown in Table 6. No significant alteration, except an initial decrease in tensile strength already documented in the literature [2], was revealed.

Gel permeation chromatography (GPC) analysis on two explanted specimens revealed no significant differences in molecular weight (Mw) from control specimens (Mw control 375 000, Mw explant 376 800 relative to polystyrene, Table 3).

Histological evaluation showed minimal local tissue response during the first 28 days after implant. Except for a slight local tissue response from 28 to 181 days, no other effects were noted on the host, Table 3.

Our first indication of unexpected polymer behavior resulted from SEM examination of microtensile specimens explanted after three months in rats. Surface cracking was observed on the gage length of the microtensile specimens (Fig. 2). The observation of cracks raised concerns about the long-term integrity of thin-wall pacing leads. No surface changes were observed on specimens examined prior to this time. The extent of surface covered by the cracked areas increased with time, and by nine months had spread to the grip region of the

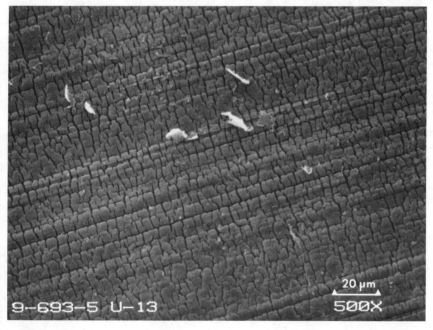

FIG. 2—*First observation of surface cracking, three-month rat explant. SEM magnifications are from originals.*

FIG. 3—*Control microtensile specimen illustrating a very thin section (flash) of polymer prior to implant. Example of microtensile specimen nine-months postimplant shows cracking and loss of entire flash area structure.*

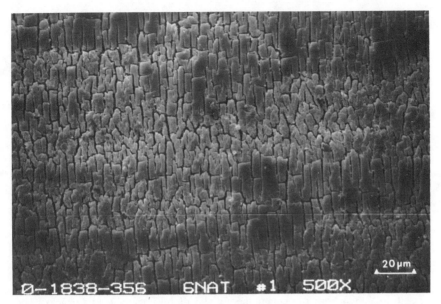

FIG. 4—*Commercial tubing, three-month subcutaneous rat implant.*

microtensile specimen as well. Even more significant was the observation that polymer flash had been completely removed at some locations (Fig. 3). (Flash is the fine ridge of polymer formed during injection molding when the melted polymer flows between the two mold halves.) The observations of cracked and missing flash confirmed that cracking was not a protein coating. Ambient controls and *in vitro* specimens (stored at 37°C in Locke-Ringer's solution) did not develop surface cracks.

A six-month rat implant study of microtensile and commercial tubing specimens verified the original observations. Cracks appeared at approximately three months onset time. A regular cracking pattern was observed on both types of specimens (Fig. 4).

Significant molecular weight changes were not detected. The molecular weight (Mw) of the implanted polymer specimens ranged from values greater than 636 000 to approximately 250 000, Table 7. Some specimens of polymer were only partially soluble and accurate Mw analysis was not possible. Insoluble polymer specimens had molecular weights greater than 636 000, solubility of the polymer being to a first approximation inversely proportional to the square root of the Mw. The manufacturer's recommended post cure for stress relief did not prevent the formation of surface cracking on specimens.

Review of the biological data showed no significant symptoms of systemic toxicity or local tissue response.

On canine implants, the appearance of surface cracking on microtensile specimens at six months was a longer induction period than observed for rats. The

TABLE 7—*Weight Average (molecular weight in thousands).*[a]

	Tubing Specimens		Microtensile Specimens	
	Post Cured	Untreated	Post Cured	Untreated
Control	260	300	636	partially soluble
1 month	256	297	444	partially soluble
2 months	255	315	540	partially soluble
3 months	268	314	539	partially soluble

[a]Based upon polystyrene.

cracking was not as extensive after six months in canines as that observed in the rat implants, but surface coverage did increase with time. Unexplained surface irregularities were observed (Fig. 5) on both specimen types at three months. The significance of this morphology change is unknown. No such change appeared in rat implant specimens.

Canine implanted leads (Fig. 6), exhibited cracking after six months. Cracking was observed on both the subcutaneous and blood contact lead surfaces. The highly stressed regions of commercial leads, suture tie-down point, tubing overlap, and molded tines exhibited extensive cracking, more pronounced than that seen on the other surfaces of the same leads (Figs. 7, 8, and 9).

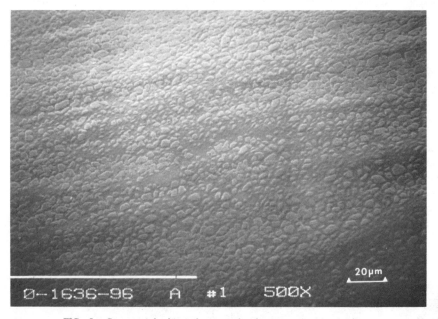

FIG. 5—*Commercial tubing, three-month subcutaneous canine implant.*

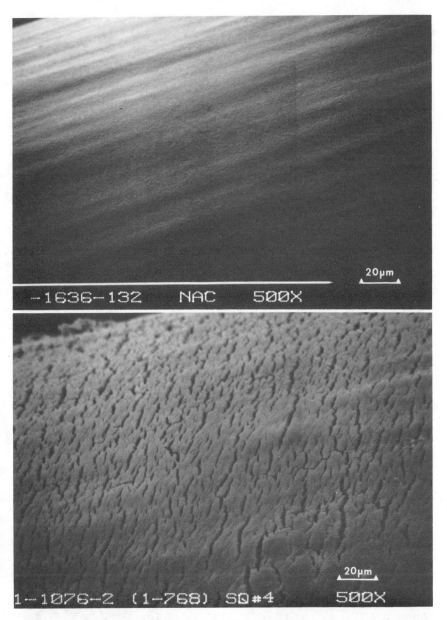

FIG. 6—*Typical commercial tubing control, Pellethane 2363-80A tubing has smooth extrusion lines with a slight surface texture. Surface of commercial pacing lead after six months subcutaneous canine implant.*

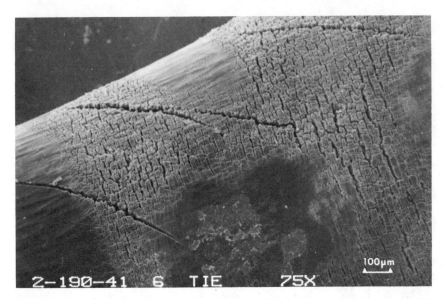

FIG. 7—*Suture tie down—the smooth section is where the suture was in direct contact with the tubing.*

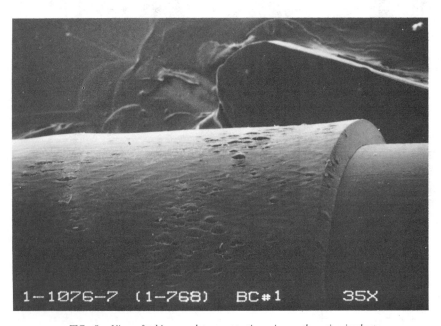

FIG. 8—*View of tubing overlap construction, six-month canine implant.*

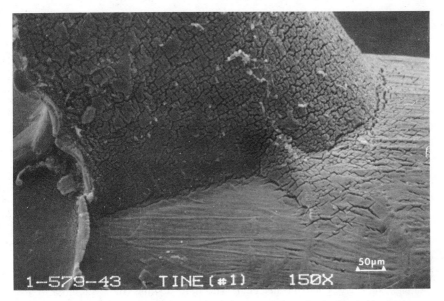

FIG. 9—*Tine section of commercial pacing lead after six-months canine implant in blood contact.*

Within the experimental time frame (12 months), no evidence of lead insulation failure was detected. Mechanical analysis revealed a large variance in tensile strength, Table 6.

Discussion

The initial observation of surface cracking on implanted Pellethane 2363-80A was unexpected. Additional studies were performed to help identify the causes of the cracking phenomenon and to assess the safety and efficacy of long-term implanted pacemaker leads fabricated from this polymer. Correlations between experimental parameters and surface cracking were examined.

Mechanical property testing of explanted polyurethane specimens did not reveal any correlation with material changes observed. Tension testing of tubing inherently yields high variances. Implanting thin-film microtensile specimens with a greater surface-to-volume ratio may yield more meaningful results.

The acute toxicological program utilized in this study consisted of test methods, commonly used by the medical device industry to screen materials for possible adverse reactions. The Pellethane 2363-80A specimens investigated in this study did not produce toxicological evidence contraindicating the consideration of this material for implant applications. This study evaluated a limited set of material parameters for toxicological effects. Changes in any of the following factors may alter the results: material formulation or processing, the types of tissues exposed, the amount of polymer implanted, and duration of implant. The hematological or toxicological significance of the cracked lead surface remains to be investigated. Additional work would be necessary for detecting adverse host responses that require long onset time periods. Because of manufacturing variations, comprehensive long-term implant studies utilizing finished product provide the best information of the appropriateness of using Pellethane 2363-80A for treating cardiovascular disease.

The appearance of surface cracks on implanted polymer specimens was independent of the polymer weight average molecular weight. The molecular weights were obtained by GPC on cross sections of the explanted and control specimens. It would have been desirable to produce surface slices, but the resilient nature of the elastomer prevented the reproducible preparation of thin slices. Anticipated changes in surface Mw would therefore be diluted by the bulk polymer Mw. Any chemical degradation of the polymer backbone would not alter the Mw until it affected a significant portion of the bulk polymer. GPC analysis of polyurethanes is difficult. The polymer solutions in DMF (N, N- dimethylformamide) required stabilization to reduce the tendency for autodegradation during the analysis. No consistent correlation between implant time, surface changes, and Mw could be detected. The implant time (three months) of the samples analyzed was too brief in comparison to the time required for the cracking to affect a substantial portion of the polymer. Evaluation of implanted thin polymer specimens should provide more meaningful results.

SEM has proven to be a valuable tool for early detection of polymer degradation and surface defects. SEM has been used to detect defects in polyurethane catheters [9,10] and to investigate calcification of the polyurethane Total Artificial Heart [11,12].

The surface cracking of implant specimens exhibited a consistent morphology. Initiation was perpendicular to both material flow and stress boundary lines (Fig. 10), and then smaller secondary cracks developed creating a cross-hatched pattern. This pattern is consistent with that seen by other investigators [14,15].

This polyurethane surface cracking appears to be related to internal stresses. The morphology of the cracks (perpendicular to stress boundaries) and initial appearance in the highly stressed areas leave little doubt that residual stresses are an important factor in the appearance of cracking. Residual stresses from processing were localized in the gage length of the microtensile specimens and in molded tines (Figs. 2 and 7). Design related stress was evident on a commercial lead where the tubing was expanded and overlapped (Fig. 9). The effects of application-induced stress are particularly apparent at the lead suture tie-down locations (Fig. 8). The post-cure heat treatment did not significantly alter the internal stresses within the polymer specimens. Annealing at temperatures close to the melting point are now known to be more effective at reducing residual stress. Such processing should, however, be rapid and require a well-dried polymer to prevent degradation. It remains to be shown if the surface cracking can be completely eliminated through improved manufacturing process control and design.

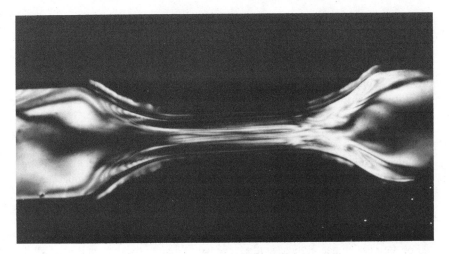

FIG. 10—*Microtensile specimen: polarized light stress patterns.*

The observation of crack propagation completely through the polymer flash (Fig. 3) illustrated that the observed surface cracking cannot be attributed to a protein coating or other unremoved biological debris. Explanted specimens, enzymatically cleaned to remove biological debris, have exhibited the same regular morphology [14]. Examples of protein and coating artifacts have been observed and present a random crack morphology.

CPI studies revealed that the initial appearance of cracking was time dependent. The relationship of surface crack depth to implant time is uncertain. Clinical results suggest a time dependency [14–15]. But others have theorized that surface cracking is self-limiting and will stop at a postulated morphology change identified as the "zero-stress boundary layer"[8,13].

Evidence that pacing leads develop surface cracking clinically has been documented by pacing manufacturers [8,13–16] and also by clinicians [17–18]. Lead insulation failure in the presence of high stress has also been acknowledged [8,13,14,18]. Pande [16] reported on a polyether polyurethane of a Shore Hardness D55. No evidence of cracking was found on specimens examined up to one year subcutaneous implantation in rabbits. A polyether polyurethane of Shore A 85 did exhibit cracking in his study. The *in vitro* experiments did not produce any observable cracking. However, the effects of lipids, blood serum, and other body fluids not present in the *in vitro* media on polymer swelling remain to be investigated.

Is the *in vivo* surface cracking of this polyurethane an example of biodegradation? If one defines biodegradation as a structural or chemical change caused by the biological environment, then it appears likely to be biodegradation. If the definition is limited to a chemical modification of the polymer, then the evidence is not conclusive at this time. Any observation of cracking resulting from an *in vitro* experiment would support the theory that the cracking is the result of solvent swelling and subsequent stress cracking [7].

Three separate animal studies have confirmed our findings that surface cracking of this polyurethane occurs *in vivo*. Secondly, surface changes occur in different animal models and implant sites, subcutaneous and blood contact. Cracking occurred in injection molded and extruded tubing specimens. The relationship between stress and surface cracking was evident in commercial pacing leads.

The rate and extent of surface degradation is influenced by such differences as physiological conditions, material processing methods and specimen configuration. The performance of Pellethane 2363-80A for long-term human implant is contingent upon proper design, fabrication, and application.

Acknowledgments

For their assistance in these studies, we would like to thank Kris Black, Molly Vomhof, Edward Kubitschek, and Nick Youker.

References

[1] Boretos, J. W., Detmer, D. E., and Donachy, J. H., *Journal of Biomedical Materials Research*, Vol. 5, 1971, p. 373.

[2] Stokes, K., Cobian, and K., Lathrop, T., *Proceedings*, VIth World Symposium on Cardiac Pacing, Claude Meere, Ed., PACESYMP, Montreal, Canada, 1979, Chapter 28-2.

[3] Devanathan T. and Sluetz, J. E., *Biomaterials, Medical Devices, and Artificial Organs*, Vol. 8, No. 4, 1980, pp. 369–379.

[4] Irnich, W., Scheuer-Leeser, M., and Kreuzer, J., "Comparison Between Silicone and Polyurethane Insulation of Leads," 4th Annual German Pacemaker Workshop, Cologne, W. Germany, Oct. 1981.

[5] Shindo, G., Mizuno, A., Furuse, A., Kotsuka, Y., Fu, G., Saigusa, M., Matsuo, H., and Takayanagi, I. H., *Artificial Organs*, Vol. 5, No. 3, Aug. 1981, p. 324.

[6] Biester, F. D., Behrend, D. M., and Klinkman, H., "Assessment of Mechanical Properties and Biocompatibility of Pellethane®," Austrian Society for Artificial Organs Second International Workshop, Innsbruck, Austria, Nov. 1981.

[7] *Technical Bulletin, 023*, Pellethane 2363-80A, Upjohn CPR, Torrance, CA, Aug. 1978.

[8] Termin, P., "Biostability of Urethane Cardiac Pacing Leads," No. 4, Medtronic, Inc. 12 April, 1982.

[9] Nachnani, G. H., Lessin, L. S., Motomiya, T. et al, *New England Journal of Medicine*, Vol. 286, 1972, pp. 139–140.

[10] Bourassa, M. G., Cantin, M., Sanborn, E. B. et al, "Scanning Electron Microscopy of Surface Irregularities and Thrombogenesis of Polyurethane and Polyethylene Coronary Catheters," *Circulation*, Vol. 53, 1976, pp. 992–996.

[11] Lemm, W., Krukenberg, T., Gerlach, K., and Bucherl, S., "Biodegradation of Some Biomaterials After Subcutaneous Implantation," European Society for Artificial Organs, Eighth Annual Meeting, Copenhagen, Denmark, 1981.

[12] Coleman, D. L., Lim, D., Kessler, T. et al, Calcification on Non-Textured Implantable Blood Pumps," *Proceedings*, American Society for Artificial Internal Organs, Vol. 10, 1981, p. 3.

[13] Stokes, K., "The Long-Term Biostability of Polyurethane Leads," *Stimucoeur Medical*, 1982.

[14] McArthur, W. A., "Long-Term Implant Effects on Polyurethane Leads in Humans," Pacesetter Systems, Inc., April 1982.

[15] Fahlstroem, U., "Polyether Polyurethane as an Insulation Material for Pacemaker Leads—Some Hesitations" Polyurethane in Medical Technics, International Colloquim, Stuttgart, W. Germany, Jan. 1983.

[*16*] Pande, G. S., "Thermoplastic Polyurethanes as Insulating Materials for Long Life Cardiac Pacing Leads," Cordis Corp., May, 1982.

[*17*] Timmis, G. C., "Scanning Electron Microscopy of Surface Defects in Chronically Implanted Polyurethane Cardiac Pacing Leads," North American Society for Pacing and Electrophysiology, 3rd Annual Session, Vol. 47, 1982 (abstract).

[*18*] Godin, J. F. and Welti, J. J., *The Stimarec Bulletin*, 31 July, 1982.

[*19*] *Aspects of Degradation and Stabilization of Polymers*, Jellinek, H. H. G., Ed., Elsevier, New York, 1978.

[*20*] Yau, W. W. and Kirkland, J. J., *Modern Size Exclusion Liquid Chromatography*, John Wiley Interscience, New York, 1979.

[*21*] Billmeyer, I., *Textbook of Polymer Science*, "Polymer Solutions," 2nd ed., Wiley, New York, 1971 Chapter 2.

[22] Flory, J., *Principles of Polymer Chemistry*, Cornell University Press, 1978, Chapter X.I.

DISCUSSION

Todd Smith[1] (*written discussion*)—What other candidate materials are you considering for this application?

D. J. Parins, K. D. McCoy, N. Horvath, and R. W. Olson (*authors' closure*)— The other candidate materials being considered for this application are: B. F. Goodrich's Estane, Mobay's Texin, Quinn's Q-Thane, Thermo Electron's Tecoflex, and Upjohn's 55D Pellethane.

[1]DePuy, Inc., Warsaw, IN 46580.

Robert A. Casper,[1] Richard L. Dunn,[1] and Donald R. Cowsar[1]

Biodegradable Fracture Fixation Plates for Use in Maxillofacial Surgery

REFERENCE: Casper, R. A., Dunn, R. L., and Cowsar, D. R., **"Biodegradable Fracture Fixation Plates for Use in Maxillofacial Surgery,"** *Corrosion and Degradation of Implant Materials: Second Symposium, ASTM STP 859*, A. C. Fraker and C. D. Griffin, Eds., American Society for Testing and Materials, Philadelphia, 1985, pp. 340–348.

ABSTRACT: Partially biodegradable composites were prepared from a biodegradable polyester, reinforced with carbon or ceramic fibers, and evaluated for their potential use in fixing mandibular fractures resulting from maxillofacial injuries. The composites, made using a lamination technique, were found to possess the strength and rigidity of bone. Their incubation in phosphate-buffered saline at 37°C resulted in the gradual reduction of both flexural strength and flexural modulus of the samples. Reductions in strength and rigidity were paralleled by commensurate reductions in the viscosity of the matrix resin. Evaluation of the model system demonstrated the potential for improved management of fracture wounds using biodegradable materials.

KEY WORDS: polymer, ceramic fibers, carbon fibers, reinforced plastics, fiber composites, wound healing, mandible, reinforcement, implant materials

The need for improved methods and materials to manage severe maxillofacial injuries is well recognized. In such cases, the surgeon faces the dual problem of restoring function and appearance. The patient suffering from an extensive maxillofacial injury is typically confronted with disfigurement, impaired speech, and eating difficulties as well as the psychological trauma resulting from the injury. Although such injuries happen to individuals in all walks of life, a substantial number of these injuries occur to military personnel in the line of duty. In recent years, groups such as the Army Institute of Dental Research have investigated numerous approaches to improving surgical methods and patient care. As a part of these studies, they have constantly sought more satisfactory materials for implantation and fracture fixation.

Implant materials used over the years belong to three traditional classes: metals, ceramics, and polymers. The choice of material for a particular application depends on the type and magnitude of applied loads that the implant is expected

[1]Head, Polymer Division; head, Biomaterials Section; and director, Applied Science Department; respectively, Southern Research Institute, Birmingham, AL 35255.

to experience *in vivo* and whether the implant is to be a permanent or a temporary augmentation. When trying to make repairs to the skeletal systems, surgeons and engineers must attempt to replicate the static and dynamic responses of bone. Bone consists of a framework of collagenous fibers, a mineral matrix consisting primarily of calcium hydroxyapatite, and a small amount of polysaccharides. Bone exhibits a higher flexural strength and flexural modulus than polymeric materials but is weaker and more deformable than metals. Currently, high-strength graphites demonstrate the combination of stiffness and strength exhibited by bone, but these materials are generally unsatisfactory as substitutes because they are substantially less ductile than bone.

Historically, metals have received wide application for the construction of devices for fixing fractures. Metals exhibit high values of tensile strength and compressive modulus; they can be fabricated into fixation hardware by a variety of conventional techniques; and they provide excellent resistance to the *in vivo* environment. Metals and alloys now used as surgical implants include Type 316 stainless steel, several cobalt-chromium alloys, titanium, zirconium alloys, and tantalum.

In mandibular fracture repair, the major disadvantages with metal implants are the need for follow-up surgery to remove the implant and atrophy of the healing bone as a result of the stress-protection effect of the metal plate.[2] The Army Institute of Dental Research demonstrated that the need for surgical removal of the implants after fracture union could be alleviated by using biodegradable polymers for fixation plate construction.[3] Aliphatic polyesters such as poly(DL-lactide) (DL-PLA) and poly(glycolide) (PGA) were made into plates of the same geometry and dimensions as conventionally used steel plates. Screws made from the same polymers were used to attach the plates to the mandibles of dogs. The animals were sacrificed at selected intervals up to 40 weeks, and they showed no signs of rejection, edema, or discoloration. These experiments demonstrated that mandibular fixation with resorbable implants was possible, that the devices could be customized, that degradation of the device was slow and uneventful, and that a second operation to remove them was not required.

Although the results of these pilot studies demonstrated that the biodegradable polymers offered new possibilities in oral surgery, the experiments revealed deficiencies with the implants. Primarily, the implants used in the preliminary studies did not exhibit sufficient flexural strength and rigidity during the first four to six weeks post-implantation. Lack of strength and rigidity during the early stage of healing contributes to mobility of the fracture site, results in improper alignment, and presents difficulty in the formation of the initial osseous union.

[2]McKenna, G. G., Bradley, G. N., Dunn, H. K., and Statton, W. O., *Journal of Biomedical Materials Research*, Vol. 13, 1979, pp. 783–798.

[3]Getter, L., Cutright, D. E., Bhaskai, S. N., and Augsburg, J. K., *Journal of Oral Surgery*, Vol. 30, May 1972, pp. 344–348.

TABLE 1—*Mechanical properties of reinforcing fibers.*

Fiber	Type	Tensile Strength, GPa	Tensile Modulus, GPa
Thornel 50 (Union Carbide)	carbon	2.2	393
Nextel AB 312 (3M)	ceramic	1.7	152
Fiber FP (Du Pont)	ceramic	1.4	379

In our study, we investigated the possibilities for obtaining the strength and rigidity necessary for initial fracture union through using biodegradable polymer modified by the incorporation of reinforcing fibers. Although our work relates to the study of model systems that are not totally biodegradable, the ultimate goal of our investigation is to develop a completely absorbable system adequate for initially fixing fractures. And the system should be capable of transferring an increasing portion of the mastication forces to the newly formed bone as healing progresses. This transfer results in a fixation site that, after healing is complete, exhibits the same strength and rigidity as virgin bone.

Biodegradable Resins with Fiber Reinforcement

Homogenous fixation plates fabricated from biodegradable polymers were shown to possess insufficient strength and rigidity for initial fracture fixation. Thus, plate reinforcement with high-strength, high-modulus, nonbiodegradable fibers was examined as a means of achieving the elevated mechanical properties needed in these devices as part of a model system study. We examined both carbon and ceramic fibers as potential reinforcing media. The composition and manufacturer's data on the fibers we selected for evaluation are shown in Table 1. The diameter of fibers used for reinforcement ranged from 7 to 21 μm. All fibers were stripped of sizing prior to use in composite fabrication. We prepared and laminated both reinforced and nonreinforced films of poly(DL-lactide) to give fixation plates of the dimensions required.

Preparation of Films

We prepared films of DL-PLA from purified polymers having inherent viscosities from 0.058 to 0.127 M^3/Kg measured in chloroform at 30°C. We made nonreinforced films by injecting approximately 97 cm^3 of a 13% solution of DL-PLA in *p*-dioxane into a 7.62 by 22.85 cm spin-casting cup rotating at 216 000 rps. The spin cup was lined with a sheet of Mylar film to prevent the polymer from sticking to the interior of the cup. After the solvent evaporated, we removed the film from the cup and dried it at 60°C in vacuo for 86 400 s (24 h) to remove all traces of solvent. We then placed each film in a hydraulic press at 70°C and 275 MPa to remove bubbles caused by solvent evaporation. We prepared reinforced films by wrapping either carbon or ceramic fiber around a Mylar-coated mandril and spraying the wrapped mandril with a solution of DL-PLA in *p*-dioxane. After the polymer dried, we repeated the spraying to obtain a composite film of the desired thickness. When the appropriate thickness was obtained, we removed the film from the mandril, cut it, and pressed it with a hydraulic press maintained at 70°C and 200 MPa to provide flat, bubble-free sheets.

Preparation of Laminates

We prepared test samples of the partially biodegradable composites by laminating reinforced and nonreinforced film sections (7.62 by 1.27 cm) in alternating layers under heat (70°C) and pressure (200 MPa) to obtain composites

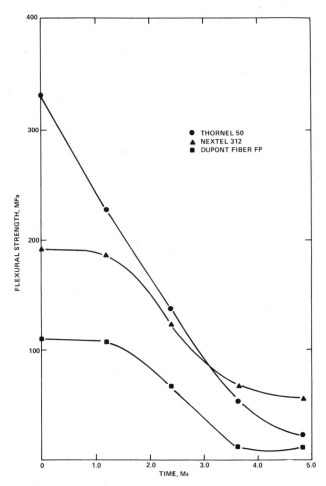

FIG. 1—*Effect of* in vitro *exposure on flexural strength of fiber-reinforced low-viscosity DL-PLA laminates.*

approximately 0.32-cm thick. In this fashion, we prepared laminates containing 55% carbon fiber by weight and laminates comprising 55% ceramic fiber by weight. For all samples, the reinforcing fibers were oriented parallel to the long axis of the specimen.

Testing and Evaluation

We evaluated the fiber-reinforced DL-PLA laminates for flexural strength and flexural modulus using ASTM Tests for Flexural Properties of Plastics and Electrical Insulating Materials (D 790–71). We made property determinations on samples conditioned for 86 400 s and on samples incubated in phosphate-buffered saline (0.2 M, pH 7.4) at 37°C for periods ranging up to 5 Ms. Incubated samples were removed from solution, blotted dry, and conditioned in a constant temperature and humidity room for 86 400 s prior to testing. All testing was done on an Instron universal testing machine.

We determined inherent viscosity on the matrix portion of incubated composite samples to ascertain the extent of chain degradation resulting from hydrolytic cleavage. We dissolved the composite samples in dichloromethane, and we filtered the resulting solution to remove the reinforcing fibers. Then we dried the filtrate under heat and vacuum to remove the dichloromethane. We dissolved the dried polymer in chloroform and determined the inherent viscosity at 30°C. We determined flexural strength, flexural modulus, and inherent viscosity on incubated and nonincubated control samples that did not contain reinforcing fibers.

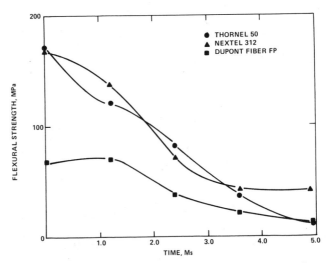

FIG. 2—*Effect of* in vitro *exposure on flexural strength of fiber-reinforced medium-viscosity DL-PLA laminates.*

Results

Flexural Strength

Composites fabricated using all three types of reinforcing fiber and using both low and medium molecular-weight DL-PLA were far superior in flexural strength to nonreinforced control samples. (The data are presented in Figs. 1 and 2.) Composites fabricated with Thornel 50 fiber and Nextel 312 fiber were well in excess of 100 MPa for initial flexural strength. The use of DuPont Fiber FP yielded composites of lower strength for DL-PLA resins with both low and

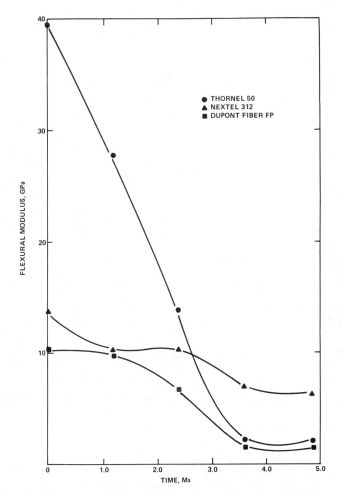

FIG. 3—*Effect of* in vitro *exposure on flexural modulus of fiber-reinforced low-viscosity DL-PLA laminates.*

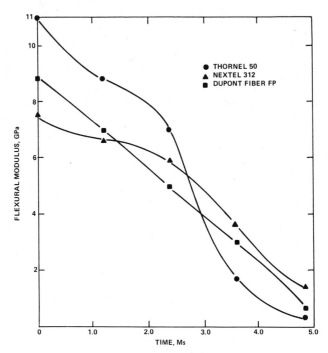

FIG. 4—*Effect of* in vitro *exposure on flexural modulus of fiber-reinforced medium-viscosity DL-PLA laminates.*

medium molecular weight. Composites made from both resin grades declined steadily in flexural strength as a result of incubation in phosphate-buffered saline.

Flexural Modulus

The effect of incubation time and choice of reinforcement fiber on composite flexural modulus paralleled the effect of these parameters on flexural strength. The incorporation of Thornel 50, Nextel 312, and DuPont Fiber FP resulted in composites with initial flexural moduli in excess of 8 GPa for both low and medium-viscosity resins. (The data are presented in Figs. 3 and 4.) Composites from low-viscosity resins were stiffer than composites made with medium-viscosity resins. In composites using both resin grades as a matrix material, we observed a steady decline in stiffness with incubation in phosphate-buffered saline.

Inherent Viscosity of DL-PLA Resins

Both low and medium-viscosity DL-PLA resins demonstrated marked decreases in viscosity during incubation in phosphate-buffered saline. (The data are presented in Figs. 5 and 6.) Viscosity decreases paralleled the declines in both flexural modulus and flexural strength.

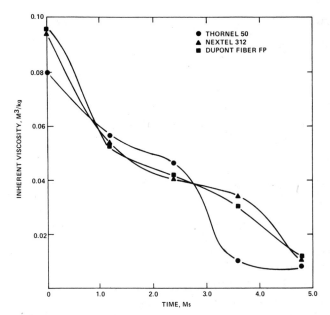

FIG. 5—*Effect of* in vitro *exposure on inherent viscosity of low-molecular-weight DL-PLA laminates.*

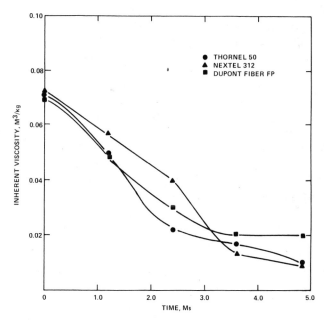

FIG. 6—*Effect of* in vitro *exposure on inherent viscosity of medium-molecular-weight DL-PLA laminates.*

Discussion

The development of fracture fixation plates that are bioabsorbable and that do not produce stress-protection atrophy of the healing bone is a significant benefit to the patient suffering from maxillofacial injuries in both treatment cost and the formation of a healthy fracture union. Such a fixation system should be rigid enough initially to immobilize the fracture site and gradually shift increasing portions of the mastication forces to the healing bone. The ideal fixation system should be completely absorbed when healing is complete.

Our model study demonstrates that partially absorbable fixation plates with the initial strength and stiffness of bone can be fabricated from a biodegradable polyester using high-strength high-modulus fibers as a reinforcement material. We further demonstrate that, in a model environment, composites of this type exhibit gradual declines in both strength and rigidity that is commensurate with a decrease in matrix-resin viscosity. This decline in mechanical properties affords the mechanism for gradually shifting mastication forces to the healing bone and alleviating the stress-protection atrophy effect encountered with the use of metal fixation plates. Although we used only one resin material for this study, numerous other biodegradable homopolymers and copolymers are available for potential use in the construction of composite fixation plates. Selection of resin material is one means of tailoring time-dependent load transfer between plate and bone. As reflected by viscosity measurements, control of molecular weight distribution of the polymer provides a second method of controlling load transfer.

We used nonbiodegradable fibers as the reinforcement material in this study. The development of absorbable, nontoxic, high-strength, high-modulus fibers would allow the construction on a totally biodegradable fixation system, and our continuing research is directed to solving this problem.

Conclusions

The use of biodegradable polyesters with fiber reinforcement allowed us to construct fracture fixation plates comparable in strength and rigidity to human bone. The development of the appropriate, biodegradable reinforcing materials used in conjunction with the resin or other degradable resins allows the development of devices to cover a wide range of surgical and patient needs in maxillofacial and orthopedic repair as well as other areas.

Acknowledgment

This study was supported by the U.S. Army Institute of Dental Research under contract DAMD17-78-C8059.

Ceramics, Polymers, and Metals: *In Vivo* and *In Vitro*

Harry A. McKellop[1] and Ian C. Clarke[2]

Degradation and Wear of Ultra-High-Molecular-Weight Polyethylene

REFERENCE: McKellop, H. A. and Clarke, I. C., **"Degradation and Wear of Ultra-High-Molecular-Weight Polyethylene,"** *Corrosion and Degradation of Implant Materials: Second Symposium, ASTM STP 859*, A. C. Fraker and C. D. Griffin, Eds., American Society for Testing and Materials, Philadelphia, 1985, pp. 351–368.

ABSTRACT: A ten-station joint simulator was used to examine the wear properties of 18 total hip prostheses. The wear rates of polyethylene acetabular cups bearing against titanium alloy femoral components were compared to those with either Type 316 stainless steel or cobalt-chrome alloy controls. Three titanium alloy prostheses and three controls were tested from three different manufacturers. Wear was determined by weighing the acetabular cups, using soak-controls to correct for fluid absorption. One million cycles were run under physiological loading, with bovine serum lubrication. For two of the three sets of prostheses, the titanium alloy generated higher mean polyethylene wear rates than the controls. However, the difference was not statistically significant ($p > 0.05$). The titanium alloy components showed more extensive surface abrasion than either the stainless steel or cobalt-chrome components. This appeared to be an artifact of joint simulator testing, since such extensive surface abrasion was not observed on numerous prostheses removed from patients after several years of use. The third set of titanium alloy prostheses, which had been subjected to a special surface hardening process, showed virtually no surface abrasion, and the mean polyethylene wear rate was identical to that with the cobalt-chrome controls. All of the polyethylene wear rates were in the range generally reported for total hip prostheses in clinical use.

KEY WORDS: prosthetic hips, joint simulator wear tests, clinical wear, titanium alloy, cobalt-chromium-molybdenum implant materials, fatigue (materials), biological degradation

In previous laboratory studies of the wear of polyethylene bearing against Ti-6Al-4V (titanium alloy), this material combination has shown both very good and very poor wear properties, depending on the particular wear test apparatus and experimental protocol. We previously published a detailed review of such experiments [1],[3] which noted that the poor wear resistance exhibited by Ti-6Al-4V in some laboratory studies was contradicted by the excellent clinical results that have been obtained with this alloy [2]. This suggested that the

[1]Instructor in Research Orthopaedics, Orthopaedic Biomechanics Laboratory, Division of Orthopaedics, University of Southern California, Los Angeles, CA 90007.
[2]Director, Bioengineering Research Institute, Los Angeles, CA 90025.
[3]The italic numbers in brackets refer to the list of references appended to this paper.

TABLE 1—*Femoral components.*

Type	Ball Diameter, mm	Alloy	Passivation	Manufacturer
T-28	28	316 Stainless steel (ASTM F 138-82)[a]	Hot nitric acid (ASTM F 86-76)[d]	Zimmer, Inc. Warsaw, IN
STH	28	Ti-6Al-4V (ASTM F 136-79 and F 620-79)[b]	Hot nitric acid (ASTM F 86-76)	Zimmer, Inc.
Anitomic	32	Cobalt-chromium-molybdenum (ASTM F 75-82)[c]	Hot nitric acid (ASTM F 86-76)	Hexcel, Inc., Dublin, CA
Anitomic	32	Ti-6Al-4V (ASTM F 136-79 and F 620-79)	Hot nitric acid (ASTM F 86-76)	Hexcel, Inc.
Link	32	Cobalt-chromium-molybdenum (ASTM F 75-82)	Hot nitric acid (ASTM F 86-76)	Waldemar Link GmbH & Co., Hamburg, West Germany
Link	32	Ti-6Al-4V (ASTM F 136-79 and F 620-79)	Diffusion hardened by immersion in molten salt bath	Waldemar Link GmbH & Co.

[a]ASTM Specification for Stainless Steel Bars and Wire for Surgical Implants (Special Quality) (F 138-82).
[b]ASTM Specification for Titanium 6Al-4V ELI Alloy for Surgical Implant Applications (F 136-79) and ASTM Specification for Titanium 6Al-4V ELI Alloy Forgings for Surgical Implants (F 620-79).
[c]ASTM Specification for Cast Cobalt-Chromium-Molybdenum Alloy for Surgical Implant Applications (F 75-82).
[d]ASTM Recommended Practice for Surface Preparation and Marking of Metallic Surgical Implants (ASTM F 86-76).

laboratory test conditions differed in some fundamental way from typical clinical use. The study reported here was an in-depth comparison of the laboratory and clinical wear properties of the various implant alloys currently in use.

Materials

Total hip prostheses from three manufacturers were tested (Tables 1 and 2). With the exception of the Link (Contour Link-SP) titanium alloy components (Specimens L4, L5, and L6), all of the metal femoral components had been passivated in hot nitric acid according to ASTM F 86–76. The Link titanium alloy components had been surface-hardened by the manufacturer by immersion in a cyanide-based molten salt bath at 800°C for 2 h. This process has been shown to produce a carbon-bearing titanium nitride surface zone, 40 μm to 60 μm deep, with a surface hardness about 55 HRC (compared to about 35 HRC below 50 μm) [3].

The metal components were polished by the manufacturers to typical prosthesis-quality mirror finish. Although we did not measure the surface finish, similarly polished specimens typically have a surface roughness of about 0.25 μm arithmatic average (A.A.) or less [4]. Prior to wear testing, each prosthesis was carefully inspected for any anomalous surface scratches or other defects.

Since the purpose of these experiments was to compare the polyethylene wear rate generated by titanium alloy femoral heads to that with the control alloys (either Type 316 stainless steel or cobalt-chrome), acetabular cups from a *single batch* from each manufacturer were used, such that the metal alloy was the only systematic variable. Both the Trapezoidal 28 (T-28) and STH femoral components were run against 28-mm-diameter STH acetabular cups. The Anitomic and Link femoral components were run against 32-mm-diameter cups from single batches from their respective manufacturers. The acetabular cups had all been sterilized with approximately 2.5 Mrad gamma radiation.

Bovine blood serum was used for pre-soaking of the cups and for lubrication during the wear test. This was membrane-filtered calf serum obtained from Sterile Systems, Inc., Logan, Utah. One-tenth percent of sodium azide was added to the serum to retard bacterial degradation.

Procedure

The wear test procedure (Table 3) and the ten-station joint simulator (Fig. 1) have been described in detail in a previous publication [5]. The acetabular cups were pre-soaked in bovine serum prior to starting the wear test. In the case of the STH cups (Set Z), the pre-soak extended through a period of 13 months while the joint simulator apparatus was being developed. The cups were cleaned and weighed several times per week for the first three weeks, and then weekly as part of a concurrent study documenting the long-term fluid absorption rate of ultra-high-molecular-weight (UHMW) polyethylene [6].

TABLE 2—*Acetabular cups.*

Type	Polymer	Preparation	Manufacturer
STH	UHMW Polyethylene Hercules Hi Fax 1900 (ASTM F 648-80)[a]	Compression molded directly from polymer flake. Sterilized with 2.5 Mrad gamma radiation.	Zimmer, Inc.
Anitomic	UHMW Polyethylene, Hoechst RCH 1000 (ASTM F 648-80)	Machined from compression molded blocks of polymer (90 cm by 60 cm by 6 cm). Sterilized with 2.5 Mrad gamma radiation.	Hexcel, Inc.
Link	UHMW Polyethylene, Hoechst RCH 1000 (ASTM F 648-80)	Machined from compression molded blocks. Sterilized with 2.5 Mrad gamma radiation.	Waldemar Link Gmbh & Co.

[a]ASTM Specification for Ultra-High-Molecular-Weight Polyethylene Powder and Fabricated Form for Surgical Implants (F 648-80).

TABLE 3—*Outline of the Joint Simulator Wear Test Procedure.*

1. Inspect femoral components and acetabular cups for obvious unusual damage (scratches, etc.).
2. Measure cup inside diameter and femoral outside diameter.
3. Pair components largest-ball-in-largest-cup, etc., to provide nearly uniform ball-cup clearance.
4. Replicate all contact surfaces.
5. Ultrasonically clean all components.
6. Dessicate and weigh acetabular cups to establish initial weights.
7. Pre-soak acetabular cups in bovine serum for approximately four weeks to eliminate early rapid fluid absorption.
8. Cut off femoral components, pot necks into mounts with pot-metal.
9. Re-clean and weigh acetabular cups (wear specimens and soak controls).
10. Mount wear specimens and soak controls in urethane holders. Place controls in soak chambers.
11. Mount femoral components and cups in joint simulator with serum lubricant bath, begin wear cycling.
12. At approximately 200 thousand cycle intervals, remove balls and cups. Clean, dessicate, and weigh the wear and control cups *simultaneously*. Inspect and replicate[a] balls and cups. Replace in fresh lubricant and resume cycling.
13. Since any fluid absorption by the wear cups tends to mask (that is, cancel) the weight loss due to wear, the magnitude of the apparent wear (weight loss) for each cup is *increased* by an amount equal to the average weight gain of the soak controls. The reverse holds if the soak controls show an average net loss in weight.

[a]Surface replication for microscopic evaluation of the wear process can be done with Xantopren (Unitek, Inc., Monrovia, Calif.) or other suitable micro-replicating compound.

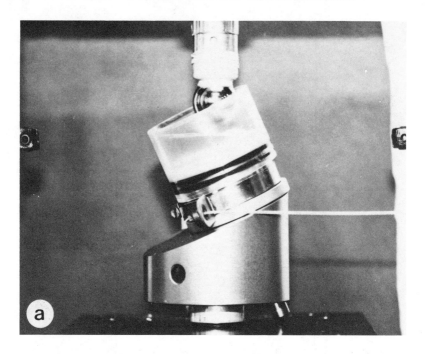

FIG. 1—(a) *Arrangement of test specimens in wear chambers. The cup is enclosed in a polyurethane holder mounted at a 23° angle to the vertical load axis. The ball is mounted above the cup on the load axis.* (b) *Ten-station joint simulator and control system.*

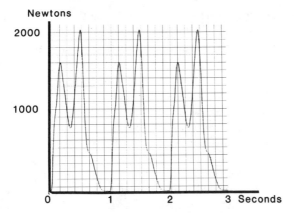

FIG. 2—*Three cycles of the load curve used on the joint simulator.*

The Anitomic cups (Set A) and the Link cups (Set L) were pre-soaked for a total of six months each. These long soak periods were primarily due to delays caused by modifications of the wear-test apparatus.

During the wear test, the soak control cups were placed in polyurethane holders identical to those used with the wear specimens, with dummy femoral heads inserted in the sockets. In this manner, the soak controls had the same surface area in contact with the serum as on the actual test specimens in the joint

TABLE 4—*Summary of Wear Test Cleaning Procedure.*

1. Scrub cups with nylon brush to remove *ALL* serum particles. Check under magnifying glass.
2. Rise under stream of deionized water.
. . . . .Protein Digestion[a]. . . . .
3. Mix 0.25 gram of pronase enzyme with 500 mL phosphate buffer solution.
4. Place all cups in beaker, fill with enzyme solution, cover with Teflon plate, seal in autoclave bag, and place in oven at 55°C for 12 to 16 h.
. . . .
5. Rinse with deionized water.
6. Clean in ultrasonic cleaner:
      (1) 5 Min in deionized water
      (2) Rinse in deionized water
      (3) 10 Min in soap solution (10 mL hydralox plus 500 mL water)
      (4) Rinse
      (5) 10 min in deionized water
      (6) Rinse
      (7) 3 min in deionized water
      (8) Rinse
7. Dry with air jet.
8. Soak in alcohol for 5 min.
9. Dry with air jet.
10. Dry in vacuum jar for 30 min.
11. Weigh on micro-balance.

[a]Protein digestion was used only with Set A to removed serum material embedded in the polyethylene.

simulator. The wear specimens were run on the joint simulator at 68 cpm, under a physiological hip load curve with 2030 N peak value (Fig. 2), for one million cycles. At intervals of approximately 200 thousand cycles, the cups were removed, cleaned, vacuum dessicated, and weighed, along with the soak controls, (Table 4). The weight loss of each wear cup was corrected by the average change of three soak controls. Wear rates for the individual cups were calculated by least-squares linear regression applied to the weight-loss data. The *volumetric* wear rates were calculated by dividing the weight loss of the cups by the density of the polyethylene (0.94 mg/mm^3). Multiple linear regression was also run for the combined groups of three, that is, with three values of weight loss for each value of total wear cycles.

The prostheses were mounted in the machine with the cup below the ball, that is, inverted relative to anatomical positioning. The primary reason for this was to eliminate the chance of air pockets forming in the cups, which would essentially be a dry-running test. The inverted positioning had the non-physiological feature that wear debris tended to collect in the cup, rather than to fall out as probably occurs *in vivo*. However, we felt that this source of error was partially offset by the fact that we cleaned the specimens and replaced the serum several times during the test. This represented a flushing-out of the wear debris that does not occur *in vivo*, where the debris accumulates in the joint space for many years, and may be drawn between the bearing surfaces by mechanical action. The joint simulator thus represented two trade-offs relative to the *in vivo* condition that, hopefully, partially cancelled each other.

Results

Fluid Absorption

During the long pre-soak period, when the cups were fully immersed in serum, the fluid absorption showed two distinct phases. For example, with Set Z, the rate of absorption over the first 28 days was about 95 μg per day, with a 95% confidence interval of ±65 μg per day (determined by linear regression). Beyond 28 days, the absorption rate was very constant at 37 ± 2 μg per day [6].

During the wear test, when the control cups were soaked in urethane holders with dummy femoral heads inserted, there was little systematic fluid absorption. At some intervals the control cups showed a slight *loss* in weight relative to the start of the wear test. The lower amount of fluid absorption during the wear test was presumably a result of less surface area being in contact with the fluid, combined with more frequent cleaning and drying, compared to the pre-soak period. The net change of the soak controls was typically less than ±1.0 mg between weighings. In comparison, as will be shown, the wear specimen weight loss was on the order of 10 mg over the same period. Thus, the weight change due to fluid absorption was typically less than 10% of that due to wear. Furthermore, since the wear weight loss was corrected by the average change of the soak controls, the potential *error* in the wear measurements due to fluid absorption was, at most, a few percent on any one weighing.

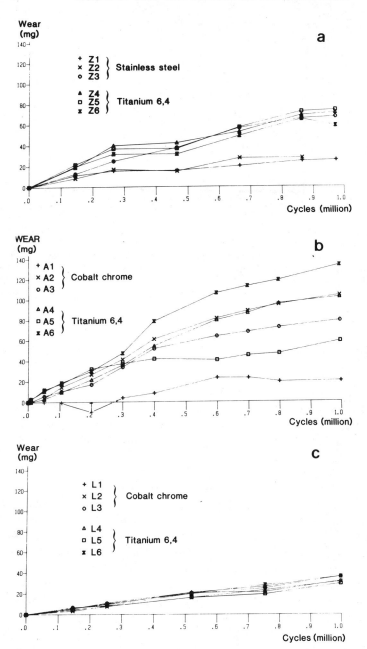

FIG. 3—*Weight loss of acetabular cups for* (a) *Set Z,* (b) *Set A, and* (c) *Set L. Our specimen (A1) showed negative wear, that is, a slight weight gain, early in the test due to embedded material (see discussion in text).*

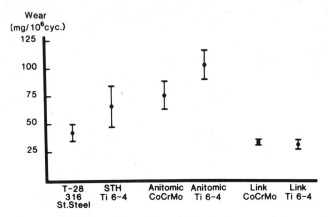

FIG. 4—*Wear rates with 95% confidence intervals for the six groups of three prostheses. Since three sets of acetabular cups were used (one set from each manufacturer), the effect of the metal alloy is determined by comparing the wear rates for the two alloys from a given manufacturer. On this basis, none of the wear rates against the titanium alloys were significantly different from the controls from the same manufacturer. The wear rates should not be compared between different manufacturers, since the batch-to-batch variation in polyethylene from a given manufacturer is unknown.*

Polyethylene Wear Rates

The weight loss for the three sets of six specimens are plotted in Fig. 3. Figure 4 and Table 5 show the wear *rates* as determined by linear regression.

The *mean* wear rate for polyethylene cups bearing against the T-28 stainless steel components averaged 34% less than that against the STH titanium alloy. Two of the three stainless steel components generated lower wear than the three titanium alloy components (Fig. 3a). However, wear with the third stainless steel head (Z3) was very close to that with titanium alloy, such that the difference between the two sets of results was not statistically significant ($p > 0.05$).

With the Anitomic prostheses, the polyethylene wear rate against cobalt-chrome alloy was 26% lower than with titanium alloy. However, the results of these tests showed the most scatter of the three groups (Figs. 3b and 4). This variation was apparently due to unusually low ball-cup clearances with the Anitomic prostheses. In the experiments with the T-28, STH, and Anitomic components, the diameters of the balls and cups were not measured before testing. We subsequently found that some of the Anitomic components had zero or even slightly negative clearances (ball larger than cup). On the joint simulator, this resulted in unusually high frictional torque, which in turn generated excessive heat. In such cases, the serum trapped between the ball and cup sometimes dried out and became embedded in the surface of the polymer. It was clear that this was a non-physiological form of wear damage that contributed to the greater amount of scatter in the wear data. We attempted to remove the embedded serum material by digestion in potassium hydroxide solution before cleaning and weigh-

TABLE 5—*Individual wear rates for polyethylene cups.*

Specimen	Ball Material	Ball Diameter, mm	Polymer Wear Rate, mg/10⁶ cycles	Regression Coefficient, R^2	Depth Wear Rate,[a] mm/year
Z1	Type 316 Stainless Steel	28	22.6	0.88	0.04
Z2		28	29.0	0.90	0.05
Z3		28	72.7	0.98	0.13
Z4	Ti-6A1-4V	28	69.1	0.94	0.12
Z5		28	73.2	0.95	0.13
Z6		28	60.6	0.93	0.10
A1	Cobalt Chrome Alloy	32	29.4	0.76	0.04
A2		32	115.6	0.97	0.15
A3		32	88.4	0.95	0.12
A4	Ti-6A1-4V	32	116.4	0.98	0.15
A5		32	52.1	0.85	0.07
A6		32	145.9	0.97	0.19
L1	Cobalt Chrome Alloy	32	36.4	0.99	0.05
L2		32	32.4	1.00	0.04
L3		32	35.9	0.99	0.05
L4	Ti-6A1-4V	32	31.1	0.97	0.04
L5		32	29.0	0.99	0.09
L6		32	37.0	1.00	0.05

[a]Depth wear rate was calculated by converting the gravimetric wear rate in milligrams per million cycles to cubic millimetres per million cycles, using a density of 0.94 mg/mm³, and then dividing by the cross-sectional area of the ball (see text for explanation).

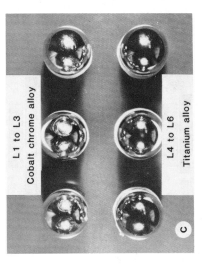

FIG. 5—Femoral components after one million wear cycles on the joint simulator (a) T-28 stainless steel (top) compared to STH titanium alloy (bottom), (b) Anitomic cobalt-chrome (top) compared to Anitomic titanium alloy (bottom), and (c) Link cobalt-chrome (top) compared to Link titanium alloy (bottom). The Link titanium alloy was surface hardened by a diffusion process. The surface abrasion generated in the joint simulator was much more extensive than has been observed on

ing (Table 4). While this was partially effective, the unusual damage to the polyethylene rendered these tests a less reliable indicator of *in vivo* wear performance. The difference in mean wear rates was not statistically significant ($p > 0.1$).

The Anitomic prostheses used in this study were prototypes of a design then in development by Hexcel, Inc. Partially based on the results from this study, the current manufacturer, Kirschner Medical Corp., specifies a minimum *positive* ball-cup clearance of 0.13 mm (0.005 in.) [7].

In the tests with the Link prostheses, the balls and cups were measured and paired largest-ball-in-largest-cup, etc., to establish a nearly uniform ball-cup clearance of about 0.3 mm. As shown in Fig. 3c, the polyethylene wear rates were very linear with wear cycles, and were very reproducible. Although the mean polyethylene wear rate was 8% lower for the titanium alloy (Table 5), this difference was clearly not statistically significant ($p > 0.25$).

Surface Appearance of the Femoral Balls

There was no transfer of polymer to the metal surfaces visible to the eye. This was consistent with previous pin-on-disk wear testing, where we found that heavy polyethylene transfer occurred using distilled water or saline lubrication, but not with blood serum. We also have not observed transferred polymer on any of the femoral components removed from patients. Apparently, the proteins present in physiological fluids such as serum or synovial fluid lubricate metal surfaces in a manner that prevents the adhesion necessary for polymer to transfer to the metal [4]. It is for this reason that we have emphasized the need to use protein-containing lubricants in laboratory wear tests of materials for joint replacements [4,5].

All of the femoral components run in the joint simulator showed a pattern of surface scratches covering the contact area that accumulated as the test progressed. This surface abrasion appeared to be more severe on the STH and Anitomic titanium alloy components than on their stainless steel and cobalt-chrome alloy controls (Fig. 5a and b). In contrast, the Link titanium alloy components, which had been surface hardened by diffusion, showed only very slight surface abrasion that was less than their cobalt-chrome controls (Fig. 5c).

Discussion

Comparison of Joint Simulator Wear Rates with Different Prostheses

It is important to note that the results of these wear tests represent three separate experiments that should not be directly compared to each other. For the six prostheses from each manufacturer, the only systematic variable was the metal alloy of the femoral component, since the acetabular cups were obtained from a single batch. The data thus obtained represents the difference in polymer wear rate for titanium alloy compared to the control alloy for that manufacturer. The results with Set Z cannot be compared to those with Set L, for example, to determine the effect of ball diameter on wear, since there were numerous other possible variables between the two sets. These include ball-cup clearance, polymer characteristics (density, molecular weight, percent crystallinity, etc.), exact polymer sterilization dose and rate, and others. An evaluation of wear as a function of ball diameter would require testing several identical specimens of each ball diameter bearing against acetabular cups manufactured from the *same batch of polymer*.

For the preceding reasons, the results should also not be interpreted as a comparison of wear rates for the different manufacturers. Such a comparison would require testing acetabular cups from several lots from each manufacturer to include the batch-to-batch variation.

The physical properties of the 18 polymer cups run in this study, along with the soak controls, are currently being analyzed. When this data is available, we may be able to correlate the wear rates of the individual specimens to their physical properties. In addition, the effect of the mechanical wear process on physical properties will be assessed (through comparison to the soak controls).

Given the difficulty in controlling the aforementioned experimental variables, it would, of course, be desirable to test a much larger sample of specimens to more firmly establish the statistical significance of the results. Unfortunately, even with a ten-station joint simulator, this is simply not practical in a reasonable time period. Each of the three experiments reported here involved two to three months to complete, including down-time for specimen cleaning and weighing. Although three replicate tests represent a bare minimum for statistical assessment, the very good reproducibility obtained in our third set of experiments (Set L), where the test variables were most carefully controlled, suggests that the joint simulator methodology will be capable of establishing small differences in wear rates with this number of specimens.

Comparison of Surfaces on Recovered Implants to Those Tested in the Joint-Simulator

The accumulation of surface scratches on the metal components run in the simulator probably affected the wear rate of the mating acetabular cups, and may explain the slightly higher group wear rates for the two titanium alloy designs that had ordinary nitric acid passivation. However, the clinical significance of this surface abrasion is not clear. Such extensive surface scratching appears to be an artifact of laboratory wear testing, since we have not observed this type of damage on any of the numerous femoral components that we have recovered from patients after revision surgery. The surfaces of recovered cobalt-chrome and stainless steel components are typically very smooth and scratch-free (Fig. 6a). Some of the titanium alloy components that we have removed from patients have had light burnished patches in portions of the contact zone (Fig. 6b). This type of wear has been described previously in the literature [8]. However, we have never observed surface abrasion on removed components as extensive as that generated by the joint simulator.

FIG. 6—(a) *Cobalt-chrome alloy hip prosthesis removed after 114 months in use. There is no visible deterioration of the surface. (b) Titanium alloy implant removed after 85 months in a patient. There is light surface burnishing. We have observed similar burnishing on approximately one third of the titanium alloy components removed from patients.*

The more extensive surface abrasion that occurs in the joint simulator may be due to minute contaminants (dust particles, etc.) becoming entrapped between the bearing surfaces. Although extensive care is taken to minimize such contamination, it is simply not practical to reproduce the clean, sterile *in vivo* environment on a laboratory wear test machine. These findings suggest that the joint simulator probably over-emphasizes the difference in abrasion resistance between two alloys and should therefore be considered a somewhat conservative test relative to *in vivo*.

Although the surface hardening treatment used on the Link titanium alloy components apparently increased their resistance to surface abrasion, the clinical significance of this result is difficult to establish since this type of abrasion is

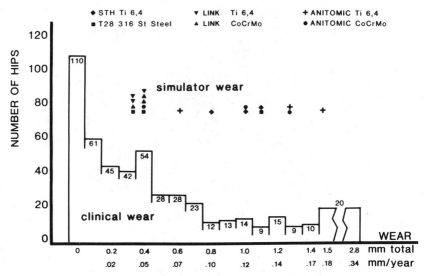

FIG. 7—*Wear rates for Charnley-type prostheses measured on radiographs of patients over an average of 8.5 years [10]. For comparison, the wear rates generated on our joint simulator are also plotted as wear depth per year.*

much less apparent *in vivo.* Nevertheless, as a general rule it is desirable to minimize the wear processes *in vivo,* since even a small amount of wear debris may generate local tissue reactions that contribute to component failure through interface loosening. On this basis, such surface hardening of titanium alloy implants would be a desirable design feature.

Comparison of Joint Simulator Wear Rates to Clinical Performance of Hip Prostheses

Table 5 lists the wear rates for the acetabular cups in millimetres of wear depth per year, assuming one million cycles to be the equivalent of one year's active use of a prosthesis. The wear depth rates were calculated by dividing the volumetric rates by the cross-sectional area of the balls. This would be a worst-case situation, that is, when the ball wears a tunnel equal to its own diameter into the cup wall. (This type of wear did occur with the early high-wear PTFE prostheses used by Charnley [9].) In this case, the volumetric wear rates obtained with the joint simulator would be the equivalent of 0.04 to 0.19 mm wear depth per year.

Griffith and colleagues [10] measured the wear rates for 493 total hip prostheses followed in patients for an average of 8.3 years. Wear was determined by measuring the distance between the ball center and a concentric metal wire embedded in the acetabular cup, as visible on successive radiographs taken over several years. Unfortunately, this method did not distinguish between cup-thinning due to actual wear and that due to polymer creep or cold-flow, which may cause more dimensional change than is due to wear. Nevertheless, Griffith's data offer a good indication of polyethylene wear rates *in vivo*.

Figure 7 shows the wear rates determined by Griffith in comparison to those measured on our joint simulator (plotted as millimetres per year). Although all of the joint simulator rates were well within the clinical range, about 50% of the *in vivo* cups wore less than the simulator-tested specimens. The higher joint simulator wear rates may have been in part due to the metal surface abrasion just discussed. In addition, it is possible that a typical patient's activity may have been less than the assumed one million walking cycles per year, resulting in less total wear. Nevertheless, the two sets of data span the same order of magnitude, which strongly suggests that the simulator induces the same wear mechanisms that are occuring *in vivo* with total hip prostheses.

Conclusions

This study has shown that the joint simulator is capable of generating wear rates in the range observed clinically. The high degree of repeatability achieved in the tests where ball-cup clearance was kept constant (Set L) indicated that the simulator will be valuable in determining very small changes in wear rates due to material processing variations or other prosthesis design parameters.

Under these laboratory conditions, titanium alloy passivated in hot nitric acid generated slightly more polyethylene wear than similarly passivated stainless steel or cobalt-chrome controls. However, the differences were not statistically significant. Comparison to prostheses removed from patients suggested that the differences in wear rates would be even less *in vivo*. Titanium alloy that had been surface hardened by diffusion was resistant to surface abrasion and produced polyethylene wear equal to the cobalt-chrome controls.

References

[1] Clarke, I., McKellop, H., McGuire, P., Okuda, R., and Sarmiento, A. in *Titanium Alloys in Surgical Implants, ASTM STP 796*, H. Luckey and F. Kubli, Eds., American Society for Testing and Materials, 1983, pp. 136–147.
[2] Gruen, T. A. and Sarmiento, A., "Roentgenographic Analysis of 323 STH Total Hip Protheses with Low-Modulus Titanium Alloy Femoral Components. A Two to Six Year Followup," *Journal of Bone and Joint Surgery,* in press.
[3] Waldemar Link GmbH & Co. Hamburg, West Germany, unpublished technical data.
[4] McKellop, H., Clarke, I., Markolf, K., and Amstutz, H., *Journal of Biomedical Materials Research,* Vol. 15, 1981, pp. 619–653.
[5] McKellop, H. and Clarke, I., "Evolution and Evaluation of Materials-Screening Machines and Joint Simulators in Predicting *in vivo* Wear Phenomena," *Functional Behavior of Orthopaedic Biomaterials,* P. Ducheyne and G. Hastings, Eds., CRC Press, Boca Raton, 1984.
[6] Clarke, I., Starkebaum, W., Hosseinian, A., McGuire, P., Okuda, R., Salovey, R., and Young, R., "Fluid Sorption Phenomena in Sterilized Polyethylene Acetabular Prostheses," submitted to *Biomaterials,* 1984.

[7] Kirschner Medical Corporation, Timonium, Md., technical data.
[8] Zych, G., Latta, L., and Mnaymneh, W. in *Titanium Alloys in Surgical Implants, ASTM STP 796,* H. Luckey and F. Kubli, Eds., American Society for Testing and Materials 1983, pp. 151–172.
[9] Charnley, J. in *Lubrication and Wear in Living and Artificial Human Joints,* Institution of Mechanical Engineers, London, 1967, pp. 104–111.
[10] Griffith, M., Seidenstein, M., Williams, D., and Charnley, J., *Clinical Orthopaedics,* Vol. 137, 1978, pp. 24–36.

Jack E. Lemons[1]

Tricalcium Phosphate Biodegradable Ceramic

REFERENCE: Lemons, J. E., "Tricalcium Phosphate Biodegradable Ceramic," *Corrosion and Degradation of Implant Materials: Second Symposium, ASTM STP 859*, A. C. Fraker and C. D. Griffin, Eds., American Society for Testing and Materials, Philadelphia, 1985, pp. 369–372.

ABSTRACT: Particulate and porous scaffolds of tricalcium phosphate ceramic (TCP) have been evaluated in laboratory, laboratory animal, and human investigations. Selected applications of the particulate biodegradable ceramic have shown favorable results, especially when used with autogeneous bone. However, long-term studies of large porous scaffolds need to be expanded prior to routine human applications. Continued research on biodegradation characteristics are recommended.

KEY WORDS: implant materials, biological degradation, fatigue (materials), tricalcium phosphate ceramic, laboratory animal implants, particulate, porous scaffolds

The need for an "off the shelf" biomaterial for the treatment of lesions in bone continues to attract the interests of clinicians and biomaterials researchers. A substance that would act as an osteoinductive or possible osteogenic substrate and would eventually biodegrade offers very special opportunities for many surgical procedures. Autogeneous bone transplant substance usually requires a second surgical site and its associated morbidity. Also, limited quantities of autogeneous substance are available that places some restrictions on the treatment of major lesions, for example, non-unions, alveolar ridge augmentations, spinal fusions, and procedures in children. Tricalcium phosphate ceramic (TCP) is recognized as one substance that provides opportunities for the treatment of many types of bony lesions. The overall objective of this short paper is to review some of the historical aspects of TCP, consider results from short and long-term laboratory animal implants, and briefly discuss some of the human clinical applications that have been initiated.

Tricalcium Phosphate Ceramic Material

The early development of the TCP materials was a part of the U.S. Army Medical and Materials Research and Development Command (USAMMRC),

[1]Professor and chairman, Department of Biomaterials, University of Alabama in Birmingham, Birmingham, AL 35294.

Army Institute of Dental Research (USAIDR) research and development programs conducted with Battelle Memorial Institute. The first materials were fabricated as macro and microporous solids that were then reduced to form microporous particulate or porous scaffolds for tissue ingrowth. The material was made from spectrochemically pure powders to produce a relatively pure form of $Ca_3(PO_4)_2$. The crystalline form, by X-ray diffraction analyses, was beta whitlockite [1].[2]

This substance was investigated to develop an extensive *in vitro* characterization profile including physical, mechanical, and chemical properties [2]. The mechanical properties of TCP are somewhat limiting, showing maximum strength properties for porous materials approximately one-third the strength of compact bone. Limited laboratory animal *in vivo* studies suggested possible applications for the particulate, porous scaffolds, and solid block forms of TCP.

Proposed Clinical Applications for TCP

As a biodegradable substance for the treatment of bone lesions, multiple applications of TCP were suggested at the outset. These included dental and medical considerations such as periodontal lesions, endodontic root canal therapy, alveolar ridge augmentation for edentulous patients, on-lay grafts for plastic and orthopaedic surgery procedures, non and delayed unions, filling spaces adjacent to total joint prostheses, etc. A number of these applications have been investigated, and these will be discussed in the human clinical trials section of this paper.

Laboratory *In Vitro* and *In Vivo* Investigations

Laboratory studies of TCP have usually been oriented towards a specific clinical application for this biomaterial. Most have emphasized basic information for the evaluation of biocompatibility and biodegradation characteristics.

Biocompatibility and Biodegradation Studies

The early studies on particulate TCP included local and systemic tissue responses in rats, and the fate of labeled materials *in vivo* [3]. In general, these studies showed minimal tissue reactions (local biocompatibility) and the excretion of the basic components after *in vivo* biodegradation. Particulate TCP was incorporated into bone and no adverse responses were noted.

Studies in rabbits [4], dogs [5], and primates [6], showed minimal tissue reactions, ingrowth into porous scaffolds with a maturation of bone within the porosities, healing of bone lesions, but a lack of consistent biodegradation profiles for the larger porous scaffold replacements. Questions were raised about the TCP chemical analysis, biomechanical aspects of force transfer and *in vivo* fracture, implant design, local vascularity of the anatomical site, *in vivo* and *in*

[2]The italic numbers in brackets refer to the list of references appended to this paper.

vitro changes in the materials (formation of apatite *in vivo* and unacceptable autoclaving prior to placement), contamination at the time of surgical placement, size fractions for particulate and degree of packing at the time of placement, and so on. Most of these questions lead to continued research projects. One such project was the long-term studies of porous scaffolds as segmental replacements of the mid portion of dog radii. This program extended to five years, being completed during 1983 [7]. Additional studies to produce oriented porous TCP biomaterials with greater mechanical strengths also evolved following the reporting of earlier results [8].

The long-term dog implants of porous scaffolds (1 by 1 cm rods) showed adverse reactions associated with residual TCP within the implant sites. These long-term (five-year) implants showed local granulation tissue and a fistulous track from the bone to the skin. Therefore, the use of large porous blocks of TCP should be carefully evaluated with respect to the material and the design of the implant. Hollow cylinders of porous TCP were not studied in this program. In general, the particulate TCP when used in limited quantities, and if used in a 50% ratio by weight with autogeneous bone, showed incorporation into the bone and good biocompatibility profiles.

Human Clinical Applications

The first major area for human investigations were periodontal lesions in dentistry [9]. The results of these studies have resulted in a general utilization of particulate TCP (usually $-40 + 100$ mesh) for the treatment of periodontally involved teeth. Some difficulties with respect to closure of the surgical site and bone regeneration and long-term biodegradation have been reported. However, most reports have been favorable. Studies on alveolar ridge augmentation have been extended to include a general availability of the TCP particulate for this application. Limited long-term data have been reported in this area. An application of a porous cylindrical scaffold replacement in orthopaedic surgery has been published recently [10]. Long-term follow-up is not yet available for these types of applications. The cylindrical design could influence long-term biodegradation characteristics for the porous TCP scaffolds. However, long-term results should be evaluated prior to extensive applications.

Summary

In summary, a significant amount is known about the basic chemistry, mechanical strengths, biocompatibility profiles of various forms, biodegradation characteristics, and clinical applications of tricalcium phosphate beta whitlockite $Ca_3(PO_4)_2$ ceramics. The previous basic science, laboratory animal, and clinical investigations have resulted in general and specific recommended practices for the manufacture, handling, placement, and clinical follow-up procedures. This biomaterial offers opportunities for the treatment of bony lesions not available with other types of synthetic compounds. However, since laboratory studies show some limitations, a careful review of the literature available on this bioceramic is strongly recommended, prior to new clinical applications.

References

[1] Driskell, T. D., O'hara, M. J., and Green, G. W., Jr., "Management of Hard Tissue Avulsive Wounds and Management of Orafacial Fractures," Report No. 1, Contract DADA 17–69–C–9118, U.S. Army Medical Research and Development Command, 1971.

[2] Driskell, T. D., O'hara, M. J., Sheets, H. D., Greene, G. W., Jr., Natiella, J. R., and Armitage, J., *Journal of Biomedical Materials Research Symposium*, Vol. 2, 1972, pp. 345–361.

[3] Bhaskar, S. M., Cutright, D. E., Knapp, M. J., Beasley, J. D., and Perez, B., *Oral Surgery, Oral Medicine, and Oral Pathology*, Vol. 31, 1971, pp. 282–289; also Vol. 32, 1971, pp. 336–346.

[4] Lemons, J. E., "Response of Combined Electrical Stimulation and Biodegradable Ceramics," Annual Reports 1–4, USAMMRC Contract No. DAMD–17–75–C–5044, U.S. Army Medical and Materials Research and Development Command, 1975–1979.

[5] Koster, K., Heide, H., Karbe, E., Konig, R., and Kramer, H., "Investigation of Biodegradable Calcium Phosphate Ceramics for Bone Replacement," Society for Biomaterials, Vol. 82, 1976.

[6] Heimke, G. and Griss, P., "Tissue Interactions to Bone Replacement Materials," Chapter 4, *Bioceramics of Calcium Phosphate*, K. deGroot, Ed., CRC Press, Boca Raton, FL, 1983.

[7] Lemons, J. E., "Response of Combined Electrical Stimulation, Applied Laboratory and Clinical Studies on Biodegradable Ceramic," USAMMRC Report No. 8, Contract No. DAMD17–79–C–9173, U.S. Army Medical and Materials Research and Development Command, 1984.

[8] Hassler, C. R. and McCoy, L. G., *Journal of Dental Research Abstracts*, S–42, Vol. 61, 1982, p. 177.

[9] Synthograft®, MITER, Inc. Worthington, OH.

[10] Goldstrohm, G. L., Roberts, J. M., and Meers, D. C., *Transactions*, Orthopedic Research Society, Vol. 8, 1983, p. 235.

Edward Ebramzadeh,[1] Massoud Mina-Araghi,[1] Ian C. Clarke,[2] and Roy Ashford[3]

Loosening of Well-Cemented Total-Hip Femoral Prosthesis Due to Creep of the Cement

REFERENCE: Ebramzadeh, E., Mina-Araghi, M., Clarke, I. C., and Ashford, R., "**Loosening of Well-Cemented Total-Hip Femoral Prosthesis Due to Creep of the Cement,**" *Corrosion and Degradation of Implant Materials: Second Symposium, ASTM STP 859,* A. C. Fraker and C. D. Griffin, Eds., American Society for Testing and Materials, Philadelphia, 1985, pp. 373–399.

ABSTRACT: A review of previous publications on the mechanical complications of cemented total-hip prosthesis femoral component indicates several failure mechanisms. However, the biomechanical significance of the various reported radiographic phenomena has not been clearly assessed with static testing or with computer models.

A new technique was developed to directly measure internal cement strains in a simulated proximal femur with cemented femoral total-hip prostheses. Static experimental testing of a Delrin plastic-bone-analog with a cemented metallic femoral prosthesis indicated highest cement strains in the proximal-medial aspect of the cement sheath, which correlated with highest cement stresses in the same location in a corresponding finite-element analysis computer model. The most noticeable observation from the cyclic testing was the 73% decrease of cement strains in the proximal-medial region after 4.1 million cycles. This time-dependent behavior, that is, creep, may have significance in the failure mechanisms of cemented femoral total-hip replacement components and should be considered in future static or cyclic experimental testing and computer modeling of such prosthetic components.

KEY WORDS: implant materials, biological degradation, fatigue (materials), total-hip replacement, femoral prosthesis, cement stresses, radiolucent zones, loosening, stainless steel, titanium alloys, cobalt-chrome alloys

Today, after the first cemented total-hip implantation by Haboush in 1951, fixation by acrylic bone cement is still the state of the art in the United States. It is currently estimated that there are approximately 70 000 cemented total-hip

[1]Research engineer and research assistant, respectively, Orthopaedic Biomechanics Laboratory, Orthopaedic Hospital-University of Southern California, Los Angeles, CA 90007.
[2]Associate professor, Departments of Orthopaedics and Mechanical Engineering, University of Southern California, and, director, Orthopaedic Biomechanics Laboratory, Orthopaedic Hospital-University of Southern California, Los Angeles, CA 90007.
[3]Research assistant, Medical School, Orthopaedic Hospital-University of Southern California, Los Angeles, CA 90007.

TABLE 1—Comparison of clinical reports describing phenomena of stem loosening radiographically by means of (A) cement cracks or loosening at (B) the cement-bone interface or (C) the stem cement interface.

Study	Number of Hips	Radiographic Follow-up, years	Interface Loosening			SAG Initiation Time
			Fractured Cement, (A)	Cement-Bone, (B)	Cement SAG, (C)	
Weber [41]	6449	1 to 10	1.5%	NS	NS	½ to 11½ years
Nicholson [42]	250	0 to 5	NS	0.4%	3.6%	NS
Charnley [16]	17[a]	1.5 to 6	(40%)[b]	NS	(40%)[b]	NS
Amstutz et al [18]	389	2 to 5	4.4%	11.1%	10.3%	1 to 3½ years
DeSmet et al [43]	101	1.5 to 5	NS	NS	14.9%	<1 year
Collis [44]	200	0 to 5	7%	NS	6%	½ to 2 years
Bocco et al [45]	119[c]	4 to 8	20%	NS	13%	NS
	97[d]	1.6 to 2.6	3%	NS	0	NS
Beckenbaugh and Ilstrup [46]	255	4 to 7.5	8%	24%(16%)[e]	(4.5%)[f]	NS
Gruen et al [47]	389	0.5 to 6	4.4%	11.1%	10.3%	NS
Carlsson and Gentz [5]	288	0 to 8.5	16%	NS	39%	<1 year[g]
Loudon and Charnley [40]	100[h]	0 to 8	26%	NS	NS	NS
	75[i]	0 to 2	0	NS	NS	NS
Moreland et al [17]	444	2 to 9	NS	40%	14%	NS
Gruen [19]	378	0 to 4.5	1.6%	2.9%	6.9%	1 to 6 years
Salvati et al [48]	97	5 years avg	5%	72%	17%	NS

[a] Subset contained only cases with stem fractures.
[b] Contained both A and C loosening criteria.
[c] Hips replaced before 1970 and only males selected.
[d] Hips replaced after 1974 and selecting only males with more than 10 mm wedge of calcar cement.
[e] Calcar resorption.
[f] Seen at one year follow-up.
[g] Increasing in size and frequency with follow-up.
[h] Conventional Charnley THR.
[i] Charnley-Cobra THR.
NOTE—NS = not specified.

replacements and 30 000 total-knee replacements performed in the United States every year [1].[4] Assuming that three packs of acrylic cement were used on average per procedure, the total amount of acrylic cement used for total-hip and total-knee procedures in the United States amounts to 18 000 kg per year. Despite this extensive and enthusiastic clinical usage of acrylic bone cement, it is generally described as the weak link responsible for loosening (failure) of the cemented femoral prosthesis [2–5].

While there have been many descriptions of clinical and radiographic loosening (Table 1) as well as laboratory stress studies of the femoral prosthesis stem, cement, and bone [6–11], there does not appear to have been any investigation of the changes occurring in the cement *during* the loosening process. If we can understand the significance of the stem-cement loosening phenomena, we should then have the necessary criteria for optimizing stem design and the surgical procedure. Therefore in this paper, we explore the role of acrylic bone cement support for the femoral stem, evaluating published clinical and biomechanical data. We shall then describe a new investigation of the time-dependent mechanical behavior of cement. The experimental study starts with basic characterization of bone cement properties using bending of cement beams, taking the factor of time into account.

The objective of this study is to measure and evaluate cement strains as a function of time under cyclic loading, which may promote a stem-cement interface failure, that is, stem loosening. This entails a more detailed experimental investigation of cement conditions around an initially well-fixed stem as it loosens under repetitive functional loading, plus monitoring of any stem-loosening phenomenon observed during cyclic loading.

Clinical Background

Although there have been many investigations on the modes of failure of the cemented femoral stem over the last 15 years, there is little understanding as to what actually initiates the loosening process and how this is affected by the implant design [2,12–14]. There are three possible sites of failure initiation: either at the cement-bone interface, at the stem-cement interface or, of course, within the cement itself. We hypothesize that loosening at the stem-cement interface involves mainly a mechanical process whereas loosening of the cement-bone junction combines both biological and mechanical mechanisms, that is, biomechanical. In this regard, it is important to note that there are *three* quite different concepts of fixation to consider: the Exeter stem, the straight Mueller stem, and the femoral stem with collar. The collarless Exeter stem from England (Fig. 1a), is designed to shear at the stem-cement interface in a "subsiding" mode, that is, to split the cement sheath intentionally. As the stem settles, it is anticipated that the cement is wedged between the stem and bone, thereby maintaining fixation [15]. The current Mueller Straight stem (also collarless) is

---

[4]The italic numbers in brackets refer to the list of references appended to this paper.

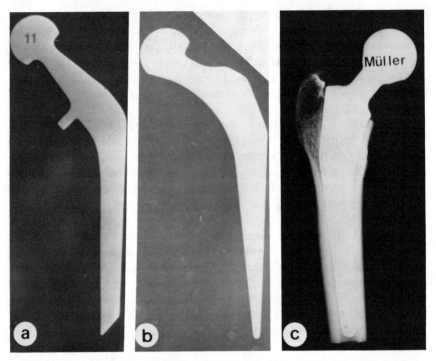

FIG. 1—*Antero-posterior outlines of* (a) *a cemented stem with large collar from the United States,* (b) *collarless Exeter stem from England, and* (c) *collarless straight Muller stem from Switzerland.*

designed for stem-bone contact on lateral and medial cortices. Therefore, this design can only experience shear at the cement-bone interface when fitted correctly (Fig. 1*b*). In contrast to these two designs, the femoral stems used in the United States generally have a collar and are intended to be used with a complete and usually *thick* cement sheath (Fig. 1*c*). No slip or shear between the cement and bone is required or even considered desirable. In other words, the perfect implantation would show neither cement cracks nor gaps between stem and cement or cement and bone. When gaps can be demonstrated radiographically, then the stem is considered loose to some degree, even though clinically it may be asymptomatic at that time.

In virtually every clinical study, the initial radiographic evidence of stem loosening (that is, varus or distal drift) predominantly involved changes in the proximal aspects of the femur [16]. The resulting gaps seen on radiographs are known as radiolucent zones (RLZ) and can occur either at the cement-bone interface or at the stem-cement interface (Fig. 2). Cement failure denoted by cement cracks visible on X-ray films can also occur, but much less frequently (Table 1).

Considerable variation has been reported for the incidence of cement-bone gaps (Table 1). However, it is hard to ascribe the clinical significance of all

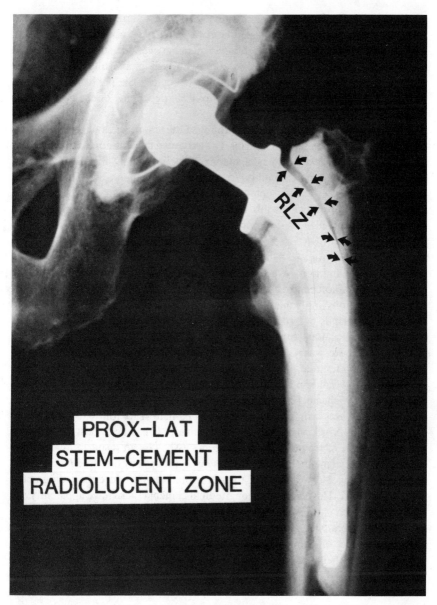

FIG. 2—*Radiograph of a femoral stem showing radiolucent zone (RLZ) between the proximal, lateral, stem surface, and the lateral cement-sheath (arrows).*

such zones. Many may be present immediately after surgery and may remain essentially nonprogressive and asymptomatic. The most obvious and dramatic sign of progressive interface loosening appears between the stem and cement at the *proximal-lateral* site (Fig. 2). A zone seen radiographically in this region shall be denoted by the term SAG (stem-acrylic-gap). The published incidence of this SAG loosening has varied from 1 to 40%, depending on the criteria used (Table 1).

All things considered, it is difficult to relate the significance of the SAG form of radiographic loosening to the diffusely distributed cement-bone demarcations. Charnley, from his vast clinical experience, believed that this SAG form of loosening was much more common [5] than generally appreciated. In a detailed review, Moreland et al [17] noted that "a very thin radiolucency between the cement and the prosthesis in this area often preceded more apparent loosening signs, such as subsidence of the prosthesis and cracks in the acrylic cement."

The first onset of SAG radiolucency for the stainless steel Charnley T-28 hip prosthesis ranged from less than one to more than three years [5,18]. For the titanium, 6%-aluminum, 4%-vanadium (Ti-6Al-4V) STH prosthesis, the first signs of SAG radiolucency were observed from one to as late as six years [19]. These observations imply that SAG loosening may be due to a mechanical process that takes thousands or even millions of cycles to develop.

Examination of total hips removed post-mortem has provided further confirmation of the stem-acrylic loosening phenomenon. One study of five Charnley-Muller total-hip femoral prostheses (up to five years follow-up) reported a *layer of fibrous tissue* adhering to the stem in each case, that is, between the stem and the cement sheath [20]. Similar findings were described by Ling [15] in three autopsy cases with five to seven and one half years follow-up.

How can the SAG phenomenon be explained? It has been postulated that a pistoning action of the stem inside the cement sheath can result in longitudinal cracking of the cement (Fig. 3) due to tensile (hoop) stresses [5,21–23]. Subsequent subsidence or varus tilt of the stem could then produce the radiographic evidence of *SAG*. However, such longitudinal cement cracks have apparently *not* been documented at revision surgery. How else could one explain the occurrence of proximal-lateral stem-cement gaps?

Fatigue studies of stems partially embedded in cement have demonstrated that gaps can occur between stem and cement without evidence of radial cement cracks [24] (Fig. 4). These gaps may have resulted from creep or from plastic deformation of the cement that was loaded beyond yield point. Pal and Saha [25] have demonstrated that acrylic cement specimens can undergo 0.2% creep strain, corresponding to approximately 10 to 20 μm (Fig. 5). Atkinson et al [26] monitored the creep of the acetabular cement bed under both static and on-off loading (2100 N peak load). For their dynamic tests, the total compression value was 106 μm with a residual (unloaded) creep value of 70 μm. This was attained after only seven such cycles (seven days), and appeared essentially completed at that time. The creep was therefore of the same magnitude as that measured

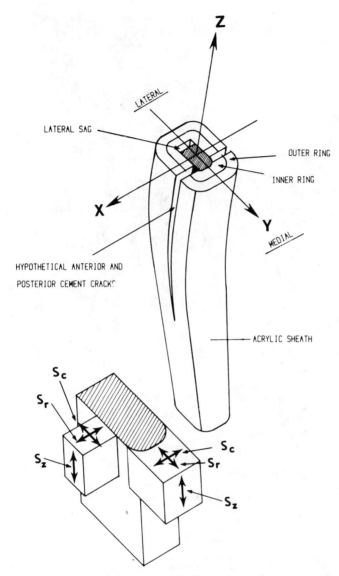

FIG. 3—*Schematic illustration of hypothetical cracks in proximal anterior and posterior facets of the acrylic cement sheath around the femoral stem. The insert shows the manner in which the circumferential* (S_c) *and radial stress* (S_y) *directions vary in direction with regard to position of cement elements on medial or posterior sites. In this study, we will focus on the mediolateral directed stresses* (S_y) *and axial* (S_z) *stress directions, on the lateral, posterior, medial, and anterior sites. Note that on the posterior and anterior sites, S_y is circumferential (that is, S_c direction and not radial S_r).*

FIG. 4—*Large gaps apparent between stem and cement on the medial side without any visible longitudinal cement cracks. These gaps were the results of fatigue tests by Semlitsch* [24].

radiographically in a clinical series where the progressive increase in stem-acrylic-gap widths was assessed at 20 to 50 μm/month on average [5].

Therefore, it may well be that the SAG can indeed result from either creep (continuous flow) or sequential plastic deformation (overloading) of the proximal-medial cement sheath, particularly when there is little or no bonding of the stem-cement interface. However, if there were either poor medial cement support or inadequate medial bone support or both, cement splitting could possibly be anticipated (Fig. 6).

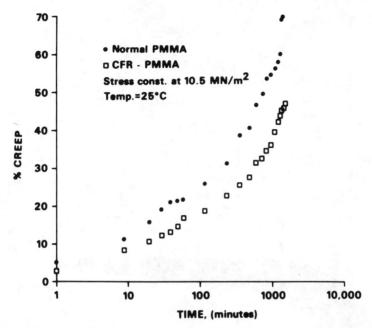

FIG. 5—*Description of creep phenomenon in acrylic cement specimens* [25].

Biomechanics Background

Data on the state of stress within the acrylic cement surrounding the femoral stem comes primarily from two-dimensional and three-dimensional linear finite-element analysis (FEA) models (Fig. 7) [2,3,11,13,14]. From such studies, we know that the stress in the proximal-lateral cement region can be of the order of 3 MPa tension, and the peak compressive stress on the corresponding proximal-

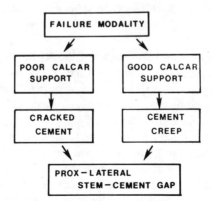

FIG. 6—*Comparison of two possible modes of stem loosening, either cement cracking initiated by poor medial support in a sub-optimal cement filling or by cement creep in a well-cemented procedure.*

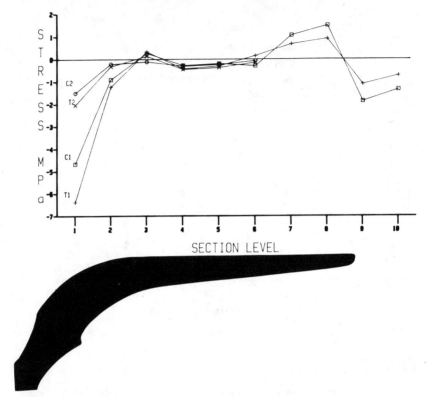

FIG. 7—*Longitudinal distribution of S_y cement stresses on the medial side of femoral stems for assumed load of 3000 N inclined in 30° load axis [10]. Collar is on the left, stem tip on the right. Ti-6Al-4V Charnley-type stem: T1-inner cement layer and T2-outer cement layer. Stainless steel Charnley-type stems: C1-inner cement layer and C2-outer cement layer.*

medial site can be 10 MPa [*3*]. Only one study that we know of has provided a nonlinear two-dimensional model of the stem-acrylic interface [*27,28*]. This study illustrated the great variation in stresses when certain interface conditions change.

A number of investigators have measured the strains on the external bone surfaces using strain gages. However, such measurements are not sensitive to the degree of loosening between the stem and cement. Several experimental studies have investigated the strains in prosthetic stem surfaces both "ideally-fixed" and "loose" in the cement using strain gages [*9,29–31*]. Crowninshield and Tolbert [*32*] appear to have provided the first experimental data on *in vitro* strain *within* the acrylic cement sheath surrounding a femoral stem. They measured strains (e_y) in the proximal-medial sheath of -0.22, and -0.28 for "well-fixed" and "unbonded" (loose) stems, respectively, that is, 27% increase in strain following loosening. (See Fig. 3 for stress directions in x, y, and z co-ordinates used in this paper. The mediolateral direction is denoted by S_y, which is radial on the medial and lateral sites. On the posterior and anterior sites, the mediolateral direction is circumferential). The axial (longitudinal) strains (e_z) rose from 0.14 to 0.17 (21% increase) from the "fixed" to the "unbonded" case. The anterior e_y strains were minimal, approximately 3 to 26 times lower than the axial strains. Possibly the largest relative increase was for the e_x strains, rising from 0.09 "well-fixed" to 0.16 for the "unbonded" case (that is, an increase of 78%). Therefore, this study investigated the "ideally-fixed" stem and compared that to a "fixed as loose," "unbonded" situation. However, there was no analysis of cement stress changes with respect to time and progressive loosening.

As a starting point, a technique was developed to measure strains in desired directions within the cement, which was verified to be of reasonable accuracy. An STH prosthesis was cemented into a cylindrical-shaped Delrin tube representing the midshaft of the human femur, and using the same strain-gage technique, *mediolateral* strains were measured with respect to progressive cyclic loading. A finite-element model of this Delrin bone analog was developed and comparison was made also to an anatomical model developed previously by Tarr et al [10]. Both FEA models assumed linear elastic homogeneous properties and displacement continuity along the interfaces. Since stresses and strains are strongly affected by choice of element geometry and boundary conditions, this evaluation of FEA data provided validation criteria for theoretical models and assisted in the development of more accurate and clinically relevant modeling schemes.

Methods

Strain-Gage Modules (SGM)

An important part of this project was to obtain data from strain gages embedded in bone cement. Draganich [33] evaluated strain gages embedded in acrylic cement beams using cantilever bend tests. In the present study, ASTM Test Methods for Flexural Properties of Unreinforced and Reinforced Plastics and Electrical Insulating Materials (D 790-81) was used as the basis for three- and four-point bend tests to verify proper functioning of the strain gages. Eleven specimens were loaded to failure in order to characterize the cement used for our study. Strain gages were then embedded in beams, and the beams were loaded in three- and four-point bending. Measured strains for these tests were compared to that predicted by Bernoulli-Euler beam theory, using the elastic modulus obtained from initial bend tests.

Dental acrylic cement was used for the beam specimens (150 g polymer: 75 cm³ monomer) and poured after 1 min of mixing. The mechanical properties of this material lie well within the range of the properties of most orthopaedic bone cements (Table 2). The cast and machined beams (final overall size: 140 mm by 18 mm by 5 mm) were X-rayed and then fitted (Fig. 8) with strain gages on the surfaces (Micro-measurements, EA-06-062AP-120) using cyanoacrylate cement (Sicomet) and waterproof coating (M-coat-A Polyurethane). After condi-

TABLE 2—Comparison of published elastic modulii and failure criteria for acrylic cement in flexure using different cement types and test methods.

Author	Method	Cement	Flexural Data				
			Minimum Time to Test, h	Number of Specimens	Youngs Modulus, GPa	Ultimate Load, N	Ultimate Stress, MPa
Haas et al [49]	ADA12	Simplex	NA	NA	2.1 to 2.2	42 to 48	NA
Lautenschlager et al [50]	ASTM D 790-71	Simplex	24+	a	2.1 to 2.7	NA	36 to 43
Holm [51]	Custom	b	8+	376	2.3 to 2.6	NA	54 to 67
Lee et al [52]	Custom	c	48	NA	1.8 to 2.4	NA	50 to 70
Lee et al [53]	Custom	d	48	18	2.9 to 3.0	NA	73 to 75
Present study	ASTM D 790-81e	Dental	24+	11	1.8 to 2.3	89 to 109e	27.4 to 31

NOTE—NA = not available.
a Three or more specimens per group.
b CMW, Simplex, Palacos.
c AKZ, Simplex, Palacos, Sulfix.
d AKZ, Simplex, RO.
e Data for beams 5-mm thick.

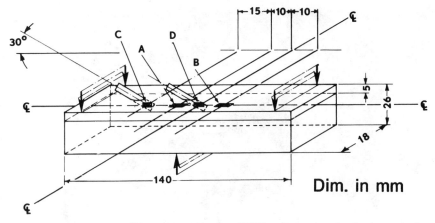

FIG. 8—*Sketch of cement-beam showing location of SGMs as pre-cemented and as inserted using the SGM concept.*

tioning at room temperature for at least 24 h, the beams were mechanically tested to check the gages' performance.

Upon completion of the load tests, an additional 21-mm layer of cement was cast on top of the beams, giving them an effective thickness of 26 mm (Fig. 8). After a minimum of 24 h, the beams were again tested mechanically to check the performance of the embedded gages.

In order to embed gages into the cement sheath around a femoral prosthesis in a certain position, the gage was first attached to a machined pre-polymerized hemicylinder of cement. A hole slightly larger than the diameter of the hemi-cylinder was drilled, and the strain gage module was placed in the hole after injecting cement in the hole and held until fully polymerized. In order to verify proper functioning of the gages embedded in cement using this method, two 4-mm diameter holes were then drilled into the beams at 30° with respect to the long axis of the beam for the strain-gage-modules. Vent holes 1-mm diameter were drilled into each of the SGM holes to permit optimal cement filling.

For the SGMs, dental acrylic cement was cast in cylindrical polytetrafluoro-ethylene (PTFE) molds (ten holes, each 6-mm diameter by 40-mm deep) using the 2:1 ratio (40 g:20 cm³) and 1-min mixing time. These cement modules were then machined to hemicylinders and the strain-gages mounted in the required orientations (Fig. 9). The SGMs were then sanded down to the edges of the strain gage. The holes were filled with cement (using a syringe) that was mixed for less than 1 min, and the SGMs were inserted into the holes and held in place until the cement cured.

Radiographs were taken to identify and verify all strain-gage positions. Ten acrylic beams were loaded to failure to determine the ultimate flexural properties and the flexural modulus. The strain-gaged beams were then loaded in 200 N step-load increments up to 1000 N. The strain-gage data were read after 6-min

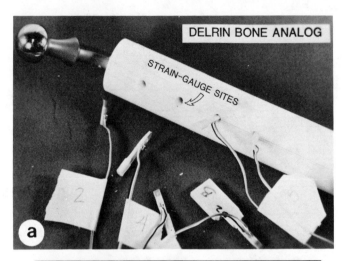

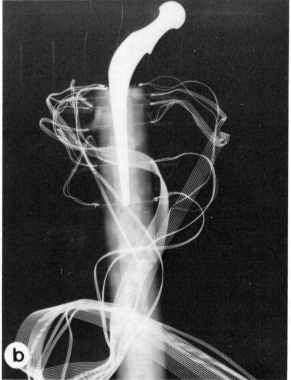

FIG. 9—(a) *Typical proportions of SGMs shown adjacent to DBA prior to insertion.* (b) *Typical radiographic appearance of completed DBA model with SGMs.*

relaxation periods using an MMED 10-channel amplifier and plotted on a Gould *X-Y* recorder.

Bone-Analog Model

The sectional modulus (EI) of the midshaft of the human femur varies in the 200 to 400 Nm² range [*34*]. This has been varyingly simulated by tabular models in numerous published femoral stem studies [*29,34–36*].

A 260-mm long Delrin (polyacetal) tube was fabricated as an analog for the femoral shaft of the femur to provide a *reproducible* mechanical model into which prostheses could be cemented (Fig. 9). The diameters (20-mm inside and 40-mm outside) of the Delrin-bone-analog (DBA) were chosen to provide a sectional modulus (EI) of 330 Nm² comparable to that of the human femur that ranges from 200 to 400 Nm² [*34*]. Stainless-steel Charnley and Ti-6Al-4V STH femoral stems were marked in 10-mm levels from the collar down to the stem tips. These values corresponded to the stem-performance-factor (SPF) analysis sections [*37,38*]. The stems were aligned using cement-blocks as spacers attached to the lateral stem surfaces. The medial and lateral cement thicknesses as seen by X-ray varied from 2 to 6 mm with distance from the proximal collar to the distal tip.

The three most proximal element layers in the anatomical femur model correspond to 17% of the total length of the stem [*10*]. Therefore 17% of the embedded part of the DBA stem (that is, the three most proximal embedded layers, each 3-mm thick), were considered representative of the first three element layers of the anatomical, even though the latter were of larger size. The stems were embedded up to section Level 04 (40 mm below collar) using acrylic cement (Fig. 10). The DBA-stem structure was X-rayed to determine the positioning dimensions for the SGMs and then 3.2-mm-diameter holes were drilled up to the stem's surface. The SGMs were cemented in place as before and left for a minimum of 24 h prior to testing (Fig. 10).

The dimensions of the DBA, the pivoting mounting-cradle, and the polyethylene loading-cup for the femoral head were arranged to provide a reproducible load axis of 9° relative to the longitudinal axis of the DBA. Static loads were applied in 250 N increments up to 1500 N, strain gage data being taken following 6 min relaxation time at each load application.

Upon satisfactory completion of the range of static tests, the DBA models were run under 500 N and 1000 N cyclic loading designed to provide clinically relevant, peak, proximal-medial cement stresses in the 5 to 10 MPa range, typical of the three-dimensional computer models [*3,10*]. Cyclic frequencies of 1 and 5 Hz were selected for the initial tests.

Finite-Element Models (FEM) of Delrin Plastic Bone Analog (PBA)

The cemented femoral stem and the DBA model were incorporated in a three-dimensional FEA model (Fig. 11). Due to its geometrical simplicity, it was very

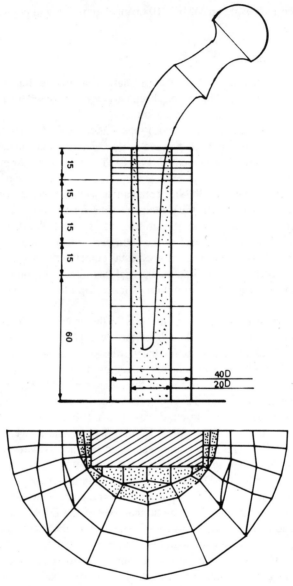

FIG. 10—*Longitudinal and cross-sectional descriptions of FEM derived for DBA.* (*Dimensions in mm.*)

straightforward to incorporate additional element density as and when required. The prototype model had approximately the same number of elements as the three-dimensional anatomical model [*10*]. There were 936 three-dimensional isoparametric elements with 1187 nodes (Table 3). The materials were modeled as homogeneous, linear, elastic media.

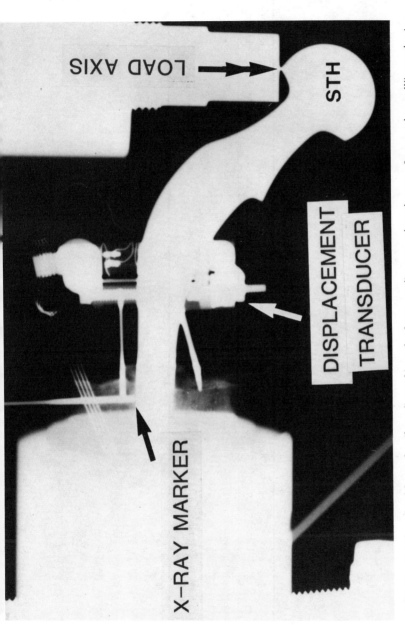

FIG. 11—*Cantilever-bend fixture for pilot radiographic study of stem-acrylic gaps on lateral aspect of cemented stem. Wire marker in cement sheath used to identify separation at stem-acrylic interface. Displacement transducer monitored relative motion between lateral stem surface and cement sheath.*

TABLE 3—Summary of three-dimensional FEM model.

	Element Type	Displacement Field	Nodes/ Element	Number of Nodes	Load	Load Axis	Symmetry	Number of Degrees of Freedom
Three-dimensional anatomical model (Tarr)	tetrahedron	quadratic	8	1215	896	3000 N	½	3645
Three-dimensional DBA model	tetrahedron	quadratic	8	1187	936	1500 N	½	3561

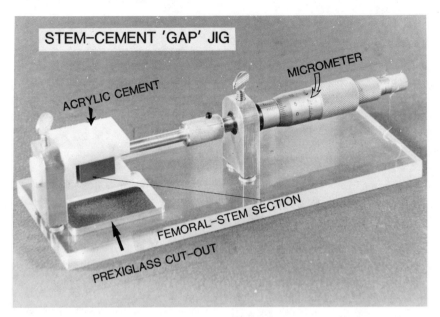

FIG. 12—*Plastic micrometer jig for radiographic assessment of SAGs.*

Quantitative X-Ray Analysis of Gaps at the Stem-Cement Interface

The objective of this study was to determine the minimum size of the SAG that could be detected by radiographic techniques. In a pilot study, we loaded a Ti-6Al-4V STH stem as a cantilever beam to examine the effect on the lateral stem-acrylic interface (Fig. 11). A metal X-ray marker was positioned through the acrylic sheath to contact the lateral stem surface prior to loading. A displacement transducer monitored relative motion between the stem and the lateral cement. Displacement of the stem with respect to the cement was monitored continuously up to 1000 N load.

For the second model, the typical dimensions of a femoral stem and cement sheath (collar region) were modeled in a micrometer set-up (Fig. 12). The displacement of the micrometer shaft moved the stem cross-section relative to the cement, thereby creating a SAG of known width. At suitable increments, the SAG size was verified by gages and by viewing in strong back-lighting. A high-resolution Faxitron (43 805 N) was used to take the radiographs centered on the stem-acrylic interface, with gaps varying from 0 to 2 mm width. The gaps on the radiographs were measured (*a*) directly from the X-ray films using digital calipers and a magnifying glass and (*b*) on photomicrographs of the X-ray films at ×40 magnification.

Results

Strain-Gage Modules (SGM)

The initial test of flat rectangular acrylic cement beams provided a flexural modulus of elasticity averaging 2055 ± 250 MPa and ultimate flexural strength values of 29.5 ± 2 MPa. The strains measured using embedded strain gages were within 6% of the values calculated from beam theory. This modulus value was used in subsequent studies with the DBA models.

Three-Dimensional FEM Bone Model Comparisons

For comparison, we used the data published in a previous investigation of a three-dimensional anatomical model of the femur fitted with a Charnley-type femoral stem [10]. This computer model analyzed an ideally-cemented Charnley femoral prosthesis in both Ti-6Al-4V and stainless steel versions. The latero-medial (Fig. 3, S_y) and axial (S_z) cement stresses were the two largest components of stress, whereas the hoop stresses (S_x) on the medial side were the lowest. The peak radial compressive stresses occurred immediately under the collar (Fig. 13) and then decreased rapidly with distance from the collar (Fig. 7).

Comparing the cement stresses (S_y, S_z) along the medial aspect of the stem for an assumed load of 3000 N (Fig. 7), it was apparent that:

1. The cement stresses were higher proximally than at the distal tip but between 0 and 1 MPa in the mid-stem region.
2. The (S_y) cement stresses decayed rapidly just a short distance away from the collar.
3. The (S_y) cement stresses around the stem surface were three to four times higher than those adjacent to the bone at the same level.
4. The proximal-medial radial (S_y) cement stresses for the Ti-6Al-4V stem design (2 to 6.4 MPa) were 30% or more higher than the cobalt-chromium design (1.5 to 4.6 MPa).

The FEM-anatomical and FEM-DBA models represented two different experimental conditions (Table 3).

Overall, the three-dimensional anatomical and DBA-FEM models had a somewhat similar circumferential S_y distribution pattern despite the large variations in input conditions between models (Fig. 13). At the proximal-medial site, the three-dimensional anatomical model predicted cement stresses (S_y) in the 7 MPa range whereas the three-dimensional DBA model predicted 11 MPa on the medial side. On the proximal-lateral site, the three-dimensional DBA model also predicted higher tensile stresses, 13 MPa and 6 MPa for S_y and S_z, respectively. The ratio of medial to lateral S_y stresses (radial) was 0.85 and for S_z (axial) was 1.1. In contrast, the three-dimensional anatomical model predicted S_y stresses of only 0 to 1 MPa at the lateral site with a resulting medial:lateral ratio of 7.9.

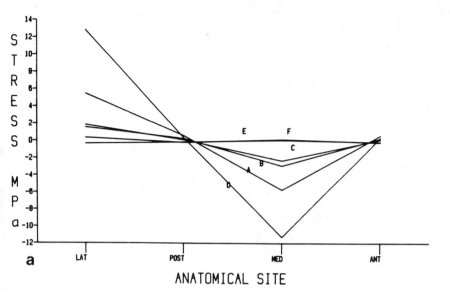

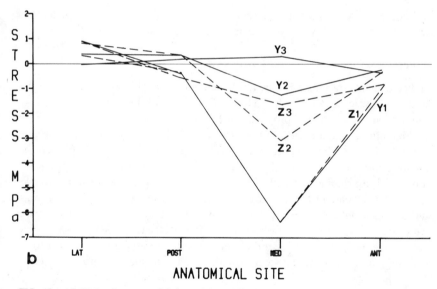

FIG. 13—(a) *Finite-element-model data of* S_y *and* S_z *stress distribution around most proximal section of the DBA (assumed load of 1500 N with 9° load axis to obtain required medial* S_y *cement stress magnitude in 5 to 10 MPa range).* (b) *Finite-element-model data of stress distribution* (S_y *and* S_z) *around three-dimensional anatomical model* [10]. (*POST* = *posterior, LAT* = *lateral, MED* = *medial, and ANT* = *anterior.*)

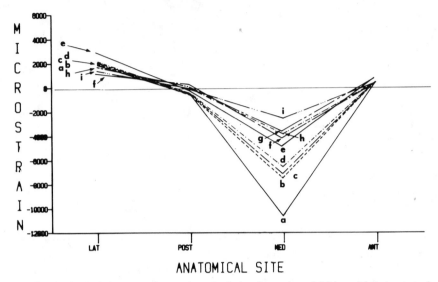

FIG. 14—*Strain* (e_y) *measured around proximal site of experimenal DBA model during test of total duration 4.1 million load cycles (1500 N load, 9° load axis).*

For the slightly more distal element rows, both models showed a large decrease in stress, the medial S_y stresses decreasing faster than the S_z stresses.

Experimental Delrin-Bone-Analog Studies

The cement strain data for the DBA were recorded following each of the first four load cycles and then after 5400, 110 000, 310 000, 2.1 million, and 4.1 million cycles. The initial proximal-medial cement strains (e_y) of 11 000 με fell to 7000 after the first cycle, then to 5440 and 3000 με on completion of second, 5400, and 4.1 million cycles, respectively (Fig. 14). This time-dependent behavior corresponded to a 73% decrease in proximal-medial cement strain over a nominal test duration of 800 h. Although the proximal-medial strains showed a dramatic decrease just below the collar, there was a slight but consistent increase more distally with time (Fig. 15). On the posterior and anterior sites, the stresses were smaller as expected, between 0 and 1 MPa and 0 and −1 MPa, respectively. During the cyclic tests, the stem was observed to be flexing with respect to the cement bed but there was neither visual nor radiographic evidence of SAGs.

X-Ray Jig Studies of SAG Magnitudes

The radiographic analyses of SAGs, demonstrated that gaps were not clearly identified until 0.1 mm or larger. In the 0.05 to 0.1-mm range, some investigators noted (somewhat hesitantly) that there appeared to be a faint "indication" of the gap, however marginal. It was observed that, when the investigators first focused on the interfaces, small gaps (0.08 to 0.12 mm) could just be detected.

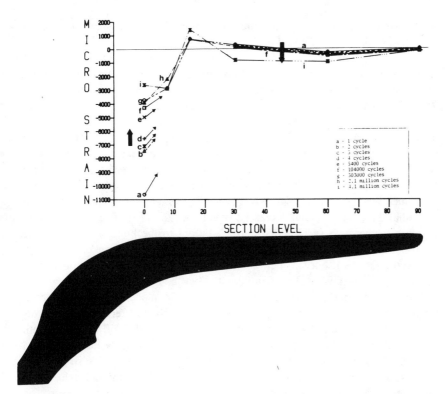

FIG. 15—*Longitudinal strain distribution* (ϵ_y) *along medial cement column adjacent to stem in an experimental DBA during cyclic tests of 4 million cycles duration. Tests* a *to* g *demonstrated progressively proximally decreasing strains with a slight increase distally. For Tests* i *and* h, *two additional SGMs were placed at 10-mm and 15-mm section levels. Note that Data Points* a *to* g *are not shown connected between Level 0 and Level 30 since the intervening cement strains were not initially recorded.*

However, a short time later, the faint gap was sometimes noted by some observers to apparently disappear.

Discussion

There have now been two experimental studies of strain measurements *within* cement. Draganich [33] studied the performance of gages embedded inside cement beams using cantilever bend tests and Crowninshield et al [32] embedded multiple strain gages in the cement sheath around femoral stems. Both studies agreed that the cement strains were accurate and reliable to within ±10%. We can confirm that the accuracy of embedded gages is within 6%. However, the advantage of the SGM technique used in this study is that it provides a practical method to position gages precisely within the cement without much effort, and broken gages may be replaced by simply using a new SGM without destroying the model. In fact, the SGMs were replaced repeatedly seven to eight times for the proximal sites during the dynamic tests when either the gages became inoperative or data was required at different sites.

The observations from the SAG calibrations by X-ray procedures showed that basically the various investigators underestimated the gap size, that is, the average measurement error was -0.2 mm. We may conclude that conventional X-ray techniques may not detect gaps smaller than 0.1 mm.

The stress patterns calculated with the three-dimensional DBA and the three-dimensional anatomical models were surprisingly similar despite differences in material properties and stem-support conditions in the models. One exception was the high tensile cement stress on the lateral side of the three-dimensional DBA model compared to the three-dimensional anatomical model. Both models confirmed the presence of high proximal-medial compressive cement stresses in the radial direction (S_y) that decreased rapidly distally [13]. These were higher adjacent to the stem surface than those adjacent to the bone surface. On the proximal-medial cement surface, the radial stress (S_y) was as large or larger than the axial stress (S_z), but still below failure strength of cement in compression. The lowest stresses were the circumferential or "hoop" stresses.

The strain distribution from the experimental DBA model appeared similar to the mathematical models. However, the ratio of medial to lateral strain (S_y) was 6.9 initially compared to 0.85 for the three-dimensional DBA-FEM model. With cyclic loading, the lateral tensile strains (S_y) increased slightly then decreased to 60% of their initial value. However, the medial compressive strains (S_y) decreased continuously with cyclic loading to 37% of their initial value. The final medial:lateral S_y ratio then had fallen to 1.8. This therefore represented a dramatic time-dependant change in the stress measured in cement supporting the stem proximal-medially.

Upon completion of 4 million cycles, the pistoning of the stem against the cement sheath was more evident. However, radiographically there was no discernible gap on the proximal-lateral interfaces.

It may well be that the stem in the experimental DBA model was already marginally loose as a result of either a creep mechanism or fatigue-collapse acting on the medial cement column. The radiographic analyses of the interface gaps demonstrated that it is very difficult to identify gaps smaller than 0.1 mm (100 μm). Under clinical conditions, the minimum detection level may easily be as large as 0.15 to 0.2 mm due to the overlay of bone and soft tissues, and problems involved in positioning relative to the X-ray beam. Thus, many of the clinical hip stems apparently in contact with the cement on the radiographs may, in fact, have gaps as large as 1 mm. Clearly, the radiographic techniques in clinical use cannot visualize such narrow gaps. It was also significant that 60 to 70% of the cases were reported as having no "observable" radiolucent zones. However, due to the masking effects of (a) X-ray gap resolution and (b) stem rotation artifacts [39], there can be no certainty that such gaps were, in fact, absent. Even measurement of such gaps must now be considered only a minimal indication of their actual width.

This is the first investigation of the time-dependency of the stress distribution in the cement supporting femoral stems, under cyclic loading. In this pilot study, the ratio of medial:lateral radial strains (e_y) fell from 6.9 to 1.8 with cyclic loading that still contrasted with the corresponding values of 0.85 for the FEM-DBA model and 7.9 for the three-dimensional FEM anatomical model.

We are continuing to investigate the role of stem design features in stressing the surrounding cement. Since the cement appears to be undergoing continuous stress relaxation, the eventual stress pattern may be quite different from what we first anticipated. There is also an obvious need for more clarification of the observed differences between mathematical and experimental models.

The initiation of a failure mechanism at the stem-acrylic interface could possibly be controlled by stem surfaces incorporating mechanical interlocking as well as pre-coated acrylic-surfaced stems. The effect of the cobra type of flange [40] or its successors [7] on limiting plastic flow of the medial cement sheath may also be significant.

Summary

1. A review of the clinical literature revealed up to a 40% incidence of stem loosening within the acrylic sheath of total-hip prostheses.

2. Radiographs of a model stem with known cement gap widths demonstrated that gaps smaller than 100 µm are not detectable on typical clinical radiographs. Therefore, many of the stems that appear well-fixed on X-ray may, in fact, have gaps as large as 100 µm. This agrees with the finding that several stems examined post-mortem have had 100-µm-thick membranes between stem and cement sheath.

3. Computer models and static experimental tests of cemented femoral stems demonstrated that the highest cement stresses were on the most proximal site adjacent to the medial stem surface. These stresses were shown to be high enough to result in significant cement creep.

4. Cyclic tests of cemented femoral stems demonstrated that by 4 million cycles there was a 73% decrease in the proximal-medial cement strains, and a slight increase more distally. There was a decrease in the ratio of medial:lateral strains (e_y) from 6.9 to 1.8 with time.

5. This first evidence of stress relaxation of the cement surrounding the proximal stem may have profound implications in our understanding of the effects of stem design on the cement and bone stresses. Prevention of loosening by rough-surfaces, pre-bonded acrylic surfaces, or lateral stem-flanges may have an important clinical effect.

Acknowledgments

This study was partially supported by the Orthopaedic Foundation of Los Angeles. Thanks are due to Elena Herrera and Cheryl Piazzola for their diligence in typing the manuscript.

References

[1] Lewis, J. L. and Hori, R. in *Proceedings*, Workshop on Internal Joint Replacement, C. L. Compere and J. L. Lewis, Eds., Northwestern University, Chicago, 1977, pp. 1–9.

[2] Crowninshield, R. D., Brand, R. A., Johnston, R. C., and Milroy, J. C., *Clinical Orthopaedics*, Vol. 146, 1980, pp. 71–77.

[3] Crowninshield, R. D., Pedersen, D. R., and Brand, R. A. in *Transactions*, American Society of Mechanical Engineers, Vol. 102, 1980, pp. 230–233.

[4] Savino, A. W., Andersson, G. B. J., Andriacchi, T. P., Hampton, S., and Galante, J. O., *Acta Orthopaedica Scandinavica*, Vol. 53, 1982, pp. 23–27.

[5] Carlsson, A. S. and Gentz, C-F., *Clinical Orthopaedics*, Vol. 147, 1980, pp. 262–270.

[6] McBeath, A. A., Schopler, S. A., and Narenchania, R. G., *Clinical Orthopaedics*, Vol. 150, 1980, pp. 301–305.

[7] Oh, I., Treharne, R. W., and Sander, T. W. in *Transactions*, Society for Biomaterials, Vol. 6, 1983, p. 95.

[8] Oh, I. in *Transactions*, Society for Biomaterials, Vol. 6, 1983, p. 94.

[9] Markolf, K. L. and Amstutz, H. C., *Journal of Biomechanics*, Vol. 9, 1976, pp. 73–79.

[10] Tarr, R. R., Clarke, I. C., Gruen, A., and Sarmiento, A. in *Finite Elements in Biomechanics*, R. H. Gallagher, B. R. Simon, P. C. Johnson, and J. F. Gross, Eds., University of Arizona, Tucson, 1982, pp. 345–359.

[11] Andriacchi, T. P., Galante, J. O., Belytschko, T. B., and Hampton, S., *Journal of Bone and Joint Surgery*, Vol. 58A, 1976, pp. 616–624.

[12] Clarke, I. C., Gruen, T. A. W., Tarr, R. R., and Sarmiento, A. in *Proceedings*, International Conference on Finite Elements in Biomechanics, B. R. Simon, Ed., University of Arizona, Tucson, Vol. 2, 1980, pp. 487–510.

[13] Huiskes, R., *Acta Orthopaedica Scandinavica*, Supplementum No. 185, 1979.

[14] Huiskes, R. and Slooff, T. J. J. H. in *Transactions*, Orthopaedic Research Society, Vol. 3, 1978, p. 148.

[15] Ling, R. S. M. in *The Hip*, C. V. Mosby, St. Louis, 1980, pp. 82–111.

[16] Charnley, J., *Clinical Orthopaedics*, Vol. 111, 1975, pp. 105–120.

[17] Moreland, J. R., Gruen, T. A., Mai, L., and Amstutz, H. C. in *The Hip*, C. V. Mosby, St. Louis, 1980, pp. 281–291.

[18] Amstutz, H. C., Markolf, K. L., McNiece, G. M., and Gruen, T. A. in *The Hip*, C. V. Mosby, St. Louis, 1980, pp. 102–116.

[19] Gruen, T. A., "Clinical Reality of a Titanium Alloy Total Hip Replacement," presented at the 48th Annual Meeting of the American Academy of Orthopaedic Surgeons, Las Vegas, 1981.

[20] Fornasier, V. L. and Cameron, H. U., *Clinical Orthopaedics*, Vol. 116, 1976, pp. 248–252.

[21] Stauffer, R. N., *Journal of Bone and Joint Surgery*, Vol. 64A, 1982, pp. 983–990.

[22] Ahmed, A. M., Miller, J., Stachiewicz, J. W., D'Souza, N., and Burke, D. L. in *Transactions*, Orthopaedic Research Society, Vol. 2, 1977, p. 29.

[23] Gruen, T. A., "Criteria for Radiographic Evaluation of Total Hip Replacement," Scientific Exhibit (#6005), at the 50th Annual Meeting of The American Academy of Orthopaedic Surgeons, Anaheim, 1983.

[24] Semlitsch, M. and Panic, B., *Biomedizinische Technik*, Vol. 28, No. 4, 1983, pp. 66–78.

[25] Pal, S. and Saha, S., *Biomaterials*, Vol. 3, 1982, pp. 93–96.

[26] Atkinson, J. R., Dowling, J. M., and Cicek, R. Z., *Biomaterials*, Vol. 2, 1980, pp. 89–99.

[27] Hampton, S. J., "A Nonlinear Finite Element Model of Adhesive Bond Failure and Application to Total Hip Replacement Analysis," Ph.D. thesis, University of Illinois, Chicago, 1981.

[28] Hampton, S. J., Andriacchi, T. P., Urban, R. M., and Galante, J. O. in *Transactions*, Orthopaedic Research Society, Vol. 8, 1983, p. 64.

[29] Svensson, N. L., Valliappan, S., and McMahon, R., *Engineering in Medicine*, Vol. 9, 1980, pp. 143–146.

[30] Collier, J. P., Kennedy, F., Mayor, M., Townley, C., and Amoakuh, E. in *Transactions*, Society for Biomaterials, Vol. 6, 1983, p. 96.

[31] Pilliar, R. M. and Bratina, W. J., *Journal of Biomedical Engineering*, Vol. 2, 1980, pp. 49–53.

[32] Crowninshield, R. D. and Tolbert, J. R., *Journal of Biomedical Materials Research*, Vol. 17, 1983, pp. 819–828.

[33] Draganich, L. F., Andriacchi, T. P., and Galante, J. O., *Journal of Biomechanics*, Vol. 15, 1982, pp. 789–790.

[34] Dobbs, H. S. and Chaplin, C. R., *Journal of Biomedical Engineering*, Vol. 3, 1981, pp. 225–234.

[35] Townsend, P. and Diamond, R. in *Transactions*, Orthopaedic Research Society, Vol. 4, 1979, p. 196.

[36] Rohlmann, A., Bergmann, G., and Koelbel, R., *Zeitschrift für Orthopaedie*, Vol. 118, 1978, pp. 122–131.

[37] Espiritu, E. T., Rao, S., Sew Hoy, A. L., Clarke, I. C., and Sarmiento, A. in *Titanium Alloys in Surgical Implants, ASTM STP 796*, H. A. Luckey and F. Kubli, Jr., Eds., American Society for Testing and Materials, 1983, pp. 74–87.

[38] Clarke, I. C., Mina-Araghi, M., and McGuire, P., "Stem-Performance-Factor Analysis of Total Hip-Stem Design," in preparation.

[39] Ashford, R. and Clarke, I. C., "Quantification of Radiographic Shadowing Due to Rotation of the Femoral Stem," in preparation for *Clinical Orthopaedics*.

[40] Loudon, J. R. and Charnley, J., *Journal of Bone and Joint Surgery*, Vol. 62B, 1980, pp. 450–453.

[41] Weber, F. A., *Journal of Bone and Joint Surgery*, Vol. 57B, 1975, pp. 297–301.

[42] Nicholson, O. R., *Clinical Orthopaedics*, Vol. 95, 1973, pp. 217–223.

[43] DeSmet, A. A., Kramer, D., and Martel, W., *Journal of Bone and Joint Surgery*, Vol. 60A, 1978, pp. 279–313.

[44] Collis, D. K., *Journal of Bone and Joint Surgery*, Vol. 99A, 1977, pp. 1033–1041.

[45] Bocco, F., Langan, P., and Charnley, J., *Clinical Orthopaedics*, Vol. 128, 1977, pp. 287–294.

[46] Beckenbaugh, R. D. and Ilstrup, D. M., *Journal of Bone Joint Surgery*, Vol. 60A, 1978, pp. 306–313.

[47] Gruen, T. A., McNeice, G. M., and Amstutz, H. C., *Clinical Orthopaedics*, Vol. 141, 1979, pp. 17–27.

[48] Salvati, E. A., Wilson, P. D., Jr., Joelly, M. N., Vakili, F., Aglietti, P., and Brown, G. C., *Journal of Bone and Joint Surgery*, Vol. 63A, 1981, pp. 753–767.

[49] Haas, S. S., Brauer, G. M., and Dickson, G., *Journal of Bone and Joint Surgery*, Vol. 57A, 1975, pp. 380–391.

[50] Lautenschlager, E. P., Jacobs, J. J., and Marshall, G. W., *Journal of Biomedical Materials Research*, Vol. 10, 1976, pp. 929–938.

[51] Holm, N. J., *Acta Orthopaedica Scandinavica*, Vol. 48, 1977, pp. 436–442.

[52] Lee, A. J. C., Ling, R. S. M., and Vangala, S. S., *Journal of Medical Engineering Technology*, Vol. 1, 1977, pp. 137–140.

[53] Lee, A. J. C., Ling, R. S. M., and Vangala, S. S., *Archives of Orthopaedic and Traumatic Surgery*, Vol. 92, 1978, pp. 1–18.

Retrieval Analyses and Standards

Roger W. Hood[1]

Retrieval Analyses of Total Joint Components from the University of Missouri-Kansas City School of Medicine Implant Retrieval Program

REFERENCE: Hood, R. W., "**Retrieval Analyses of Total Joint Components from the University of Missouri-Kansas City School of Medicine Implant Retrieval Program,**" *Corrosion and Degradation of Implant Materials: Second Symposium, ASTM STP 859*, A. C. Fraker and C. D. Griffin, Eds., American Society for Testing and Materials, Philadelphia, 1985, pp. 403–414.

ABSTRACT: The retrieval analyses of orthopaedic implants gathered at revision surgery or at autopsy provide much needed data as to the performance of such implants *in vivo*. This paper details the organization of such a retrieval program in a university medical center. The primary objective of this program is to identify specific implant design and material problem areas. Since such a variation of designs and materials of total joint implants exists, special emphasis is placed on these retrieval analyses. Femoral stem fracture in older total hip designs has been observed. Surface failure and cold flow of the ultra-high molecular weight polyethylene has been also found, particularly in some total knee designs.

KEY WORDS: retrieval analyses, orthopaedic implants, femoral stem fracture, polyethylene wear, carbon-reinforced polyethylene wear debris, implant materials

The reconstructive nature of orthopaedic surgery frequently requires the implantation of replacement or stabilization devices. Most often this is done for fracture fixation, and the devices are subsequently removed once the fracture has healed. Some implants such as total joints are placed permanently and only removed if a complication arises. For many years the implant manufacturers have been interested in the retrieval of failed implants,[2] but in the last few years several implant retrieval programs have been established at the university level to examine all retrieved implants [1–7].[3] This allows a more widespread interchange of the data obtained on both the prostheses that have performed well (and have been removed for sepsis or after fracture healing) as well as those

---

[1]Assistant clinical professor of Orthopaedic Surgery, MidAmerica Joint Replacement Institute, University of Missouri-Kansas City School of Medicine, Kansas City, MO 64108.

[2]R. Fuson, personal communication.

[3]The italic numbers in brackets refer to the list of references appended to this paper.

removed for implant failure, that is, those implants that have fractured or failed due to excessive wear. This has allowed an analysis of these implants that have functioned *in vivo* that ought to provide more reliable data than bench testing *in vitro*, which cannot fully simulate accurately all conditions to which an implant is subjected in actual use. It also removes the failure analysis connotation by also examining the implants that have functioned well.

Our primary interest in implant retrieval at the University of Missouri-Kansas City (UMKC) concerns total joint prostheses. We are particularly interested in the problem of polyethylene wear and deformation in the various retrieved implants. This provides realistic data for the selection of implant design and material based on estimated patient usage for a particular clinical situation. As more designs are examined over time, we will accumulate enough numbers of retrievals to eliminate implant geometry factors and concentrate on material factors influencing long-term results.

This paper presents a method for setting up an implant retrieval program at a large university hospital with community participation in the retrievals by metropolitan area orthopaedic surgeons. Examples of interesting cases are presented with particular emphasis on the problem of polyethylene wear in total joint implants.

Materials

All orthopaedic implants removed in the operating room at the University of Missouri-Kansas City/Truman Medical Center and, in most instances, implants removed at any of the four affiliated hospitals in the UMKC Orthopaedic Surgery Residency Program, are collected. The retrieval rate is approximately 150 implants per year. The program was begun in July 1981, and 207 implants have been collected in the first 18 months of operation. Total joint prostheses account for 40% of this total (81/207). Total knees slightly outnumber total hips (45:36).

Some of the private orthopaedic surgeons in the community have been resistant to the idea of having their prostheses included in the University Retrieval Program. It has been a routine in this area that the patient would receive any implant that was removed (if they so desired, once the surgeon's collection was complete), since they had paid for the implant. However, the opinion of hospital legal counsel that any removed implant should be treated as a pathologic specimen and retained at least for the duration of the statute of limitations has brought only slightly more cooperation from area orthopaedists. Additionally, we have emphasized our interest in just the total joint retrievals to the private surgeons in the hopes of obtaining those implants perhaps at the expense of obtaining fracture fixation devices.

A subset of the implant retrieval protocol has been developed for handling those implants retrieved at autopsy or amputation. These specimens are removed and stored *en bloc* in a freezer. We have obtained one hip and four knee prostheses in the first 18 months of program operation. These will be used for various mechanical testing protocols once enough are obtained to run a series. The fact that the morgue for Kansas City, Missouri, is located in our hospital should allow this type of retrieval to become more plentiful in the future as we identify those patients with total joints who are undergoing postmortem examination.

Methods

Implant Retrieval Procedure

All orthopaedic implants removed at surgery are put through the ultrasonic instrument washer in the surgery equipment room. They are then placed in appropriate-sized individual boxes that are labeled with the patient's identification data (name, hospital number, date of birth). Foam rubber packing is used to separate and pad multiple-part prostheses to prevent artificial damage to the implants that might later be construed as occurring *in vivo*. This cleaning and boxing procedure is handled by the orthopaedic equipment nurse in the operating room. The implants are then taken to the orthopaedic research laboratory where they are logged into the implant retrieval book in sequential order of receipt. Additionally, a 3 by 5 index card is prepared for each implant and filed under a category designation—total hip, total knee, fracture plate, compression hip screw, intramedullary rod, etc. This allows rapid access to a specific group of prostheses. The file number in the master log is placed on the box containing the implant and on the index card.

An implant inspection report request form for the removed implant is completed at the time of removal by the orthopaedic resident or staff surgeon responsible for the patient. This form contains demographic information as well as diagnosis at implant insertion, reason for removal, dates of each, patient's height and weight, and activity level (see Fig. 1). The form then accompanies the implant to the research lab to await cataloging, inspection, and storage. For all total joint retrievals, a representative specimen of synovium and of granulation membrane at the bone/cement interface is sent to Pathology. The pathologists are instructed to carefully report any cement or polyethylene debris observed in the tissue. A copy of the pathology report is attached to the retrieval report. The bottom half of the triplicate request form is for the inspection report on the implant. Various code numbers, size, and manufacturer are recorded as well as a visual description of the device. Space is available for more specific analysis if warranted or requested. The completed form is then signed by the director of the retrieval program and becomes an official document to be inserted in the patient's permanent hospital record. One copy is sent to the surgeon's office for inclusion in his/her files, and the third copy is kept in the orthopaedic department implant retrieval files.

It should be noted that we make no attempt to retrieve percutaneous Steinmann pins, Kirschner wires, or external fixator pins. We could not imagine any real value in retrieval analyses of such commonly used devices and felt the added paperwork and expense for several hundred such devices per year could easily be avoided for all concerned.

School of Medicine

UNIVERSITY OF MISSOURI - KANSAS CITY

2411 Holmes Street
Kansas City, Missouri 64108
Telephone 816-474-4100

IMPLANT RETRIEVAL PROGRAM

Patient Name _____ Hospital Record No. _____

Address _____ Phone _____

_____ Birth Date _____

Date of Removal _____ Removal Surgeon _____

Hospital of Removal (circle): TMC   St.L   VAH   CMH   BMH   RMC   MMC   SMMC

Other _____

Date of Implantation _____ Hospital of Implantation _____

Site of Implantation _____ / _____
                      side R or L   bone or joint

Patient Weight _____ lbs.   Patient Height _____ ft. _____ in.

Activity Level While Implant in Place: ☐ very active   ☐ moderately active
                                        ☐ sedentary   ☐ bed or wheelchair confine

Diagnosis at Implantation _____

Reason for Removal: ☐ routine post fracture   ☐ sepsis
                     ☐ implant failure   ☐ other _____

*Send Sample of Synovium for Microscopic Analysis from All Total Joint Implants.

RETRIEVED IMPLANT INSPECTION REPORT

Date Received _____

Implant _____ Catalog No. _____

Manufacturer _____

Code Numbers _____ Size _____

Gross Condition:

Detailed Analysis (if indicated):

_____
Roger W. Hood, M.D.
Director, Implant Retrieval Program

an equal opportunity institution

FIG. 1—*Copy of Retrieved Implant Inspection Report Data form.*

En bloc *Implant Retrieval Procedure*

Those patients who are terminally ill and have a total joint prosthesis in place are identified by the medical staff at Truman Medical Center. The orthopaedic department has alerted the other medical services that we are very interested in retrieval of these prostheses at autopsy (or at amputation) so that we might examine total joint prostheses that have functioned well *in vivo* (presumably) when death occurs from an unrelated cause. Permission for autopsy is then obtained. The specimens are taken *en bloc* with joint capsule and ligaments intact. The orthopaedic residents do the actual retrieval at the postmortem examination to assure that the specimen retrieved is intact. The pathologist records this retrieval in the autopsy report. A small incision is made through the capsule and a synovial biopsy is obtained for microscopic pathological analysis (looking for polyethylene debris and tissue reaction) [8]. The joint is then wrapped in a plastic freezer bag and stored in a freezer for later examination.

Patient Data and Radiographs

A complete set of radiographs and the medical chart are available for all implants removed at the University medical center. The implant retrieval request sheet requires the most pertinent data to be completed by the physician at the time of removal when the chart and radiographs are immediately available. This has virtually eliminated the very time-consuming chart reviews to obtain the necessary information. For implants from other institutions, we ask the orthopaedic resident involved in the case (or the staff physician) to complete the form at the time they deliver the prosthesis to the UMKC orthopaedic department.

The total joint retrievals are handled slightly differently in that more information is obtained and pertinent serial radiographs are copied and filed with the implant data for later ease of data analysis/retrieval. Copies of the operative note are frequently included for a description of the findings in the joint at the time of revision/removal. Numerical score systems are used to clinically follow patients with total joints, and this sequential record is obtained (or reconstructed, it possible, on referral patients) to help establish activity/performance levels prior to the problem that led to revision arthroplasty. Since a large portion of my orthopaedic practice is made up of patients referred for revision of failed total joint arthroplasties, I ask the referring orthopaedist to send me copies of pertinent information (operative notes, discharge summaries, office follow-up notes) that will be helpful in reconstructing the clinical situation that led to failure. A copy of the information is included in the retrieval record file for easy accessibility.

Examination Methods for Retrieved Implants

Each removed implant is visually examined (using straight edges, 90° guides, and vernier calipers) for evidence of gross deformation or mechanical damage sustained *in vivo*. Once one becomes familiar with such examinations, it becomes quite easy to discern which marks are from the retrieval effort and which are due to damage that occurred *in vivo*. Closer examination is then performed under light microscopy at × 10. During this examination the amount of mechanical damage present and its distribution are recorded in the manner previously described by the biomechanics department at The Hospital for Special Surgery [1,5,6,7]. Metallic components are graded for corrosion, fracture, and scratching of the polished surface in the track areas. The ultrahigh molecular weight polyethylene (UHMWPE) components' articulating surfaces are graded for scratches, burnishing, pitting, embedded polymethylmethacrylate (PMMA) debris, surface deformation, abrasion, and delamination. The system of ten surface areas for tibial components and four areas for patellar buttons and acetabular components described previously was used [6,9]. A vernier caliper is used to measure patellar button eccentricity. All implants are photographed on high-resolution color slide film from several perspectives at this time. The slides are coded with the implant retrieval number for later anonymity. The pertinent radiographs are also photographed using SO-185 (Kodak) film and coded with the identification number. The 35-mm slides are stored in the patient's folder in the retrieval analysis file. The serial radiographs are used extensively to provide temporal as well as clinical correlation as to failure mode of the implant.

A more detailed analysis is not usually performed on most components. However, in certain circumstances, the polyethylene components are subjected to scanning electron microscopy to better determine the etiology of various wear processes. We have not routinely examined the metal components radiographically, nor have we used a fluorescent penetrant to look for surface cracks. No destructive or semi-destructive testing has been performed to date.

The testing for the *en bloc* specimens follows the protocol again developed at The Hospital for Special Surgery [8]. The *in situ* mechanical testing will be performed when a larger number of specimens have been collected to complete a research project on the subject.

Storage

Once inspection and photography of the implants have been completed, they are returned to their individual boxes for storage. All implants of the same kind are stored together in large boxes. Thus, all intramedullary rods or all total knees will be in a separate large box that is labeled as to type of prostheses stored within. These are kept in a locked closet in the research lab. Finding an individual component has proven relatively easy using this system. We have agreed to keep all prostheses for a minimum of ten years and plan to keep the total joint prostheses indefinitely. Once implants are discarded from the retrieval program, they are used in the orthopaedic bioskills laboratory for application onto cadaver bones and plastic bones by residents in training to perfect their manual skills.

TABLE 1—*Total joint retrievals.*

	July-December 1981	January-December 1982	January 1983
Hips	5	25	6
Knees	2	36	7
Totals	7	61	13

TABLE 2—*Total joint retrievals, reason for removal.*

	Hips	Knees	Totals
Prosthesis failure (fracture)	3	4	7
Prosthesis loosening	22	34	56
Joint sepsis	11	7	18

Results of Total Joint Retrievals

Large amounts of data are generated by retrieval programs such as this, and the conclusions continue to change as more implants are added to the series. This section will present some of the total joint retrieval trends that have been noted in the series at the University of Missouri-Kansas City. A list of retrievals is given in Table 1.

Metal Problems with Retrieved Total Joints

Two general types of problems are noted in this category. The first involves prostheses with metal-to-metal articulations such as some of the hinge knees. Axle wear has been noted by other authors [2,5], and we have three such specimens including one with a broken axle housing. The synovial response to the large amount of metallic debris has been also well-documented by others [5,10].

The second metal problem is that of fracture—either of the femoral stems of total hips or of the metal stems or tibial retainers/stems on total knee components. Fracture of a component usually requires revision surgery before further bony destruction occurs, whereas metal debris does not of itself require revision. Its presence may, however, predispose to later sepsis or prosthetic loosening as the synovium becomes overwhelmed with the debris-related phagocytic activity. We have retrieved three fractured hip stems and two hinged-knee component stem fractures (see Table 2). Both hinged-knee stem fractures and two of the femoral stem fractures were in components made of Type 316L stainless steel. The third femoral stem fracture was a standard stem Müller-type (see Fig. 2). All three stem fractures were felt to be the result of good distal fixation in the face of lost proximal cement support. All three hip stems were placed *in varus* originally (ranging from 6° to 14° varus) and were thus technically predisposed to problems. This suggests that the stem failures are related to higher localized stress in the stems under loading situation created by the cement distribution.

Scratching and corrosion of the metal components is a relatively minor problem. Certainly, all components that have been in use show some degree of scratching of the articular surface in the direction of motion. This does not, however, seem excessive nor does it seem to be releasing large quantities of metal debris to the synovium based on the pathological analysis of the synovium. Metal debris was not noted on any pathological analysis of retrieved synovium except in the metal-on-metal hinge designs.

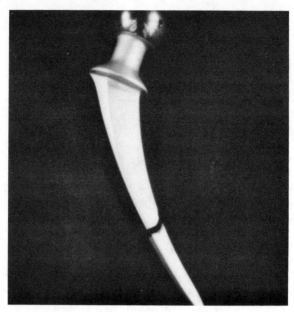

FIG. 2—*Fractured Müller-type cobalt-chromium molybdenum (ASTM Specification for Cast Co-balt-Chromium Molybdenum Alloy for Surgical Implant Applications F 75-76) femoral stem (Zimmer-Zimaloy). Fracture occurred after seven years implantation time.*

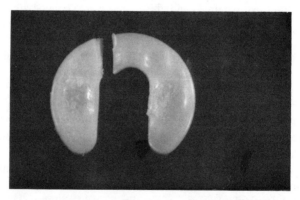

FIG. 3—*Retrieved tibial component from Biocompartmental Cruciate Retention prosthesis (How-medica 6439 design, RCH 1000 polyethylene) showing multiple failure mechanisms. The component fractured in vivo after implantation of less than three years with redevelopment of 30° varus deformity. Also, note the marked amount of surface destruction with gross pitting.*

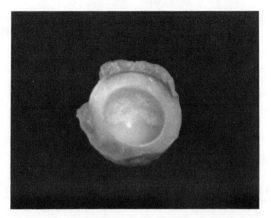

FIG. 4—*Retrieved 32-mm inside-diameter Müller cup showing fairly advanced surface failure of the polyethylene. This acetabular component was encountered at revision of a loosened femoral stem and was removed due to the marked polyethylene wear. No loose cement had escaped from the femoral canal, and all damage was felt to be in the absence of PMMA.*

Polyethylene Components

Several authors have commented on polyethylene wear from retrieved components [9,11,12]. Fracture and permanent deformation have been also noted [2,5,9]. In this series, there are four fractured polyethylene components. Two are Müller acetabular cups with the deep cement fixation grooves where the head has caused a "protrusio fracture" through the apical groove (that groove with the smallest radius). Neither of these was revised for this reason (both were loosening revisions), but the fracture was noted prior to revision on radiographs in one case. The other two polyethylene fractures are in tibial components from cruciate-retaining designs, one of which is pictured in Fig. 3. Both tibial component fractures occurred in active patients. None of these fractured polyethylene components was metal reinforced.

Polyethylene wear debris has been noted in the synovial biopsy sample of all retrieved components in the series (including a total knee patient who died from myocardial infarction three weeks postoperatively). Figures 3 and 4 show evidence of the typical polyethylene wear problems seen in retrieved components. Implantation time and body weight are correlated very well with the amount of polyethylene damage in the prostheses in this series (see Table 3). In loosening

TABLE 3—*Damage score correlation.*

Implantation time	$P < 0.001$
Activity level	$P < 0.1$
Body weight	$P < 0.001$

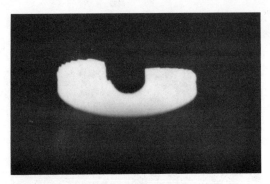

FIG. 5—*A cruciate retention tibial component from Biocompartmental design (Howmedica 6439). This is from the other knee of the patient whose fractured tibial component is pictured in Fig. 3. Both knees collapsed into recurrent varus, and this component had not yet broken but was 1 cm lower on the medial side of the knee.*

cases where PMMA has been free in the joint, obviously such damage mechanisms are vastly accelerated.

Permanent deformation of the polyethylene components was also found, particularly among the thinner (less than 10 mm) tibial components. Collapse of underlying bony support (perhaps due to excessive force transmission) under these thin components has frequently led to permanent deformation of the polyethylene. A typical example of permanent deformation is pictured in Fig. 5. Whether the current trend for metal reinforcement of tibial and acetabular components will eliminate these deformation and fracture problems remains to be seen. Additionally, all retrieved patellar buttons were found to be out of round (average 2.4% out of round with the elongation occurring in the superior-inferior direction). Once again, these are relatively thin polyethylene components without metal backing.

Wear Debris from Carbon-Reinforced Polyethylene

In an attempt to improve wear properties, some manufacturers have added carbon fibers to UHMWPE. Very few such prostheses have required revision to date; thus, retrieval data on this form of polyethylene is sparse. The UMKC Retrieval Program currently has seven such components. In general, the carbon-reinforced polyethylene appears to be functioning much the same as conventional white polyethylene. However, the debris from the wear that does occur is much more visible to the surgeon. Indeed, I have revised a total hip in which the synovium was black and visually appeared much the same as a metal-to-metal articulated total joint. Examination of the reinforced polyethylene surface does show the carbon fragments to be protruding from the surface after *in vivo* wear has occurred. Whether this effect is simply visual (that is, the black debris shows up whereas the white debris does not) or is more significant remains to be seen. The pathological analysis of the synovium removed from the aforementioned hip revealed no specific tissue reaction to the carbon fiber debris.

Discussion

It is the goal of implant retrieval programs to show persistent problems with materials or designs or both, and to suggest changes for improvement. Certainly the hip stem improvements of the late 1970s and early 1980s were the result of analysis of stem fractures. Now we are seeing the first failures of these new "supermetal" stems with reports of forged Vitallium and MP-35N stem fractures [13][4].

Metal reinforcement of tibial trays and acetabular cups has resulted from retrieval analyses. Not enough of these types of new designs have yet been retrieved to predict the success of this design/material change. The carbon-fiber-reinforced polyethylene may give improved wear results, although the yield point of the reinforced polyethylene may be exceeded by some (if not most) contemporary knee designs [14]. The color of the wear products is certainly a cause for concern, particularly in the more peripheral subcutaneous joints, even if the tissue reactivity to such carbon fibers proves to be benign with long-term follow-up.

Conclusion

A program for the retrieval analysis of orthopaedic implants has been established at the University of Missouri-Kansas City. The mechanism for setting up such a program is outlined. The early results of this program have further supported results reported by other authors. Additionally, retrieval results from carbon-reinforced polyethylene components are discussed along with the problem of visible staining of the synovium by the wear-liberated carbon particles.

The cost of establishing such a program and paying for such analysis of implants has to date been self-funded. The University has donated storage space, and I have been able to get summer students in engineering fields to do the chart reviews, photography, etc. However, were it not for generous self support, the program would not be able to do anything but collect implants. It is hoped that funding from other sources will be forthcoming.

References

[1] Wright, T. M. and Burstein, A. H. in *NBS-Implant Retrieval: Material and Biological Analysis*, U.S. Department of Commerce, Washington, DC, 1981, pp. 559–570.

[2] Hori, R. Y., Lewis, J. L., Wixon, R. L., and Kramer, G. M. in *NBS-Implant Retrieval: Material and Biological Analysis*, U.S. Department of Commerce, Washington, DC, 1981, pp. 509–559.

[3] Cook, S. D. and Weinstein, A. M. in *NBS-Implant Retrieval: Material and Biological Analysis*, U.S. Department of Commerce, Washington, DC, 1981, pp. 573–614.

[4] Schurman, D. J., Burton, D. S., Nagel, D. A., and McShane, D.: in *NBS-Implant Retrieval: Material and Biological Analysis*, U.S. Department of Commerce, Washington, DC, 1981, pp. 615–642.

[5] Hood, R. W., Wright, T. M., and Burstein, A. H., *Orthopaedic Transactions*, Vol. 6, 1982, p. 368.

[4]T. M. Wright, personal communication.

[6]   Hood, R. W., Wright, T. M., and Burstein, A. H., *Orthopaedic Transactions*, Vol. 5, 1981, pp. 319–320.
[7]   Hood, R. W., Wright, T. M., and Burstein, A. H., *Orthopaedic Transactions*, Vol. 5, 1981, pp. 291–292.
[8]   Wright, T. M., Hughes, P. W., Torzilli, P. A., and Wilson, P. D., Jr., *Journal of Bone and Joint Surgery*, Vol. 61-A, 1979, pp. 661–668.
[9]   Hood, R. W., Wright, T. M., and Burstein, A. H., *Journal of Biomedical Materials Research*, Vol. 17, 1983, pp. 829–842.
[10] Uchida, S., Yoshino, S., Doi, M., and Kudo, H., *International Orthopaedics*, Vol. 3, 1980, pp. 285–291.
[11] Dowling, J. M., Atkinson, J. R., Dowson, D., and Charnley, J., *Journal of Bone and Joint Surgery*, Vol. 60-B, 1978, pp. 375–382.
[12] Rose, R. M. et al, *Clinical Orthopaedics*, Vol. 145, 1979, pp. 277–286.
[13] Miller, E. H., Shastri, R., and Shih, C., *Journal of Bone and Joint Surgery*, Vol. 64-A, 1982, pp. 1359–1362.
[14] Hood, R. W., Wright, T. M., Fukubayashi, T., and Burstein, A. H. in *Transactions*, 27th Orthopaedic Research Society, Vol. 6, 1981, p. 181.

DISCUSSION

Warren Starkebaum[1] *(written discussion)*—Did you find bone cement contamination in acetabular cups and tibial knee components in equal proportion?

R. W. Hood (author's closure)—We found bone cement much more often in the tibial components, probably because the acetabular components face downward and are not debris-retaining as are the upward-facing tibial components.

Manfred Semlitsch[2] *(written discussion)*—Together with Prof. Willert, in the last 15 years, more than 500 artificial joints have been reinvestigated after reoperation. This is a negative selection. Some 70 to 80% of implanted joints are in function for 5 to 15 years. Would it not be of great interest to look at these positive cases? Do you have the impression than many of the revisions are often caused by poor surgical technique?

R. W. Hood (author's closure)—I certainly agree that retrieval analysis of only the reoperated cases is basically failure analysis. Obtaining the postmortem specimens is very difficult in most cases, but certainly the information obtained from those retrievals is much more informative in terms of how the vast majority of implants are doing clinically.

I think the most common cause for revision is poor surgical technique followed closely by an excessive activity level.

[1]Kirschner Medical, Dublin, CA 94568.
[2]Sulzer Bros. Ltd., Department R & D, CH-8401 Winterthur, Switzerland.

Timothy M. Wright,[1] Albert H. Burstein,[1] and Donald L. Bartel[2]

Retrieval Analysis of Total Joint Replacement Components: A Six-Year Experience

REFERENCE: Wright, T. M., Burstein, A. H., and Bartel, D. L., "**Retrieval Analysis of Total Joint Replacement Components: A Six-Year Experience,**" *Corrosion and Degradation of Implant Materials: Second Symposium, ASTM STP 859*, A. C. Fraker and C. D. Griffin, Eds., American Society for Testing and Materials, Philadelphia, 1985, pp. 415–428.

ABSTRACT: Total joint replacement components have been examined as part of an ongoing implant retrieval analysis program. Examination of 65 fractured metal femoral components from total hip replacements revealed that all were caused by a fatigue mechanism and that factors such as stem position, material defects, cement technique, patient weight and activity, and stem design played an important role. Metal components from total knee replacements faired better, except those from metal-on-metal constrained designs, which were subject to fracture and considerable wear. Ultrahigh molecular weight polyethylene components showed several modes of damage, including surface wear, permanent deformation, and fracture. Subjective damage grading of 75 acetabular and 57 tibial components revealed significant correlations with patient weight and implantation time.

KEY WORDS: joint replacement, retrieval analysis, fracture, wear, implant materials, damage, fatigue (materials)

An implant retrieval analysis program has been in existence at the Hospital for Special Surgery since 1977. Over this period, all implants obtained at revision or removal surgery and, where permission is given, at autopsy have been collected and examined. This collection includes components from hemiarthroplasties and from total joint replacements, as well as fixation devices such as plates, screws, rods, pins, and wire.

From the inception of the retrieval program, the primary purpose has been to examine the influence of material and design on the mechanical performance of implants. The initial emphasis has been on total joint replacement components, and this will be the focus of this paper.

[1] Associate professor of Biomechanics and professor of Biomechanics, respectively, The Hospital for Special Surgery (affiliated with New York Hospital and Cornell University Medical College), New York, NY 10021.

[2] Associate professor, The Sibley School of Mechanical & Aerospace Engineering, Cornell University, Ithaca, NY 14853.

TABLE 1—*Total joint replacement components retrieved at the Hospital for Special Surgery (through 31 Dec. 1982).*

Total Hip (including surface replacements)	
Femoral	314
Acetabular	203
Total Knee	
Femoral	218
Tibial	238
Patellar	65
Total Shoulder	
Humeral	1
Glenoid	1
Total Elbow	
Humeral	13
Ulnar or radial	11
Total Wrist	
Radial	1
Metacarpal	1
Articulated	2
Total Finger	
Phalangeal	5
Articulated	7
Total Ankle	
Talar	1
Tibial	1
	Total 1082

TABLE 2—*Summary of 65 fractured femoral components from total hip replacements.*

Design Type	Distributor	Material, ASTM Standard	Number of Components
Trapezoidal-28	Zimmer USA[a]	316L steel (F138)	21
Mueller	Howmedica[b]	Co-Cr-Mo alloy (F75)	17
	DePuy[c]	Co-Cr-Mo alloy (F75)	1
	DePuy	Co-Ni-Cr-Mo alloy (F90)	1
Charnley	Howmedica	Co-Cr-Mo alloy (F75)	14
	Johnson & Johnson[d]	316L steel (F138)	3
Bechtol	Richards[e]	316L steel (F138)	5
Aufranc-Turner	Howmedica	Co-Cr-Mo alloy (F75)	2
"I" Beam	Howmedica	Co-Cr-Mo alloy (F75)	1

[a]Zimmer, Inc., Warsaw, IN.
[b]Howmedica, Inc., Rutherford, NJ.
[c]DePuy, Division of Bio-Dynamics, Inc., Warsaw, IN.
[d]Johnson and Johnson Products, Inc., Orthopaedic Div., Braintree, MA.
[e]Richards Medical Co., Inc., Memphis, TN.

Methods and Materials

The detailed procedures for the routine examination of removed total joint replacement components have been described elsewhere [1][3]. Briefly, they are as follows: ultrasonic cleaning, visual and light microscopy ($\times$ 10) examination, scanning electron microscopy examination where appropriate (for example, fracture and wear surfaces), and a review of the patient's medical record and radiographs. In addition, articulating surfaces of ultrahigh molecular weight polyethylene (UHMWPE) are subjectively graded for the presence and severity of several surface damage modes (burnishing, abrasion, delamination, pitting, surface deformation, scratching, and poly(methyl methacrylate) debris) [2]. The resulting damage scores can be compared between design types or against clinical variables.

Over the six years the program has been operating, 1082 components from total joint replacements have been collected. Table 1 shows the anatomical location of these components.

Results and Discussion

Total Hip Replacement

Of importance in any retrieval analysis of femoral components from total hip replacements is the occurrence of *in vivo* fracture. Of the 314 components retrieved, 65 components had fractured and their design type and origin are summarized in Table 2. The mean length of time the components were implanted was 70 (s = 23) months with a range of from 13 to 107 months. The mean patient weight was 79.7 (s = 16.0) kg with a range of from 35.4 to 112.5 kg, and the mean age of implantation was 64 (s = 12) years and ranged from 26 to 84 years. The original diagnosis was osteoarthritis in 81% of the cases, avascular necrosis in 9%, congenital dislocation of the hip in 4%, and rheumatoid arthritis, revision of hemiarthroplasty, and femoral fracture in 2% each.

The mechanism of failure in all cases was fatigue as verified from macroscopic and microscopic features on the fracture surfaces. Included was one case of partial fatigue fracture (Fig. 1) noted in a Mueller type stem removed after 71 months for femoral component loosening. The location of fatigue failure generally varied according to design, with Mueller type components fracturing in the distal third and Charnley type and Trapezoidal-28 (T-28) type components fracturing in the proximal and middle thirds of the stem portion of the component. There were three notable exceptions (Fig. 2). One case involved a component that fractured in two places in a 78-kg man after eight years implantation [3]. The other two cases were fractures through the neck of the femoral component. Both these cases involved large-stem T-28 type components that fractured after six and eight years implantation in heavy, active males. In both cases, fracture occurred through the base of the neck. Stress analysis of the trapezoidally shaped

---

[3]The italic numbers in brackets refer to the list of references appended to this paper.

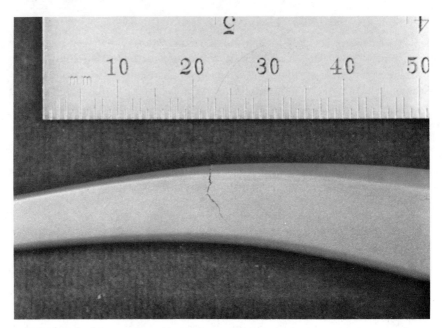

FIG. 1—*Partial fracture through a Mueller type femoral component. Component was removed for loosening from a 69-year-old; 102-kg male, 71 months after implantation.*

neck in this component design demonstrated that stresses can reach the published fatigue strength for Type 316L stainless steel [4]. One of the two components with fracture through the neck also had a well-developed fatigue crack in the proximal third portion of the stem. Such secondary fatigue cracks were quite common in the T-28 type design, occurring in 11 of the 21 fractured components, including 7 of 12 small stems and 4 of 9 large stems. These secondary cracks initiated on the medial side of the stem and were probably the result of super-position of residual stresses and stresses resulting from bending moments applied to the trapezoidal cross-section [1].

There was also a component fracture in a prosthesis made from Co-Ni-Cr-Mo alloy, one of the higher fatigue strength alloys used in implants. This was a long (305-mm) stem component implanted in a 23-year-old female who had been involved in an automobile accident in which she suffered a fracture in the proximal femur. The component fractured at the site of a non-union approximately three years after implantation.

The variables that affected the fatigue failures in this series do not differ from those already well identified in the literature [5,6]. These variables included varus positioning of the stem, poor bone cement technique, heavy (often active) patients, metallurgical defects (most notably porosity in cast stems), and a stress concentration defect (in the form of a nicked stem either caused by a drill bit at surgery or post-operatively by rubbing against a trochanteric wire).

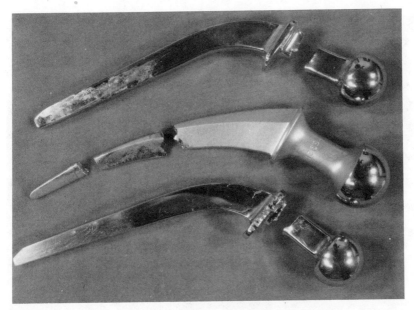

FIG. 2—*Three unusual cases of femoral component fracture. Top and bottom are neck fractures in large stem, T-28 type components occurring eight and six years after implantation, respectively. Middle is a double fracture of the stem in a Mueller type component occurring eight years after implantation.*

Attempts to correlate patient weight with implantation time using regression analysis for the fractured femoral components did not result in a statistically significant correlation, probably due to the effect of the other variables just mentioned and to the fact that a number of stem designs were included. This differs from the results of Wroblewski [7], who found a strong negative correlation between patient weight and implantation time for a large group of Charnley-type fractured femoral components.

With retrieved UHMWPE acetabular components from total hip replacements, the major concerns are wear (or surface damage) and gross, permanent deformation, either of which can compromise the performance of the total joint as a low friction arthroplasty. However, in the present series, fracture was also noted, occurring in three of the 203 retrieved components. All three components were of the Mueller design with fracture occurring at the base of the superior notch placed in the outside surface of the cup for cement fixation (Fig. 3). Examination of the fracture surfaces has revealed features indicative of a fatigue mechanism [8]. That stresses sufficient for component fracture could arise, particularly in cases in which the underlying subchondral bone is absent, has been demonstrated by analytical stress analysis of the acetabular region [9].

Wear and permanent deformation of acetabular components were determined by examination of a consecutive group of 75 components from within the 203 retrieved thus far. When the articulating surfaces of the components were sub-

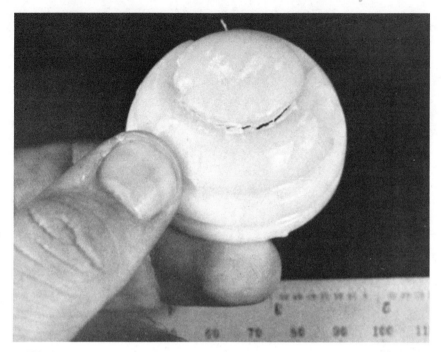

FIG. 3—*Fractured Mueller type acetabular cup. Component was removed from a 65-year-old, 79-kg male, 6½ years after implantation.*

jectively graded for surface damage, it was evident that two of the mechanisms, delamination and abrasion, seen on the articulating surfaces of tibial components from total knee replacements (see next section) were absent from the acetabular components. Delamination is a damage mode in which subsurface failure occurs and results in the loss of large "sheets" of polyethylene. It is similar in appearance to the damage modes seen on metallic ball bearings. Abrasion is a damage mode caused by contact between UHMWPE and either bone or poly(methyl methacrylate) bone cement. The absence of abrasion is understandable, since it is difficult for the articulating surface of the acetabular component to come in contact with bone or bone cement from around the femoral component. In general, the amount of damage was less for the 75 acetabular components than for the tibial components examined (see next section). The decreased surface damage and the absence of the delamination mode were probably due to the decreased contact stresses that occur in the highly conforming total hip geometry (as compared to the less conforming total knee geometry) [10].

Significant ($p < 0.05$) positive correlations were found using linear regression analysis between the total damage score and both patient weight and time of implantation for the 75 examined acetabular components. Thus, the heavier the patient and the longer the component was implanted, the more damaged the articulating surface appeared (Fig. 4a). This underscores the fact that the surface

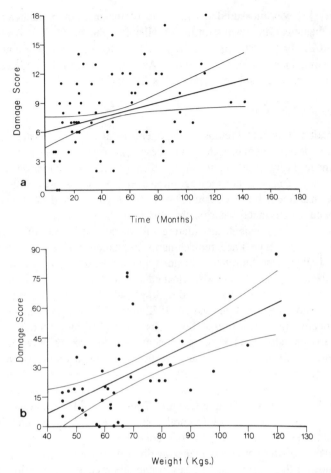

FIG. 4—(a) *Damage scores (obtained from subjective grading) plotted against amount of time implanted for 75 acetabular components. Shown with the data are the regression line and the 95% confidence intervals.* (b) *Damage scores (obtained from subjective grading) plotted against patient weight for 48 tibial components of the Total Condylar type [2]. Shown with the data are the regression line and the 95% confidence intervals.*

damage modes result from mechanisms that are dependent on both the magnitude of the applied loads and the number of cycles of loading. In addition, a number of the components were grossly deformed so that they appeared elliptical rather than circular in shape. This was found much more frequently in thin components (such as the Mueller type) than in thick components (such as the Charnley and Trapezoidal-28 types). For example, of the 75 components, 30 Mueller and 19 Charnley and Trapezoidal-28 components had been removed in one piece. Analysis by Student's t-test showed the wall thickness of the Mueller components (8.06 mm, $s = 1.47$ mm) to be significantly ($p < 0.01$) less than that of the Charnley and Trapezoidal-28 components (9.64 mm, $s = 1.49$ mm). When the eccentricity [2] was measured (as the ratio of the maximum diameter to the minimum diameter), the results for the Mueller components (1.023, $s = 0.020$) were significantly ($p < 0.05$) greater than those for the Charnley and Trapezoidal-28 components (1.012, $s = 0.04$). As in the case of fractured acetabular components, analytical studies support this finding [9,10].

Total Knee Replacement

Examination of the metal femoral components from metal-on-plastic total knee replacement designs demonstrated that these components experienced mostly minor damage during implantation in the form of mild scratching on the articulating surface. This was not the case, however, for metal-on-metal total knee designs. Constrained, hinged designs (such as the Guepar and Walldius types) often showed severe metal wear on the axle portion and on the bearings of the femoral and tibial components articulating with the axle. For example, five worn axles from Guepar type knee replacements implanted a mean of 30 months ($s = 13$ months) were thoroughly cleaned and weighed. Three new axles that had never been implanted were also cleaned and weighed. The mean weight of the worn group was 38.2314 g ($s = 0.0430$ g), which was less than that of the new group, whose mean weight was 38.3572 g ($s = 0.0135$ g). This difference was statistically significantly ($p < 0.001$) as determined by a Student's t-test.

Fracture of metallic total knee components was also more prevalent in the constrained, metal hinged type total knee designs than in the less constrained designs. For example, of the 29 hinged total knee components, seven experienced fatigue fracture. However, only two metallic components from less constrained total knee designs showed fatigue fracture. These were a Duocondylar femoral component and a Gustillo UCI femoral component, both of which fractured through the cross-bar connecting the two condyles.

Examination of the surface damage on tibial and patellar UHMWPE components showed the presence of the seven damage modes just mentioned: burnishing, abrasion, delamination, pitting, surface deformation, scratching, and poly(methyl methacrylate) debris. As with the acetabular components, a series of 57 condylar type prostheses were subjectively graded for the presence and severity of these damage modes. This series included a group of 48 tibial components of the original Total Condylar design [2]. Again, the total damage experienced by the components was found to be significantly ($p < 0.05$) correlated with patient weight and the time of implantation, using linear regression analysis between damage score and each variable. Components from heavier patients or from longer times of implantation or both showed more damage (Fig. 4*b*).

Twenty-eight patellar components from condylar type knee replacements were also graded for damage. In general, these components showed little surface damage, except in six cases that involved heavy, active patients or longer implantation times.

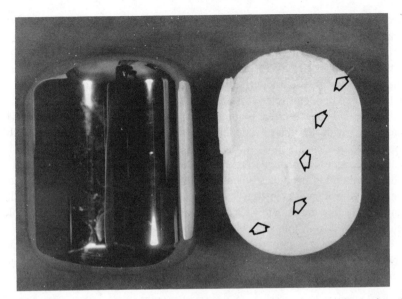

FIG. 5—*Freeman-Swanson type total knee replacement removed eight years after implantation showing gross disruption of articulating surface on tibial component.*

In addition to surface damage, permanent deformation and fracture were seen in retrieved tibial and patellar UHMWPE components. One form of gross, permanent deformation was tilting of the plateau of condylar type tibial components. In these cases, the fixation pegs remained straight and were found to be well-fixed at the time of removal surgery. Such deformation coincided with poor tracking of the femoral component. Poor tracking refers to the condyles of the femoral component not being parallel, so that as the component goes from extension into flexion, the contact areas move from the center of the tibial plateau to the extreme edges. Such eccentric loading leads to increased stresses on the underlying cancellous bone. This, in turn, can lead to failure of the supporting cancellous bone and deformation of the plateau. Since the fixation peg transfers little of the load placed on the tibial component, it remains well-fixed, even as the plateau deforms [*11*]. Another form of gross, permanent deformation was noted in the patellar components, which upon removal were found to no longer be circular in shape, but rather to be elliptical with the major axis of the ellipse being a mean of 3% larger than the minor axis. Again, such deformation coincides with the high stresses predicted for the patella [*12*].

Fractures of UHMWPE components were limited to two tibial components of the Freeman-Swanson type and one tibial component of the Duocondylar type. Small cracks were noted, though, in a number of other components, most notably emanating from the alignment holes in Unicondylar type tibial components. Both Freeman-Swanson type components showed well-developed cracks of an apparent fatigue mode running the length of the component in an anteroposterior direction and located on the inferior surface of the component. The Duocondylar component was fractured through the cross-bar connecting the two condyles.

The positive correlation between the amount of surface damage and the time of implantation in UHMWPE components deserves further comment. Though there is no evidence to support extrapolation of the correlation, which would predict that components retrieved after longer implantation could be expected to be even more damaged, there is some reason to believe that this will be so. The components with the longest implantation times among those retrieved in this series demonstrated the most severe damage (Fig. 5). Furthermore, as surface damage occurs on implanted components the contact areas can be expected to be disrupted and in many cases decreased in size, so that contact stresses could be expected to increase, in turn leading to more damage. While it is true that many of the components in this series must be considered as worst-case examples (being retrieved from joints in which loosening had occurred), the remainder of the components were removed for reasons possibly unrelated to surface damage (for example, infection) and these components were found to be just as damaged when compared on the basis of implantation time and patient weight.

Other Total Joint Replacements

As can be seen from Table 1, retrieval of components from total joint replacements other than the hip and the knee has been minimal, with the majority of components coming from total elbow replacements. Examination of these components revealed two problems. One was the fracture of the polyethylene humeral component of a Pritchard-Walker Mark I design. The fracture occurred at the base of the distal stem and appeared to be the result of a fatigue mechanism. The second problem involved dislocation of a Tri-Axial Total Elbow Replacement (Fig. 6). The Tri-Axial Elbow is a semi-constrained design that employs a "snap-fit" articulation created by a polyethylene bushing between two metallic stems [13]. Disarticulation occurred in this case in the post-operative physical therapy in a very active individual and was caused by excessive deformation of the polyethylene bushing.

Seven of the twelve finger joint components were of the articulated Swanson design made from silicone elastomer. Of these seven components, five had fractured, all at the base of one of the stems of the prosthesis. Fracture of components of this design has been identified as a significant clinical complication [14], though what rate of incidence the five fractures in this series reflects is impossible to determine because the population of total finger joint replacement patients in which the retrieved components are included is unknown.

The total ankle replacement included in this series was an interesting case in that the tibial component is made of carbon-fiber reinforced UHMWPE. The ankle replacement was of the Oregon design and had been implanted for 13 months. The tibial component was one of only five components of this material included in the series (the others being an acetabular total hip component and three tibial total knee components). Visual and light microscopy examination of

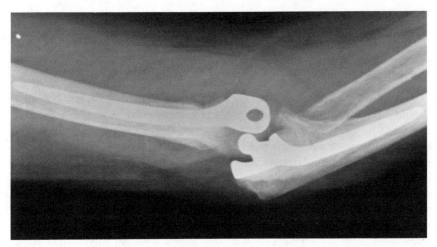

FIG. 6—*Radiograph showing disarticulated tri-axial total elbow replacement.*

the articulating surface of the tibial component revealed only minor scratching. However, histology of the surrounding tissue revealed the presence of numerous carbon fibers (Fig. 7) together with a large amount of refractile debris consistent in appearance with polyethylene. There was a marked histiocyte and giant cell reaction, but not beyond that typically seen in the presence of wear debris from prostheses with plain polyethylene. No reaction specific to the carbon debris

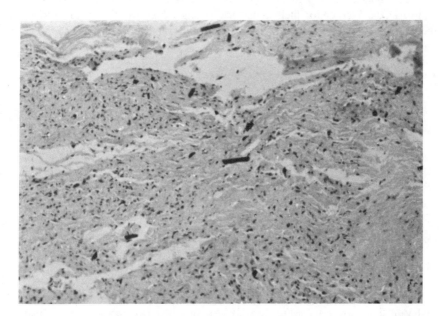

FIG. 7—*Light photomicrograph showing carbon fibers in the surrounding tissue of an Oregon type total ankle.*

was noted. A previous report of a retrieved component of this type showed similar findings [15].

Concluding Remarks

A continuing retrieval analysis program of orthopaedic implants including detailed examination of total joint replacement components has been used to assess the performance of these components and to identify material and design problems. Some of the results provided additional evidence of problems previously mentioned in the literature: for example, fracture through the stem portion of total hip femoral components and metal-on-metal wear on components from constrained total knee prostheses. Other results, such as fracture through the neck portion of femoral components, constitute new problems that necessitate reassessment of neck designs in contemporary hip replacements. Finally, the assessment of the performance of ultrahigh molecular weight polyethylene components reveals several damage modes that appear dependent on applied load (patient weight), component design, and length of time implanted. The long-term performance must be questioned based on these results. It may be possible, however, to optimize the performance by considering the effect of design variables on the resulting stresses to which the polyethylene is exposed [10,11].

Acknowledgments

The authors express their gratitude to Drs. Dethmers, Hood, and Johannson and to Diane Warner for their assistance in this work. We also wish to thank the Clark Foundation and the Sherman Fairchild Foundation for their support.

References

[1] Wright, T. M. and Burstein, A. H. in *Implant Retrieval: Material and Biological Analysis,* National Bureau of Standards Special Publication 601, A. Weinstein, D. Gibbons, S. Brown, and W. Ruff, Eds., U.S. Government Printing Office, Washington, DC, 1981, pp. 559–572.
[2] Hood, R. W., Wright, T. M., and Burstein, A. H., *Journal of Biomedical Materials Research,* Vol. 17, No. 5, 1983, pp. 829–842.
[3] Pellicci, P. M., Wright, T. M., Atkinson, R., and Salvati, E. A., *Clinical Orthopaedics and Related Research,* No. 170, Oct. 1982, pp. 248–251.
[4] Burstein, A. H., Bartel, D. L., Wright, T. M., Salvati, E. A., and Wilson, P. D., Jr., "Fracture Through the Neck of the Femoral Component After Total Hip Replacement," *Journal of Bone and Joint Surgery,* in press.
[5] Galante, J. O., *Journal of Bone and Joint Surgery,* Vol. 62–A, No. 4, June 1980, pp. 670–673.
[6] Chao, E. Y. S. and Coventry, M. B., *Journal of Bone and Joint Surgery,* Vol. 63–A, No. 7, Sept. 1981, pp. 1078–1094.
[7] Wroblewski, B. M., *Acta Orthopaedica Scandinavica,* Vol. 53, No. 2, April 1982, pp. 279–284.
[8] Salvati, E. A., Wright, T. M., Burstein, A. H., and Jacobs, B., *Journal of Bone and Joint Surgery,* Vol. 61–A, No. 8, Dec. 1979, pp. 1239–1242.
[9] Vasu, R., Carter, D. R., and Harris, W. H., *Journal of Biomechanics,* Vol. 15, No. 3, 1983, pp. 155–164.

[*10*] Bartel, D. L., Burstein, A. H., and Toda, M. D., "The Effect of Conformity and Plastic Thickness on Contact Stresses in Metal-Backed Plastic Implants," *Journal of Biomechanical Engineering*, in press.
[*11*] Bartel, D. L., Burstein, A. H., Santavicca, E. A., and Insall, J. N., *Journal of Bone and Joint Surgery*, Vol. 64-A, Sept. 1982, pp. 1026–1033.
[*12*] Hayes, W. C. and Levine, B. M. in *Transactions*, 26th Orthopaedic Research Society, Vol. 5, Feb. 1980, p. 163.
[*13*] Inglis, A. E. in *Total Joint Replacement of the Upper Extremity*, A. E. Inglis, Ed., C. V. Mosby Co., St. Louis, 1982, pp. 100–110.
[*14*] Beckenbaugh, R. D., Dobyns, J. H., Linscheid, R. L., and Bryan, R. S., *Journal of Bone and Joint Surgery*, Vol. 58-A, No. 4, June 1976, pp. 483–487.
[*15*] Groth, H. E., Shen, G., Gilbertson, L., Farling, G., and Shilling, J. M. in *Transactions*, 24th Orthopaedic Research Society, Vol. 3, Feb. 1978, p. 186.

DISCUSSION

Emanuel Horowitz[1] (*written discussion*)—You stated that you examine all joint components that come out in your hospital? How effective is your system for recording and tracking the joint components that are placed into your patients?

T. M. Wright, A. H. Burstein, and D. L. Bartel (*authors' closure*)—The type of prosthesis used and any necessary identification markings should be noted directly on the Operative Record, which becomes a permanent part of the patients medical history. The responsibility for recording this information lies with the operating room staff. Though we have not yet been involved in a prospective study in which it has been necessary to track protheses, it should be possible to do so with this mechanism.

Ann C. VanOrden[2] (*written discussion*)—Could you describe the corrosion you have found on retrieved prosthesis, and have you found corrosion a significant problem?

T. M. Wright, A. H. Burstein, and D. L. Bartel (*authors' closure*)—Corrosion on retrieved metallic components from total joint replacements has not been found to be a significant problem. We have seen no evidence of corrosion on retrieved components from cobalt alloy or titanium alloy. Corrosion, in the form of pitting, has been seen on stainless steel femoral components (of the Charnley, Bechtol, and T-28 designs). This pitting has generally been found on the distal third of the stem portion of these femoral components. Because of the location, it is possible that the corrosion is secondary to abrasion between the femoral component and the surrounding bone cement.

Manfred Semlitsch[3] (*written discussion*)—Which design (Charnley or other) had the total hip prosthesis delivered by DePuy? Which type of material (cast

[1]Johns Hopkins University, Baltimore, MD 21218.
[2]National Bureau of Standards, Washington, DC 20234.
[3]Sulzer Bros. Ltd., Department R&D, CH-8401 Winterthur, Switzerland.

of forged)? Have you determined the chemical analysis and the microstructure of the revised implants? Have you described the failure mode in more detail?

T. M. Wright, A. H. Burstein, and D. L. Bartel (authors' closure)—Referring to Table 2, there were two fractured femoral components manufactured by DePuy. Both were of the Mueller design. One stem was made from cast Protasul (cobalt-chrome alloy); the other was made from Protasul 10 (cobalt-nickel-chrome wrought alloy). Chemical analysis has only been performed in one or two of the fractured femoral components, and neither of these were from DePuy. Microstructural analysis has also only been performed on a few of the removed components. The failure mode in each case has been determined from light microscopic examination of the fracture surfaces and by fractographic analysis in the scanning electron microscope.

Michael B. Mayor[1]

Performance Standards: Flexible Tool or Straight Jacket?

REFERENCE: Mayor, M. B., **"Performance Standards: Flexible Tool or Straight Jacket?"** *Corrosion and Degradation of Implant Materials, ASTM STP 859*, A. C. Fraker and C. D. Griffin, Eds., American Society for Testing and Materials, Philadelphia, 1985, pp. 429–433.

ABSTRACT: Standards documents can be written in several ways. The distinction between design standards and performance standards is examined here, in the special context of surgical materials and devices. The work of writing performance standards is significantly different, and that difference will affect the work of the task forces and committees of ASTM Committee F-4 on Medical and Surgical Materials and Devices. If the performance documents produced by Committee F-4 are well crafted, their usefulness will extend beyond the limits of the American Society for Testing and Materials to service in many different contexts.

KEY WORDS: implant materials, standards, design standards, performance standards, medical devices, surgical devices, medical materials

Standards writers have the responsibility to produce documents that offer assurances about the properties of materials and devices, so that the documents apply in diverse situations of use. Reality forces difficult choices on the standards writer, so that the document written bears close relationship to the state of the art at the time of writing. One fundamental choice that must be faced by each of us in the business of writing standards is the choice between design standards and performance standards.[2] What is the difference? Are we moving in the desired direction in going from design standards to performance standards, and what will that mean to the work of the task forces in ASTM Committee F-4?

Design Standards

A design standard seeks to assure the performance of a device or material by designating in detail such things as composition, size, density, and other measurable basic parameters. The writing of such a standard takes advantage of

[1]Associate professor of Clinical Surgery in Orthopaedics, Dartmouth Hitchcock Medical Center, Hanover, NH 03756.

[2]Heminway, D., "Performance vs Design Standards," National Bureau of Standards/GCR 80–287, Oct. 1980.

established laboratory practices, many of which are already part of the usual and customary quality control procedures used in the manufacturing process. Such practices are reassuring in their familiarity and their often long history of application in manufacturing and laboratory science, and offer a time-tested and uniform source of common understanding about the materials and devices on which our day-to-day practices depend.

Design standards are, then, an attractive option for standards writers. They incorporate the use of accessible variables that are relatively easy to measure, and their incorporation into standards is relatively easy to agree on, in the process of consensus building. Experience with these standards, in medical device and material science, and in other fields, reveals some problems.

Design standards, in their effort to afford uniformity and predictability in materials and devices, are forced to describe the objects of the standards in exhaustive detail. For a material or a device to qualify under the written standard, it must precisely "measure up," for it is on that basis that the performance of that device/material is being assured.

A little reflection on the process reveals both the problem and the solution. The standards, in their effort to provide predictability, have restricted innovation. They have made it impossible to introduce novel solutions to achieving equal or superior performance by prescribing the methods for solving the problem, not by defining the problem to be solved. An example from current experience might help to clarify the point.

Engine lubricating oils are required to minimize wear of bearing surfaces against each other in a wide variety of environments. In writing standards to assure that the oils actually function in a predictable fashion, the authors of such standards might elect to do their job by exhaustively cataloging the polymeric and additive combinations that the large body of accumulated experience has proven will do the job. This would assure the user that the oils being chosen would measure up to other critical applications and earlier experience. The effect of such a standard on innovation and progress is predictable. Any variation from the lubricant formulation proscribed by the standard would not qualify under the standard, with chilling effect on the decision makers regarding departures from the constraints of the standard.

Performance Standards

The workable option that has been exercised in lubricant standards-writing is the performance standard.[3] Instead of crafting the standard to describe the already extant working solutions now doing the lubricating, engine oil standards describe the minimum results considered appropriate by the standards-writing consensus. The performance standard is indifferent to the constituents of the oil, and asks only that the oil does the things called for by the standard under the conditions

[3]Wright, J. R., "The Performance Approach: History and Status," *Performance Concept in Buildings*, NBS Special Publication 361, National Bureau of Standards, Vol. 2, 1972.

prescribed by the standard. Any formulation, then, that performs up to the demands of the standard will be deemed to qualify under the standard.

Performance standards thus provide the necessary assurance of predictable outcome without constraining imaginative problem-solvers in their efforts to achieve that outcome. How do standards writers produce performance standards for medical materials and devices? They do it with considerable additional effort. The work of producing a useful performance standard may be as much as an order of magnitude greater than that previously demanded. The achievement requires that the expected performance be very carefully and completely defined.[4] A Foley catheter retention standard has been crafted. It addresses the resistance felt to be appropriate in a urinary catheter to assure that it will not be inadvertently pulled from the urinary bladder. It does not address other aspects of urinary catheter function, even though other functions are clearly important. The achievement of a standard like this demands that the scope of the document be very carefully circumscribed, or the range of concerns included in the document will be so broad that the document will be unwritable. The document's authors are required to be very rigorous in deciding about the range of performance characteristics to include, as well as the description of the characteristics themselves.

In the field of medical materials and devices the situation is particularly sticky. We are dealing with human tissues and human lives where failure becomes an especially distressing prospect. Historically, producers, medical professionals, and patients have been well served by the combined efforts of all to keep to a high standard of care. In the process, problems have arisen and they have attracted attention. The regulatory agencies have responded to the inevitable expressions of concern with a demand for standards that speak to the issue of performance. How should Committee F-4 respond?

All of the things touched on previously led Committee F-4, in its considerations during the 1979 meeting in San Francisco, to draw up guidelines to the inclusion of performance language in the documents being written. The sense of that meeting was that a logical beginning could be made by addressing the question of performance of materials and devices at three discrete levels. For some materials and devices, very particular limits of performance could be called for, with upper and lower limits to the variables measured, resulting in acceptance or rejection of the tested item. For others, the state of the art would not permit such rigid responses and the standard could instead call for the measurement of certain agreed performance-related variables, without demanding a response for acceptance or rejection. In the third category, it is not possible, with the current state of the art, to discover a measurable parameter that reliably relates to the hoped-for performance of the device or the material under consideration.

The press for performance standards is expected to press the state of the art. It will require the development of test methods that do not yet exist. It will

---

[4]Brenner, F. C., "The Challenge of Developing Performance Tests," *ASTM Standardization News*, Sept. 1973.

require increased committee and task force work. It requires that each of us look at each standard developed or revised with the question: can a performance standard stand alone, supported by test methods and common experience? If not, performance language in each device standard should speak to the problems that the task force or committee were wrestling with. For help in that, it is important to use the guidelines for performance language contained in the minutes of the May 1980 meeting in Denver, Colorado. The effect of these documents will be greatly enhanced by a well-crafted rationale statement. Extra effort with the rationale behind the performance statement certainly will be well spent. There is no expectation that the document should stand ahead of the state of the art. It is equally important that the document reflect the state of the art at the time of writing, or revision.

Ultimately, the work of Committee F-4 should move toward pure performance standards via experience with hybrids as just described. It is no surprise that talk in the general case generates terrible anxiety. Work in the specific case should alleviate that, and should be supported by citations of the literature and recruitment of outside help, where necessary, to fully reflect the state of the art. Also helpful is reference to the blue book of *Form and Style for ASTM Standards* for writing rules, coupled with help from the minutes of May 1979 and May 1980 regarding performance and rationale.

Throughout this effort, it will be difficult to keep in mind that we are writing standards for ourselves, for ASTM Committee F-4, and not for any outside agency or target. We should be guided by (or driven by) our own interest in performance standards, for the advantages they offer to the *ASTM Book of Standards* for which we ultimately are responsible. If they are useful to regulatory agencies, we will enjoy added value for the work we do.

DISCUSSION

Anna Fraker[1] (*written discussion*)—Are you referring to performance standards for the device or for the implanted device and its performance?

M. B. Mayor (*author's closure*)—Dr. Anna Fraker has raised the fundamental question; am I referring to performance standards for the device, or for the implanted device and its performance? Since the effort required to write meaningful standards is grand, the effort should be aimed at a worthy target. If it were practical to do so, we should be writing documents that speak to the performance of the device in its place of ultimate function, the human body. All of us involved in implant retrieval are very much aware, however, that the real world hides implants from us, much as we struggle to find them all. The

[1]National Bureau of Standards, Washington, DC 20234.

implants we must write about, then, are the ones we can get our hands on, the implants that are on their way into the human body. The skillful means we must employ in writing these performance standards should take us as close as possible to those conditions of use the implants will need to withstand. If we are good at what we do, the standard will say things about the implant on the shelf that will be close to truth about the implant in the patient. Our standards documents can then support our confidence in the implants we offer our patients.

Henry R. Piehler[1]

Pluralistic Medical Device Risk Management: Standards, Regulation, and Litigation

REFERENCE: Piehler, H. R., **"Pluralistic Medical Device Risk Management: Standards, Regulation, and Litigation,"** *Corrosion and Degradation of Implant Materials: Second Symposium, ASTM STP 859*, A. C. Fraker and C. D. Griffin, Eds., American Society for Testing and Materials, Philadelphia, 1985, pp. 434–440.

ABSTRACT: The operation of the currently existing institutions and processes for managing risks associated with the use of medical devices are reviewed and analyzed. Questions addressed include the definition(s) of device failure, the establishment of acceptable failure rates, and the selection of the parties to be involved in these risk-management processes, both prospectively and retrospectively. Currently perceived difficulties associated with the unilateral risk-management activities of both the FDA regulatory apparatus plus the product liability and medical malpractice litigation systems are described. A pluralistic risk-management approach based upon marketplace incentives is proposed. This marketplace-incentive approach encourages the development and use of performance standards both to enhance benefits to patients and to reduce the adversarial tension between the FDA and the private sector as well.

KEY WORDS: implant materials, fatigue (materials), degradation, performance standards, medical devices, risk assessment, risk management, medical malpractice, product liability, consensus development, consensus standards, medical device regulation, social regulation, marketplace incentives

Before considering specific processes for medical device risk management, several threshold questions must be addressed. These include:

1. What constitutes device failure?
2. What is an acceptable failure rate?
3. Who should be involved in this decision?
   (a) Prospectively.
   (b) Retrospectively.

The Association for the Study of Internal Fixation defines failure as a patient's not attaining a level of recovery prospectively targeted by a team of orthopedic

[1] Professor, Department of Engineering and Public Policy and Department of Metallurgical Engineering and Materials Science, Carnegie-Mellon University, Pittsburgh, PA 15213.

surgeons and biomedical engineers. Other previous studies [1][2] have defined failure as occurring when the degree of pain or immobility experienced by the patient is sufficient to require another surgical intervention. While the former is more appropriate for assessing the performance of medical personnel, the latter is operationally more suitable for assessing and managing the risks associated with the use of medical devices.

The second and third questions are the core concerns of both the regulatory and liability systems, the latter involving claims based both upon product liability and medical malpractice. The litigation system operates retrospectively to provide compensation for injuries incurred, though it is widely thought to have a prospective deterrent effect as well. On the other hand, the Food and Drug Administration (FDA) is empowered through the Medical Device Amendments of 1976 [2] to regulate medical devices both retrospectively and prospectively. Finally, standards, whether they be voluntary or mandatory, performance or design, are used extensively in measurement, evaluation, or level setting in both the regulatory and litigation systems.

However, both the tort liability system in general [3] and medical device regulatory system in particular [4,5] have been recently subjected to increasing scrutiny and criticism. While the major concerns associated with the litigation system are in areas other than the use of medical devices, the Medical Device Amendments of 1976 are widely acknowledged not to be working as intended. The Amendments' basic approach was to provide for three levels of increasing regulatory stringency to be specified by individual classification panels comprised of experts primarily from the private sector. These three levels are Class I, General Controls, that involve control only over labeling and good manufacturing practices; Class II, Performance Standards, that are intended either to reduce risks to health and safety to a reasonable level or to provide adequate assurances of effectiveness; and Class III, Premarket Approval, that is to be imposed only if a suitable performance standard cannot be promulgated. However, a recent General Accounting Office (GAO) Report [5] concluded that

1. FDA needs (a) comprehensive medical device information system,
2. development of performance standards for over 1000 devices will be time-consuming and expensive,
3. review of older (preenactment) devices will take years, and
4. proof of safety and effectiveness should be required for all new devices (including those substantially equivalent to preenactment devices).

The first four GAO recommendations in response to these findings instructed the FDA to strengthen its database; the fifth, to collect, analyze, and disseminate information on the scope and nature of problems caused by user error and inadequate maintenance; the sixth, to identify and regulate new "risky" Class

---

[2] The italic numbers in brackets refer to the list of references appended to this paper.

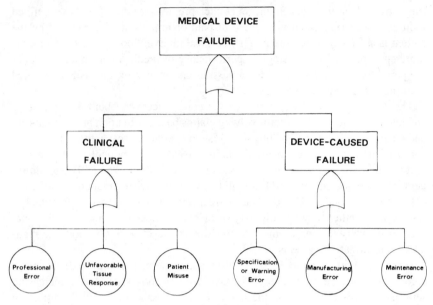

FIG. 1—*Medical device fault tree.*

II devices; and the seventh, to rely more on local institutional review board approval of significant risk devices.

Medical Device Risk Assessment

While the development and use of appropriate databases to assess risks associated with the use of medical devices is generally a worthwhile endeavor, care must be exercised not to confine their use exclusively to establishing causal inferences statistically, as one does with drugs. Given the institutional setting in which medical devices are regulated, there is a natural tendency for the "drug mentality" to carry over to medical devices as well. As has been pointed out previously [6], clinical databases can be also used to reduce risks by providing information for the development and use of responsive performance standards. An example of the development and use of such a performance standard to document the influences of materials selection and design on the corrosion-fatigue performance of hip nails is contained in this volume [7].

As mentioned previously, a substantial concern was expressed in the GAO report over the need to separate clinical from device-caused failures prior to instituting any form of regulatory control over currently existing devices. Published clinical results [8] as well as data obtained from the Association of Trial Lawyers of America [6] indicate that the overwhelming preponderance of medical device failures are clinical rather than device-caused. The various pathways that can lead to medical device failure are illustrated in Fig. 1. Figure 1 is a fault tree [9] comprised entirely of "or" gates. This means that medical device failure can result from either clinical failure or device-caused failure; clinical failure can occur as a result of professional error, unfavorable tissue response, or patient misuse; device-caused failure can occur as a result of a specification or warning error, a manufacturing error, or a maintenance error. Any of these events, either singly or in combination, can lead to medical device failure. However, only device-caused failures can be addressed through the Medical Device Amendments of 1976.

It is difficult to measure the impact of product liability litigation in retrospectively limiting the risks associated with flawed devices that have failed in service and have led to product liability suits. Responsible companies would probably have taken appropriate remedial steps as soon as the failure became known to them, which most often is prior to the filing of a lawsuit. "Bad actors" might adopt a "public-be-damned" attitude and seek refuge in the bankruptcy courts. It is clear, however, that medical device litigation is not plagued with the volume of claims and the doctrinal difficulties characteristic of toxic tort litigation [10,11], principally involving asbestos. If Senator Kasten's proposed Model Uniform Product Liability Act [12] passes, the burden of proof in design-defect cases would revert to a negligence standard, focusing on industry practice at the time of manufacture. Should this happen, there would be enhanced industrial interest in the development of voluntary consensus standards as indicators of current practice and hence defenses in design-defect litigation. Production-defect litigation would remain essentially unchanged, but this has never raised serious concern, especially not in the medical device area.

The discussion to date has focused primarily on managing risks associated with the use of currently existing medical devices, often only after discernible failure patterns have emerged. Risk management involving "risky" new devices will almost inevitably occur through Premarket Approval, which presumably will result in devices with an acceptable failure rate. But what about existing devices (principally those in Class II) for which discernible failure patterns have not emerged? It has been previously suggested that the FDA identify and deal with high priority Class II devices [4], or be given discretion as to whether devices would be subjected to performance standards [5]. A new pluralistic approach to managing risks associated with Class II devices is presented in the following section, an approach that facilitates and combines the strengths and incentives associated with standards, regulation, and litigation in minimizing risks to patients from the use of medical devices.

Pluralistic Risk Management Through the Development and Use of Performance Standards

The principal stakeholders potentially involved in medical device risk management through performance standards are patients, individual firms, voluntary consensus standardization associations, and the FDA. It is clear that the resources

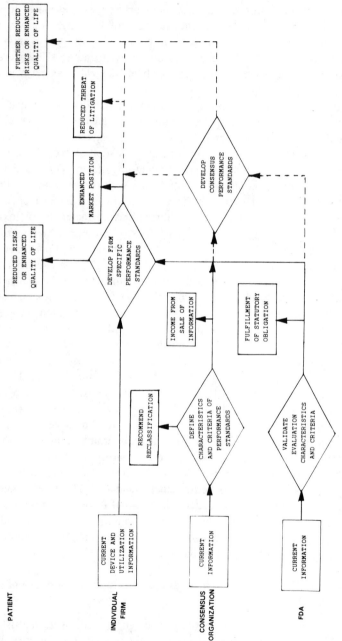

FIG. 2—*Flowchart for performance-standard development using marketplace incentives.*

and skills required to develop and implement performance standards are present to a much greater extent in the private sector than in the FDA. However, individual firms will probably not expend their sources needed to develop responsive performance standards unless:

1. assurances are given that the FDA will not preempt private sector performance standards with different standards of their own, or
2. performance standards offer individual firms a competitive edge in the marketplace, or
3. both of the above.

The risk management proposal described here allows individual firms to realize both of these benefits and at the same time offers the FDA a mechanism by which to break the Class II regulatory logjam through the use of a more broadly based consensus process.

The procedure would begin with either a consensus organization's or the FDA's defining the characteristics of, or criteria for, a medical device performance standard. This initiating role is shown for a consensus organization in Fig. 2. The next step involves validation of these criteria and characteristics by the FDA (or by a consensus organization if the FDA initiated the action).

The consensus organization would then derive income from the sale of information related to the criteria and characteristics of medical device performance standards. If such characteristics and criteria cannot indeed be identified, the device would then be recommended for reclassification into either Class I or Class III by the FDA. Individual firms would now find it attractive to develop firm-specific performance standards incorporating the characteristics and criteria identified by a voluntary consensus organization and validated by the FDA. Individual firms could enhance their market positions by showing that their devices do indeed outperform their competitor's on the basis of a credible performance standard. This process would also result in reduced risks, or an enhanced quality of life, or both, for patients while at the same time offering the FDA a pathway out of the Class II regulatory logjam. Also shown in Fig. 2 is the potential (dotted lines) for achieving further benefits through the development of voluntary consensus rather than firm-specific performance standards. Patients would benefit from further reductions in risk, or enhanced quality of life, or both. Individual firms would benefit by having a reduction in the threat of product liability litigation through conformance to a voluntary consensus performance standard.

There are, undoubtedly, refinements that can make this marketplace incentive-based performance-standard process more equitable and effective. However, the time is ripe to perform a serious experiment to see if society can indeed reap benefits from a pluralistic risk management and the concomitant breakdown of adversarial tension between the private sector and the FDA.

References

[1] Piehler, H. R., et al, *Orthopaedic Review*, Vol. VII, No. 1, 1978, pp. 75–81.

[2] The Medical Device Amendments of 1976, Public Law No. 94-295.

[3] Bok, D. C., *Harvard Magazine*, Vol. 85, No. 5, 1983, pp. 38–45, 70–71.

[4] *Medical Device Regulation: The FDA's Neglected Child*, Committee Print 98–F, Subcommittee on Oversight and Investigations, Committee on Energy and Commerce, U.S. House of Representatives, May 1983.

[5] *Federal Regulation of Medical Devices—Problems Still to be Overcome*, GAO/HRD-83-53, General Accounting Office, 30 Sept. 1983.

[6] Piehler, H. R. in *Medical Devices: Measurements, Quality Assurance, and Standards. ASTM STP 800*, C. A. Caceres, H. T. Yolken, R. J. Jones, and H. R. Piehler, Eds., American Society for Testing and Materials, 1983, pp. 244–269.

[7] Piehler, H. R., Portnoff, M. A., Sloter, L. E., Vegdahl, E. J., Gilbert, J. L., and Weber, M. J., this publication, pp. 93–104.

[8] Ahnefeld, F. W., as quoted in *Clinica*, Issue No. 30, 11 Dec. 1981, p. 2.

[9] *Fault/Failure Analysis Procedure*, ARP 926A, Society of Automotive Engineers, Inc., 1979.

[10] Maxwell, N., Hill, G. C., and Joseph, R. A., *Wall Street Journal*, 27 Aug. 1982, pp. 1, 10.

[11] Epstein, R. A., *Regulation*, Sept./Oct. 1982, pp. 14–19, 43–46.

[12] Product Liability Act (Revised), S44, introduced 14 Oct. 1983.

Summary

Summary

The papers in this book are grouped together in sections based on the topics covered. The first group deals with *in vitro* studies of metallic materials and is concerned with corrosion processes and behavior, hydrogen embrittlement, and stress corrosion cracking. The second section that covers metallic materials *in vivo* and *in vitro* deals with corrosion-fatigue, fretting polarization behavior, and electro-polishing. Aneuryism clip materials, fatigue of ball-joint rods, electro-chemical studies of dental alloys, and arc plasma sprayed coatings are topics covered in the third section. The fourth section covered biological effects of metallic implants and the fifth section deals with biocompatibility and durability of polymeric materials including silicones and polyesters. The sixth session deals with degradation of polyurethane, adsorbable, and biodegradable materials. Wear of polyethylene, tricalcium phosphate ceramic and loosening of cemented hip stems are considered in the seventh session. The final section addresses retrievel, performance standards and pluralistic risk management, regulation, and litigation. The individual papers in each section are summarized here in general terms but the papers should be read for specific information.

Basic corrosion processes for implant alloys are discussed by N. D. Greene who relates corrosion of metals in the body as being similar to corrosion of metals in aerated sodium chloride. Emphasis is placed on the appropriate selection of metals for implants. Equations for oxidation and reduction are given, and it is explained that due to the low rate of corrosion of implant metals, the best technique for measuring the rate of corrosion of implant metals is by the linear polarization method. Corrosion of implants can accelerate mechanical failure or fatigue. Hydrogen embrittlement or stress corrosion or both have not been a problem with the cobalt-chromium-molybdenum alloy, but Edwards et al found that the hydrogen that is generated at the surface when the cobalt-chromium-molybdenum is cathodic can be adsorbed and embrittle the material. Some experiments charged the specimen with hydrogen. These studies point up that if hydrogen were available, a loss of ductility could arise. The hydrogen build-up in this material could occur over time if a crack tip and suitable conditions existed. Hydrogen embrittlement, in general, is not a problem with the cobalt-chromium-molybdenum implant alloys but improper design or creviced areas could accentuate this potential problem.

Kumar et al investigated properties of cast, wrought, and powder metallurgy (P/M) processed cobalt-chromium-molybdenum alloys. The large-grained cast material had lower tensile strength and ductility than did the wrought or P/M processed small-grained materials. Wrought materials had small equiaxed grains with some twinning and with fine-grain boundary carbides. Carbides in the P/M material were fine and dispersed throughout the grain and grain boundaries. The P/M materials made from coarser powders (60/100 mesh as opposed to 250 mesh) had larger carbides and reduced mechanical strength over the other P/M material and wrought materials but greater than the cast material. All materials showed identical corrosion behavior and high corrosion resistance to Ringer's solution, but pitting tests in 0.01 M HCl resulted in the 250 mesh P/M material being the most resistant to the four types and the 60/100 mesh P/M material being the least resistant. All materials showed pitting in the HCl, but all were more resistant than Type 316L stainless steel. The fine-grained materials of the wrought and P/M types were more susceptable to stress corrosion cracking than the cast material in boiling 30% $MgCl_2$. No susceptibility to stress corrosion cracking was found in tests with Ringer's solution.

Sheehan, Moran, and Packer studied stress corrosion cracking (scc) of Type 316L stainless steel *in vitro*. Tests in this investigation were conducted using Type 316L specimens with electropolished surfaces for slow strain rate tension tests and Schneider intramedullary nails in the static bend tests. They concluded, as previous investigators had done, that scc is not a mode of failure of Type 316L stainless steel implants *in vitro* and found no indication that scc of this material would occur *in vivo*. Bundy and Desai studied scc of Type 316L stainless steel and ELI Ti-6Al-4V using fracture mechanics specimens and measuring crack propagation velocity versus stress intensity in environments of $MgCl_2$, HCl, and Ringer's solution. Crack propagation occurred in precracked Type 316L stainless steel in Ringer's solution held at a potential that disrupted the passive film. The conclusion from this investigation was that scc of Type 316L stainless steel could occur *in vivo* if these conditions existed.

Piehler et al reported on corrosion fatigue of hip nails and emphasized the importance of materials selection and design. Corrosion-fatigue tests of nail plates were conducted on a flexure fatigue machine to compare devices of identical design but different material. Large plate designs had superior corrosion-fatigue performance over small plates. Devices made of Ti-6Al-4V were superior in corrosion-fatigue performance to the Type 316L devices even though some fretting and wear occurred in the countersinks.

Brown and Merritt reported on *in vitro* fretting corrosion testing of bone plates and screws, and effects of blood serum and other proteins were investigated. Oscillatory motion experiments at a frequency of 1 Hz were run for 14 days and were followed with weight loss measurements. Results showed that protein additions significantly reduced fretting corrosion of stainless steel and the cobalt-nickel-chromium alloy, had no effect on Ti-6Al-4V, and increased fretting corrosion of pure titanium. These effects did not change the corrosion resistance ranking of the materials, and it was concluded that protein additions and associated precautions were not necessary for ranking alloys with this test involving fretting corrosion.

Ogundele and White carried out polarization measurements in Hanks' physiological solution on two stainless steels, Type 316L and SG2SA, both meeting ASTM Specification for Stainless Steel Sheet and Strip for Surgical Implants (F 56–78) composition specifications. The Type 316L stainless steel had a breakdown potential of 200 mV versus saturated calomel electrode (sce), while the SG2SA stainless steel had a breakdown potential of 300 mV versus sce. The difference was attributed partially to microstructural differences and to small variations in molybdenum and other elemental composition in the latter alloy. The corrosion potential for both materials shifts to a more negative value with the addition of Cl^- ions and to a more positive potential with the addition of HCO_3^- ions. These stainless steels can undergo active corrosion in highly acidic solutions below $-300mV$ versus sce but are not corrosively active in physiological saline solutions below the breakdown potential.

Stainless steel implants may be electropolished as one of the final surface preparation steps. The paper by Irving describes the basic electrochemical principles of electropolishing as well as the commercial process for electropolishing. The increased corrosion resistance of electropolished Type 316L stainless steels is noted and several factors are given to account for this improvement including removal of a deformed surface layer, surface roughness, and other surface variations. Caution should be exercised when using electropolishing solutions and procedures. Further studies of electrolytic solutions, surface films, and other variables are needed to maximize benefits of the electropolishing process.

Aneurysm clips, used to isolate a potentially fatal vascular aneurysm, are currently made from a variety of metallic alloys. Kossowsky, Kossovsky, and Dujovny examined commercially available clips made from 17-7ph, Type 304, both U.S. and British Type 316 stainless steels, and MP35N using a variety of metallurgical techniques including corrosion resistance measurements. Additionally, they performed failure analysis on one device made of 17-7ph which had been retrieved after failure *in vivo*. Their work suggested that some of the alloys used are too susceptible to stress corrosion to be recommended for permanent implantation. Indeed, this was believed to be the mode of failure of the retrieved clip. They suggest that clips that show a lower corrosion resistance than Type 316 stainless steel should not be used for permanent implantation.

Donald, Seligson, and Brown considered the fatigue of ball-joint rods used in external fixation devices. It is interesting that while much work has been done on fixation devices that are never used more than once, such as bone plates, very little study has been directed at the problem of fatigue in external fixation devices that may be reused. The authors concentrated on the ball-joint connectors in doing fatigue tests that indicate that the endurance limit of these connectors is below 400 N. Since some configurations in which these devices can be used result in loads higher than this value, the authors suggest that guidelines be evolved governing the reuse of such devices.

A new electrochemical technique to determine the susceptibility of dental alloys to sulfide tarnishing was developed by Marek. This technique requires careful control of experimental parameters such as temperature, sulfide solution composition, and potential. Using coulometric techniques, an index of susceptibility to tarnishing can be established. This elegant test offers an opportunity to achieve objective, quantifiable results that previously were not possible.

With increasing use in orthopedic implants, it is important to have established methods of determining the mechanical properties of the porous metals used in many devices. Hahn et al reported the results of mechanical testing of titanium coatings manufactured using the plasma spray technique. Both the result of their testing and their method of testing is of interest. Their results showed shear strengths in the 5.6 to 9.9 MPa range and tensile strengths in the 5.1 to 25 MPa range. Corrosion fatigue tests indicated no effect on endurance limit of the substrate provided that no sintering was performed. Sintering of the samples above or below the beta transus resulted in lower endurance limits.

A very interesting session on *in vivo* effects of corrosion products included papers by some of the foremost researchers in the area. Merritt and Brown have done extensive work on the biological effects of corrosion products. The paper presented here focuses on the *in vivo* responses to corrosion products of stainless steel and cobalt-chrome alloy devices. The work included study of the sensitivity responses using the LIF test, infection studies in animals, protein binding of metal salts and the distribution of metal salts in the body. The results suggest that patients sensitive to the metallic constituents of stainless steel or cobalt-chrome may be unable to defend adequately against infection due to a loss of spontaneous white blood cell migration. This question remains unanswered. However, it would appear from the animal testing that the ability to fight infection may be enhanced in the short term but reduced chronically.

Lucas, Bearden, and Lemons examined the ultrastructure of cells from rabbits and human gingiva that had been exposed to Type 316L stainless steel solutions. The cellular response to the solutions was correlated with the *in vivo* condition. The severity of the cellular response increased with increasing concentration of the stainless steel solution. This microscopic evaluation technique may be helpful in interpreting the results of macroscopic histological examinations.

Smith and Black used a novel method of estimating the corrosion rate of Type 316L stainless steel by examining the systemic transport and distribution of corrosion products in laboratory rabbits. They showed a positive correlation between implant surface area and the levels of iron and chrome circulating in the blood and accumulated in the liver. They approximated the corrosion rate of the implant by evaluating these levels and found the rate to be consistent with the reported corrosion rate of Type 316L.

Marchant et al use a cage implant system to study the biocompatibility of materials by focusing on the acute inflammatory response and the associated reactions. The cage is made of wire mesh Type 316 stainless steel, and the polymer or implant of interest is placed within this cage. Prior to tissue growth, the cage is filled with exudate resulting from the presence of the foreign materials. This exudate can be aspirated from the site using a sterile needle and syringe and then studied in terms of cell types, activities, etc., to characterize some of the complex reactions of the inflammatory response. Studies of the inflammatory response in rats were reported for the cage only, a biodegradable hydrogel, poly(2-hydroxyethyl-L-glutamine)(PHEG), and with added injections of a chemotactic tripeptide solution to increase the acute phase reaction. An acute inflammatory reaction with increased white cell concentration occurred in all cases, but the exudate from the PHEG implant system indicated increased cell formation and activity. Details of the inflammatory response of this study are discussed, and these results show that the cage system can be used as a model for studying the body's reaction to implant materials.

Swanson et al reported on biocompatibility of silicone implants in animals and humans ranging up to 12 years of implantation. Using pathological and radiological techniques, they assessed host reaction to various implants. Their results showed smooth fibrous encapsulation of the implants in dogs with no bone resorption or bursa formation. Although wear particles from the implants were evident in both the animal and human studies, the particles appeared to be well tolerated with minimal inflammatory cells and no necrosis. A single enlarged lymph node in the human showed a mild benign foreign body reaction. No evidence of particle transport was evident from pathological examination of organs in either human or animal cases.

Silicone elastomers have been used successfully in implant reconstructive surgery since the middle of the 1950's, and the finger joint prosthesis is one implant with which this material has made a great improvement. This background is provided by Frisch and Langly along with information regarding chemistry, fabrication, and biocompatibility of silicone elastomers. Their paper reports on the biodurability and high performance of a new silicone elastomer. Tests, utilizing ASTM procedures, showed the new high performance material to have a low modulus, high tear propagation strength, and high resistance to flexural fatigue crack growth. *In vivo* studies of the material in dogs over a two-year period showed no change in the physical properties. There is a 2% by weight of polydimethylsiloxane that is not chemically bound in the elastomer. Based on test data of extractions before and after implantation, the authors concluded that no significant loss of silicone to tissues occurred. Lipid absorption occurred early after implantation and was 1.5% at the end of 104 weeks, but this did not appear to affect the properties of the material.

King et al present the results of a study of the biodegradation of vascular prostheses made of poly (ethylene terephthalate), PET. Physical and chemical factors are given as contributing factors in the loss of mechanical performance and bursting strength of this material. Retrieved implants of PET material were obtained after having been implanted for periods of a few hours to 14 years. Unused control specimens of all samples, except in two cases, were used for comparative tests of bursting strength, stitch density, molecular weight, and carboxyl group content. There was a loss in bursting strength, a loss in molecular weight and an increase in carboxyl group content for the implanted PET material. It was estimated that 25% of the initial bursting strength was lost after 162 ± 23 months and that 25% of the molecular weight was lost after 120 ± 15 months of implantation. Theoretical models of degradation are discussed. The authors state that this rate of degradation does not put patients who have these prostheses at risk, but it does indicate that additional work is needed to improve this material.

Two very timely papers reported on the *in vivo* degradation of polyurethane. Szycher and McArthur compared the surface fissuring of two different polyurethanes with different hardnesses used in pacemaker lead insulation. Their examination showed surface fissures at points of high stress such as ligature sights and high stress areas introduced by the manufacturing processes used in these devices. They hypothesize that the surface fissures may be due to *in vivo* oxidation of the polyether chain. They recommend that ligature stress be lowered by use of silicone anchoring sleeves, that manufacturing processes be controlled to reduce stresses, and that higher durometer polyurethanes be used that are less susceptible to stress-induced failures.

Parins et al reported on animal testing done in support of pre-clinical investigations of polyurethane leads. During the early phase of this investigation, unexpected surface cracks were observed leading to an expanded test program. They reported no obvious correlation between surface cracking and mechanical properties, although such evidence may have been hidden by the inherent scatter in the data. Their examination confirmed the relationship between stress and cracking reported by the previous authors. They conclude similarly that the successful performance of this polyurethane depends on proper use.

In a change from most of the previous papers that reported the negative aspects of biodegradation, there were two papers presented on materials that made positive use of this phenomenon. Dunn, Casper, and Cowsar described a biodegradable composite material they have developed that will be useful in reconstructive surgery, particularly maxillofacial reconstruction. The composite is a laminate of PLA sheets with fiber-reinforced PLA sheets. The fibers used were either carbon or ceramic. The resulting composite has initial mechanical properties very similar to those of bone but degrades *in vivo* over time to prevent stress protection atrophy and alleviate the necessity for later removal of the device.

McKellop and Clarke have extensive experience at wear testing of total hip joint components. In the paper presented here, they report on the testing of titanium alloy hips from three different manufacturers and compare the results from those devices to similar tests on steel or cobalt hips from the same three sources. One of the sets of titanium hips had been given a special nitriding treatment that gave these devices a harder surface. Since previous data reported by several authors questioned the wear characteristics of titanium alloy hips, this carefully controlled and comprehensive study is of much interest. The results reported indicated that there was no significant difference in the wear rates of the polyethylene associated with any of the devices tested except the surface hardened titanium alloy devices that showed lower wear rates. All of the wear rates reported were within the ranges observed clinically.

Lemons has given an excellent review of the history, results of animal testing, and human clinical applications of tricalcium phosphate ceramics. This material is intended as a treatment for bone lesions and should biodegrade leaving a natural tissue repair.

Certainly, loosening of total hip stems within the PMMA cement bed is a clinically significant problem. Ebramzadeh et al report that the literature indicates up to 40% incidence of cemented stem loosening with more cases possibly undiagnosed due to the inability of radiographs to reveal small gaps. In the work reported here, they developed new techniques to measure bone cement strains in a model bone. An interesting result of the testing was that proximally where the cement strains were the highest the strains reduced by as much as 73% after about 4 million loading cycles. This is evidence of creep within the cement layer. Understanding this phenomenon is important in designing prostheses that reduce cement stresses.

Two papers on total joint retrieval analysis are included in the book. One paper reports on retrieved prostheses from a study by the Mid American Joint Replacement Institute, University of Missouri-Kansas City School of Medicine by Hood. The paper states that the goal of implant retrieval programs is to identify specific implant problems relating to design, material, installation, patient weight, and other factors. Cataloging this information provides a base of information for improving total joint prostheses. Guidelines are given for collecting data, examining, and storing the retrieved implants. Problems noted were wear in metal-to-metal joints, fracture of the metal stems resulting from localized stress due to cement loosening, or improper placement for carrying the load, or all. The UHMW polyethylene components are subject to wear and fracture. Implantation time and body weight were discussed.

Wright, Burstein, and Bartel of the Hospital for Special Surgery, New York, NY, reported on a six year experience of total joint retrieval and analysis. There are common elements in the approach and conclusions of the two studies. This work had as one purpose to relate material type and design to implant mechanical performance. Retrieved components were cleaned, examined, and classified according to specific procedures including a review of the patient's medical record. A total of 1082 total joint components were analyzed. These were taken from patients with weights ranging from 35.4 to 112.5 kg with a mean weight of 79 kg, and ages ranging from 26 to 84 with a mean age of 64 years. The mechanism given for metal failure was fatigue with the fracture location varying with prosthesis design. Wear and deformation of polyethylene were observed. Articulating surfaces of polyethylene exhibited damage due to delamination and abrasion.

The long-term performance of polyethylene based on these retrieved analyses is questionable, and the possibility of optimizing UHMW polyethylene performance with different designs should be investigated. There are specific descriptions of a number of components and their failures in both the retrieval papers.

The subject of developing performance standards was addressed by Mayor who described design standards as a method of assuring the performance of a material or device based on measurable parameters. These design standards provide predictability but restrict innovation. The performance standard would provide assurance of predictable use without constraining imaginative solutions to problems. Writing performance standards will be more difficult, will press the state of the art, and will require test methods not yet developed. In spite of the enormousness of the task, the author suggests that performance standards are the direction in which standards writing should be moving. An illustration of a performance standard in another field is given and additional information is provided to help the reader understand performance standards.

The paper by Piehler on pluralistic medical device risk management also cites the need for performance standards. Failure of a device is defined as the patient not being able to recover to the expected level or having to undergo additional surgery to correct the problem. Standards, either voluntary or mandatory, are used in regulation or litigation. Although a number of factors such as clinical procedure, patient misuse, unfavorable tissue response, etc. contribute to device failure, only the device itself is covered in the Medical Device Amendments of 1976. A plan is presented for the development of performance standards in which the private sector would carry out the development of the standards. The standards would meet the needs of the Food and Drug Administration (FDA) and the criteria of a voluntary concensus organization. The author discusses the marketplace incentive for the development of the performance standards and indicates that the adversarial tension between the FDA and the private sector would disappear.

Anna C. Fraker
National Bureau of Standards Washington, DC 20234; symposium chairman and editor.

Charles D. Griffin
Carbomedics, Inc. Austin, TX 78752; symposium chairman and editor.

Index